T. Yoshida R. D. Tanner (Eds.)

Bioproducts and Bioprocesses 2

Third Conference to Promote Japan/U.S. Joint
Projects and Cooperation in Biotechnology,
Honolulu, Hawaii, January 6-10, 1991

With 151 Figures and 20 Tables

Springer-Verlag

Berlin Heidelberg New York
London Paris Tokyo
Hong Kong Barcelona
Budapest

Editors

Prof. Toshiomi Yoshida

International Center of Cooperative
Research in Biotechnology
Faculty of Engineering
Osaka University
Osaka 565, Japan

Professor Dr. Robert D. Tanner

Department of Chemical Engineering,
Vanderbilt University,
Nashville, Tennessee 37235, USA

ISBN 978-3-642-49362-1   ISBN 978-3-642-49360-7 (eBook)
DOI 10.1007/978-3-642-49360-7

Library of Congress Cataloging-in-Publication Data
Conference to Promote Japan/U.S. Joint Projects and Cooperation in Biotechnology
(3rd : 1991 : Honolulu, Hawaii) Bioproducts and Bioprocesses 2 : third Conference to Promote
Japan/U.S. Joint Projects and Cooperation in Biotechnology, Honolulu, Hawaii, January 6–10,
1991 / T. Yoshida, R.D. Tanner, — (Eds.). Includes bibliographical references.
          (Berlin : acid-free paper). —   (New York : acid-free paper)
1. Biochemical   engineering—Congresses.   2. Biotechnology—Congresses.   I.   Yoshida,   T.
(Toshiomi), 1939-   . II. Tanner, R.-D. (Robert D.) III. Title.
TP248.3.C65 1991 660'.6—dc20

Typesetting: Macmillan India Ltd., Bangalore-25

51/3020-543210—Printed on acid-free paper

List of Contributors

Numbers in parentheses refer to the sections within this volume

Aikawa, J. (2.1)
Asali, E.C. (1.2)
Asama, H. (4.1)
Bailey, J.E. (2.2)
Bennett, G.N. (2.4)
Beppu, T. (2.1)
Blanch, H.W. (3.5)
Cooke, T.J. (5.5)
Dykstra, K.H. (4.4)
Endo, I. (4.1)
Frame, K.K. (5.6)
Furusaki, S. (5.4)
Furuya, T. (5.4)
Gomez, P.L. (1.2)
Hashimoto, T. (5.2)
Heineken, F.G. (1.1)
Horinouchi, S. (2.1)
Hu, W.-S. (5.5, 5.6)
Humphrey, A.E. (1.2)
Iijima, S. (5.7)
Imanaka, T. (2.5)
Karube, I. (4.3)
Kellogg, C.H. (1.6)
Kim, J.Y. (2.6)
Kimura, A. (2.3)
Kishimoto, M. (1.3)
Kita, Y. (1.3)
Klibanov, A.M. (3.3)
Kobayashi, T. (5.7)
Kokitkar, P.B. (1.6)
Konstantinov, K.B. (4.5)
Kurata, H. (5.4)
Ladisch, M.R. (4.6)
Lauffenburger, D.A. (5.3)
Li, J.-K. (1.2)

Linko, P. (4.1)
Mermelstein, L.D. (2.4)
Mukherjee, T. (1.4)
Nagamune, T. (4.1)
Nakajima, M. (1.3, 4.1)
Okada, H. (3.1, 3.6)
Omstead, D.R. (1.5)
Papoutsakis, E.T. (2.4)
Reese, J.A. (1.5)
Ryu, D.Y. (2.6)
Sakoda, H. (2.5)
Seki, M. (5.4)
Shimizu, S. (3.2)
Shuler, M.L. (5.1)
Siimes, T. (4.1)
Staba, E.J. (5.5)
Starburck, C. (5.3)
Takeuchi, S. (1.3)
Tamiya, E. (4.3)
Tanaka, A. (3.4)
Tanner, R.D. (1.6)
Ueda, M. (3.4)
Urabe, I. (3.6)
Velayudhan, A. (4.6)
Vits, H. (5.5)
Wang, D.I.C. (4.2)
Wang, H.Y. (4.4)
Wang, S.-Y. (1.5)
Wiley, H.S. (5.3)
Yamada, H. (3.2)
Yamada, Y. (5.2)
Yomo, T. (3.6)
Yoshida, T. (1.3, 4.5)
Zivin, R. (1.5)

Preface

Introduction

During the week of January 6–10, 1991, the Third U.S.–Japan Conference on Biotechnology was held at the Asian-Pacific Conference Center at the University of Hawaii in Honolulu. This book is a compilation of the papers and posters presented at the Conference. The Conference was sponsored, in part, by the National Science Foundation and U.S. pharmaceutical companies including Ortho Pharmaceutical, Merck, Genentech, SmithKline Beecham and ABEC. Its purpose was to promote information exchange between Japanese and U.S. researchers, primarily academics, in biotechnology and to seek ways to carry out collaborative research in biotechnology.

The honorary chairmen of the Conference were Professor H. Okada and me. The formal program was organized by Professors J. Bailey and T. Yoshida. Twelve invited formal presentations were given from each side. In addition, both sides were invited to bring along five observers to the Conference who were encouraged to prepare poster presentations on their research. Paper abstracts plus bibliographies were exchanged prior to the Conference in order to promote maximum technical interaction between the participants.

The presentations were selected and equally divided among four preselected topic areas. These were 1) applied genetic engineering, 2) biocatalysis, 3) bioprocess engineering, and 4) cell culture. Six papers were presented in each topical area. During lunch and coffee breaks, the participants had an opportunity to discuss their work in one-to-one situations. In general, the Japanese academic research work was strongly applications oriented while the U.S. work appeared driven by support and direction through funding sources from the National Science Foundation. There appeared to be a strong desire on both sides to do collaborative research; however, it was found that the avenues of collaboration had to be driven by individual contacts. Clearly, the more basic the research work, the easier it appeared to collaborate.

The following is a brief summary of the papers presented in each topical area.

Applied Genetic Engineering

Although the title seems internally redundant, i.e., applied and engineering, it really represented an effort on the part of the program organizers to look at research concerned with the application of recombinant DNA techniques to produce useful products or to achieve more efficient processes. For example, the first paper in this area looked at the use of secretion-signal for mucor rennin to achieve secretion of pro-urokinase and human growth hormone in a re-combinant yeast and then to use this knowledge to genetically engineer an improved rennin. The next paper discussed the use of genetic engineering techniques to improve the carbon conversion and energy metabolism in cells. In a third paper, the author discussed ways to genetically engineer the production of a tuna growth hormone in *E. coli*. The next paper focused on the usefulness of amplifying homologous genes in a bacterium in order to improve the fermenting characteristics of the solvent producing *Cl. aectobacterium*. The fifth paper was concerned with the cloning of the thermal stable alcohol dehydrogenase gene from *Bacillus stearothermophilus* NCA 1503 into a *Bacillus subtilus* as a way of shifting the pH optimum of the mutant enzyme from 7.8 to 9.0. The final paper in this series was concerned with methods to improve and/or optimize a recombinant fermentation process.

Biocatalysis

In this series of papers, such topics were discussed as the use of site directed mutagenesis to explore the structure and function of enzymes, the role of enzymes in chiral synthesis, and the mechanism of enzyme activity and function in organic solvents. In the chiral synthesis paper, nearly 100 examples of the use of enzymes in the commercial production of chemicals were reported. One process was described for producing up to 40 000 ton/year of acrylonitrile by enzymatic catalysis. Another paper discussed enzyme behavior in unusual environments such as super critical CO_2. A final paper discussed strategies for designing enzyme-like catalysis including the possibility of producing catalytic antibodies.

Bioprocess Engineering

This session provided a truly "mixed" bag. Topics covered included on-line diagnostic systems for controlling a fed batch fermentation, development of an expert system for a phenylalanine fermentation, development of micro-biosensors for biomedical research, sensors for bioprocess monitoring and control plus development of purification systems for fermentation products recovery. In the latter, paper systems included studies of 2-D gel electrophoresis and gradient elution chromatography. Of particular interest was the paper on micro-biosensors which discussed the philosophy of designing micro-biosensors and the various potential applications.

Cell Culture

Four of the six papers in this session focused on the culture of plant cells. Two of the papers looked at ways through genetic engineering to regulate the production of desired secondary metabolites. One paper was concerned with somatic embryogenesis as a method for the large scale production of transgenic plant varieties. Another emphasized the design and operation of bioreactor systems that would maximize volumetric productivity. This included individual strategies such as strain selection, medium development, cell immobilization, product secretion, in situ product removal and elicitation. The fourth paper discussed the design of a photo-bioreactor suitable for plant cell cultivation. The remaining two papers were concerned with the design and operation of bioreactors for the cultivation of animal cells. One of the presentations discussed an effort to increase the quantitative understanding of how binding and trafficking aspects of growth factor/receptor interactions influence mammalian cell proliferation.

There was general agreement among the participants that the periodic organization of these conferences promotes an understanding and greater appreciation of biotechnology activity in our respective countries. Through the informal contacts, opportunities were identified for postdoctoral and visiting scientist exchanges beyond those offered by governmental agencies. Also, these contacts catalyzed scientific friendships and enhanced the appreciation of each other's culture. Clearly, they are very valuable, particularly in view of the very low costs of the conferences. To my Japanese colleagues—dozo yoroshi-ku. To those of you who find these exchanges interesting, consider applying for observer status at the next one. To those companies that supported the Conference and to the NSF for the basic travel grant—thank you.

Director, Biotechnology Institute,
The Pennsylvania State University,
University Park, PA 16802   Arthur E. Humphrey

Contents

1 Biochemical Engineering and Biotechnology

1.1 Biochemical Engineering and Biotechnology – An NSF Perspective

Fred G. Heineken,

Program Director, Biotechnology, Division of Biological and Critical Systems, Directorate for Engineering, National Science Foundation

Abstract

Engineering research related to the life sciences is becoming more important as an increasing number of products from genetic engineering and cell fusion technology reach the market place. Novel bioprocess engineering, both upstream and downstream, is also needed to provide a fundamental engineering basis for the economical manufacturing of substances of biological origin. Research linking the expertise of engineers and life scientists is crucial to providing such a fundamental basis, and requires individuals who are broadly competent in each of their fields and who are also willing to collaborate on research projects. This presentation intends to provide an overview of how the Engineering Directorate within the National Science Foundation plans to provide support for the engineering research needed to address the problems of the economic production of products obtained from the most recent advances in the life sciences.

Bioproducts and Bioprocesses 2
Editors: Yoshida, Tanner
© Springer-Verlag Berlin Heidelberg 1993

1.2 Use of Fluorometry for On-Line Monitoring and Control of Bioreactors – Microbial Cell Concentration and Activity, Plant Cell Metabolism, Mixing Time, and Gas Hold-up

J.-K. Li[1], E.C. Asali[2], P.L. Gomez[2] and A.E. Humphrey[3]*

[1]Merck & Co., Inc, P.O. Box 600, Danville, PA 17821, USA
[2]Center for Molecular Bioscience and Biotechnology, Lehigh University, Bethlehem, PA 18015, USA
[3]Biotechnology Institute, Pennsylvania State University, University Park, PA 16802, USA

Contents

This paper discusses the potential for fluorometry in monitoring fermentations. Emphasis is on the application of multiple excitation fluorometry. This application includes monitoring of microbial concentrations and activities, monitoring metabolic response to perturbations of plant cell cultures, and determining mixing times and gas hold-up of real fermentations.

List of Symbols and Abbreviations

AFU	arbitrary fluorescence units
DO	dissolved oxygen
MEFS	multiple excitation fluorometric system
NFU	normalized fluorescence units
NADH	reduced nicotine adenine dinucleotide

C fluorescence signal
C_0 initial fluorescence signal
C_f final fluorescence signal
t_{exp} time constant which is the time when signal reaches 63.212%, i.e. $(1 - e^{-1})$ of maximum response
t_{lag} lag time which is defined as the time for first response

---

* To whom all correspondence should be addressed.

Bioproducts and Bioprocesses 2
Editors: Yoshida, Tanner
© Springer-Verlag Berlin Heidelberg 1993

1 Introduction

On-line monitoring and control of bioprocesses is one of the most challenging research areas in biochemical engineering. Sterilizable dissolved oxygen (DO) and pH probes have been available since the 1960s. No other biosensors developed since then have proved to be as good as the DO and pH probes. Dissolved oxygen, pH and temperature along with off-gas analysis are the only parameters which are monitored on-line in industrial fermentations. None of the present on-line measurement techniques involves intracellular parameters. This limitation has hampered the development of bioprocess control since our ability to control and perform on-line optimization of fermentation processes is limited by our ability to monitor what is happening in the fermentation.

In order to control a bioprocess, we need to be able to measure on-line substrate, product, and cell concentrations, as well as cellular activities within the fermentor in a non-invasive manner. Techniques currently exist to achieve this based upon enzyme, immunological and optical density probes, as well as using indirect measurements by material balancing around the bioreactor, specifically measuring oxygen uptake rates (OUR) and utilizing simple metabolic models to estimate cell growth, substrate uptake, and product formation rates. Unfortunately, none of these probes are practically robust devices. Enzyme and immunological probes cannot withstand in situ sterilization. Optical density probes are reliable only for non-mycelium cultures growing on clear, non-particle containing media, conditions that are generally uncommon in fermentation industry. The models used to estimate cell growth, substrate uptake, and product formation rates from OUR are based on conditions of defined media and single substrate limitations. As such, these models may not apply in many real situations.

Fluorometry is the technique which has great potential to be used as a bioreactor monitoring device. The reason is that there are some measurable fluorophores in growing cells such as NADH, tryptophan, pyridoxine and riboflavin. Because these fluorophores involve key vitamins, coenzymes and aromatic amino acids, their concentration changes may reflect the changes in cell concentration and cell metabolic state as well as environmental conditions. By measuring the fluorescence signals of whole broth through non-invasive optical fiber systems, one can obtain continuously on-line estimates of several fermentation parameters during a fermentation. Many of them are intracellular parameters. Fluorometry has very high sensitivity and good specificity. There is virtually no time lag in the measurements. This technique provides a means of monitoring what is going on inside the cells, rather than just the environment around them the way traditional biosensors do.

Reduced nicotinamide adenine dinucleotide (NADH) is the first cellular fluorophore which has been used to monitor bioreactors. There are two commercially available fluorescence probes (BioChem Technology and Ignold) for culture fluorescence measurements. Both of them are based on NADH fluorescence. Most of the results in previous fluorescence studies were obtained

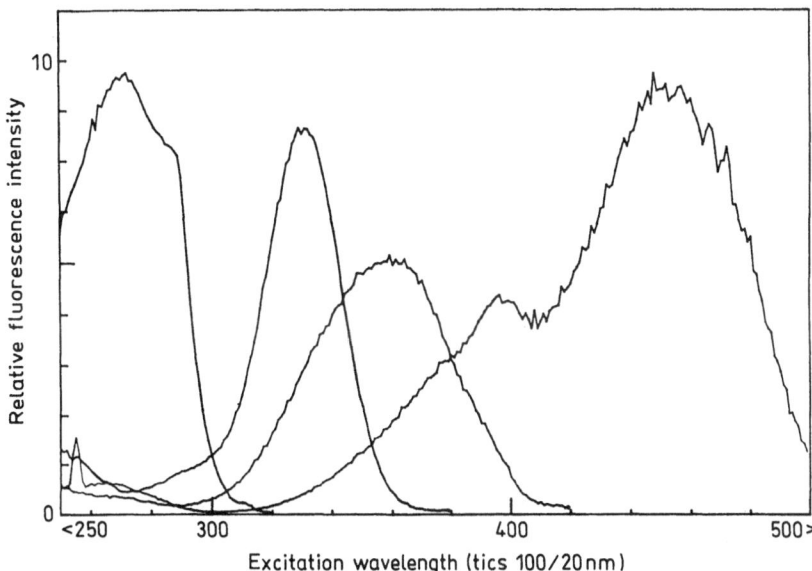

Fig. 1. Excitation spectra of the four major cellular fluorophores—excitation wavelengths vs relative fluorescence intensity. The spectra from left to right correspond to the fluorophors of tryptophan, pyridoxine, NADH and riboflavin respectively

using these NADH probes. However, using only NADH fluorescence for the monitoring of fermentations presents several problems. For example, the quantum efficiency of NADH is very low and NADH fluorescence is extremely sensitive to cellular metabolic state and environmental conditions. Since the optical filters used in the commercial NADH probes permit a relatively wide range of light wavelengths to excite the culture, the emission signal contains not only the NADH fluorescence but that of several other cellular fluorophores. Hence, it is not explicit as to what particular fluorophores are measured by commercial NADH probes. Additional useful information can be obtained by monitoring other cellular fluorophores than just NADH fluorescence [1, 2].

Based on the reasoning outlined above, we decided to build a new type of fluorometric probe titled—multiple excitation fluorometric system (MEFS)— and to monitor four major cellular fluorophores—tryptophan, pyridoxine, NADH, and riboflavin—in whole broth cultures. The reasons for selecting these particular four fluorophores are several-fold. Firstly, and most importantly, these fluorophores are key metabolic components. Secondly, they are optimally or near optimally excited by Hg arc lamp spectral lines of 289, 313, 334, 365, and 404 nm (Figs 1 and 2). Thirdly, these four fluorophores, when excited at selected wavelengths, fluoresce in separate and distinct regions with little or no overlap (Fig. 3). For example, the fluorescence emission spectra of NADH and riboflavin overlap if both of them are excited at the wavelength of 365 nm. However, NADH does not fluoresce when excited at 404 nm while riboflavin strongly

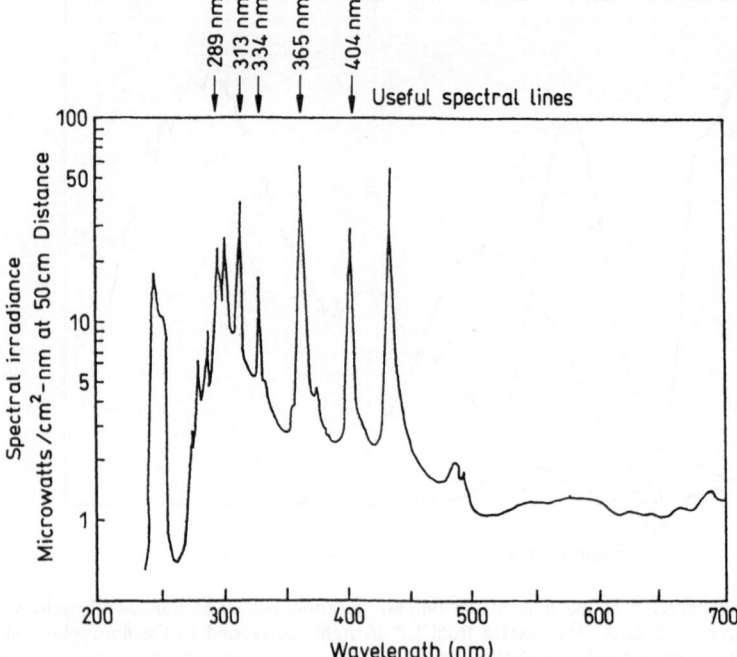

Fig. 2. Spectra of mercury arc lamp

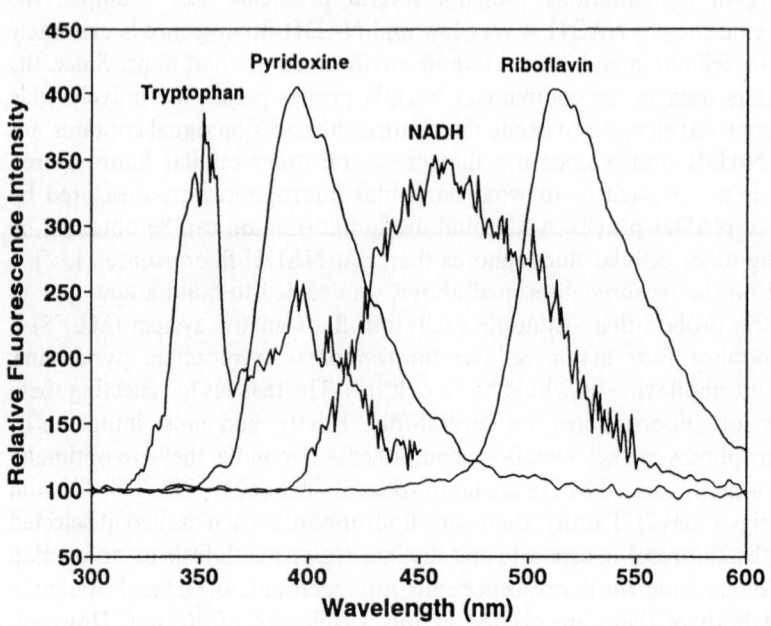

Fig. 3. Fluorescence emission spectra of cellular fluorophores excited by Hg arc lamp spectral lines

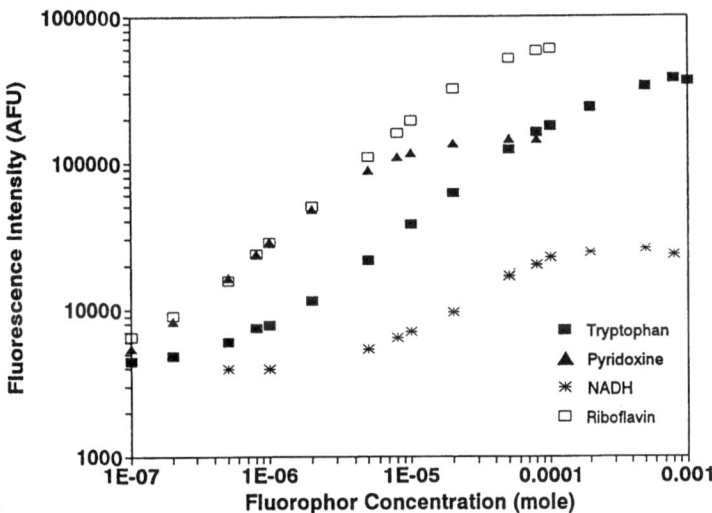

Fig. 4. Relationships between fluorescence intensities and fluorophore concentrations measured by MEFS

fluoresces. Therefore, an excitation wavelength (404 nm) can be selected where NADH does not interfere the emission spectrum of riboflavin. Similar situations exist with the other cellular fluorophores. Finally, the linear ranges between fluorescence intensities and fluorophore concentrations are very wide for all the four fluorophores and occur in the concentration regions normally encountered in most fermentations (Fig. 4).

The assembly of the MEFS is shown in Fig. 5. The light source is a 200-W mercury arc lamp (ORIEL Corporation). The light from the Hg arc lamp passes through a filter wheel containing 5 different bandpass filters to produce five different narrow band width (10 nm half-height widths) excitation light sources (289, 313, 334, 365, and 404 nm). After focusing, the excitation light is sent to a fermentor to excite the whole fermentation broth through a bifurcated optical fiber. The back scattered (180°) fluorescence signal is collected by the optical fiber and sent to a spectrum analyzer (Guided Wave, Inc.). The results are recorded by an IBM AT computer. By selecting a particular filter, a specific fluorophore, i.e., tryptophan, pyridoxine, NADH or riboflavin, can be excited. In our work fluorescence intensity was defined by the area under fluorescence peak in certain wavelength ranges and measured in arbitrary fluorescence units (AFU). The excitation wavelengths and integrated fluorescence signal ranges are shown in Table 1. For comparison purposes, a commercially available NADH probe (BioChem Technology) was also used to measure the whole broth fluorescence. Culture fluorescence was expressed in terms of normalized fluorescence units (NFU) as defined by Armiger [3].

The fermenter used in this study was a specially modified New Brunswick Scientific Inc., computer controlled, 3 L BioFlo III fermenter. The vessel was

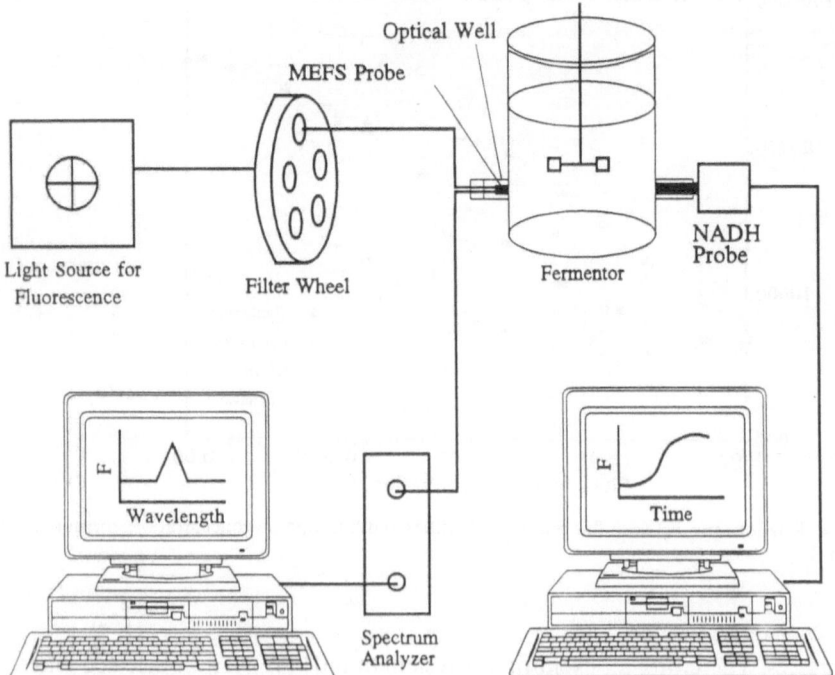

Fig. 5. Fluorometry measurement systems

Table 1. Excitation and fluorescence monitoring wavelengths (nm)

Fluorophores	Excitation	Integrated signal range
Tryptophan	289	335–365
Pyridoxine	313	385–415
NADH	365	450–480
Riboflavin	404	500–530

made of stainless steel and equipped with two sterilizable "Ingold" ports fitted with optical wells at 90° to each other. The two fluorescence probes were placed into the optical wells. A specially designed baffle was used to divide the vessel into two parts so that the light from the two fluorescence probes did not interfere each other.

Because many factors can affect fluorescence signal and the quantum yield of a fluorophore may change when it combines with other compounds, it is difficult to obtain accurate fluorophore concentrations from the back scattered culture fluorescence signal alone. However, by using fluorescence signal time profiles for the various cellular fluorophores, we believe significant information can be obtained for fermentation control purposes.

2 Applications of Fluorometry to Fermentations

2.1 Monitoring Microbial Activity

The fluorometric behavior of two model yeast fermentations have been examined in our laboratory. The first model system was a *Candida utilis* (ATCC 26387) fermentation, cultivated in non-fluorescence synthetic medium using 1% ethanol (w/v) as the carbon and energy source [10]. The cultivation conditions were 28°C, pH 5.2, and 1.5 vvm aeration. The second model system was a Bakers' yeast fermentation using glucose as the carbon and energy source. The cultivation conditions were 28°C, pH 5.5, and 1.5 vvm aeration [4].

The time course of the fluorescence signals measured by the MEFS for the *C. utilis* fermentation is shown in Fig. 6. The tryptophan fluorescence gave the strongest signal followed by pyridoxine fluorescence. Riboflavin had the weakest signal. The profile of fluorescence signal from the four cellular fluorophores were similar to the cell growth curve during exponential phase of cell growth. All four fluorescence signals can be used to monitor cell concentration during the growth phase for this particular organism and medium. Tryptophan, however, is the best fluorophore for estimating cell concentration since its fluorescence signal is the most sensitive to cell concentration.

After exponential growth phase, none of the fluorescence signals of the four fluorophores correlate linearly with cell concentration. In these growth phases, many factors may affect fluorescence intensity. Cellular metabolic activity can be one of the most important factors. Also in declining growth phase, cell autolysis can release fluorophores into broth, as well as materials which absorb light at the fluorophore sensitive wavelengths to create inner filter effects. Hence, fluorometric behavior of the declining growth phase in batch culture is complex. The fluorometric behavior of declining growth phase needs to be studied carefully if control of this phase is to be achieved by fluorometric monitoring.

Over the time course of the *Candida utilis* fermentation the cellular metabolic state was compared with the NADH fluorescence measured by MEFS. Typical results are shown in Fig. 7. When ethanol was used up and cells began to use acetate, the NADH fluorescence signal decreased. This is because more NADH is oxidized to NAD^+ when cells use acetic acid [5]. When the acetate was used up, the fluorescence signal begins to increase. Since in a fed-batch fermentation process, ethanol should be added only when acetate is used up in order to optimize the yield, the NADH fermentation offers a way to control ethanol feeding.

In the Backers' yeast (*S. cereviciae*) fermentation, pyridoxine fluorescence is the dominant signal and is a good indicator for monitoring the metabolic state

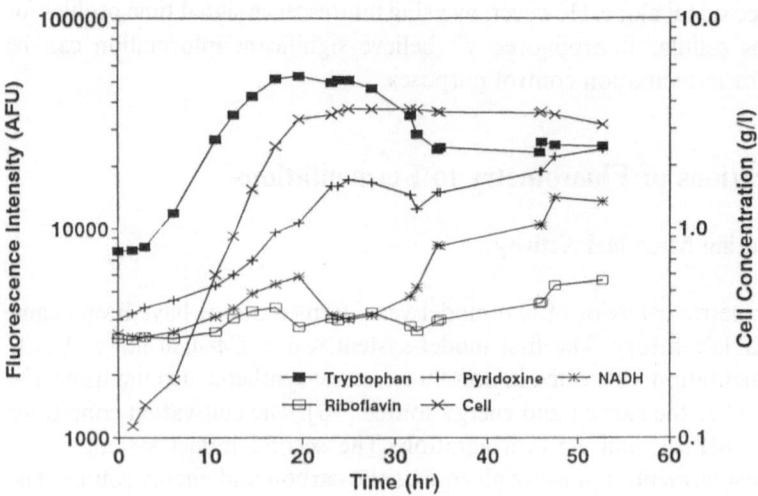

Fig. 6. Fluorescence profiles of the four cellular fluorophors for *Candida utilis* growing on ethanol

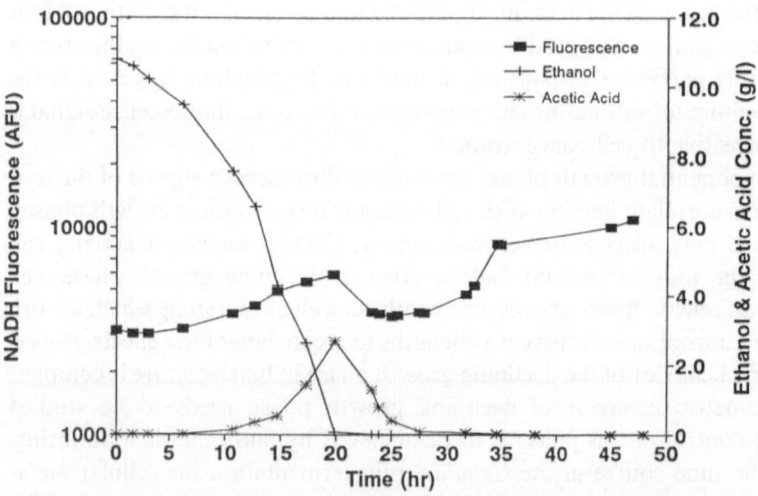

Fig. 7. Time course of *Candida utilis* fermentation in comparison to the NADH fluorescence

of the cells (Fig. 8). A change in pyridoxine fluorescence signal indicates a switch in glucose utilization. For example, it indicates the level change of key metabolic intermediates, such as ethanol and acetic acid. Again, tryptophan fluorescence is a good indicator for estimating cell concentrations for this system.

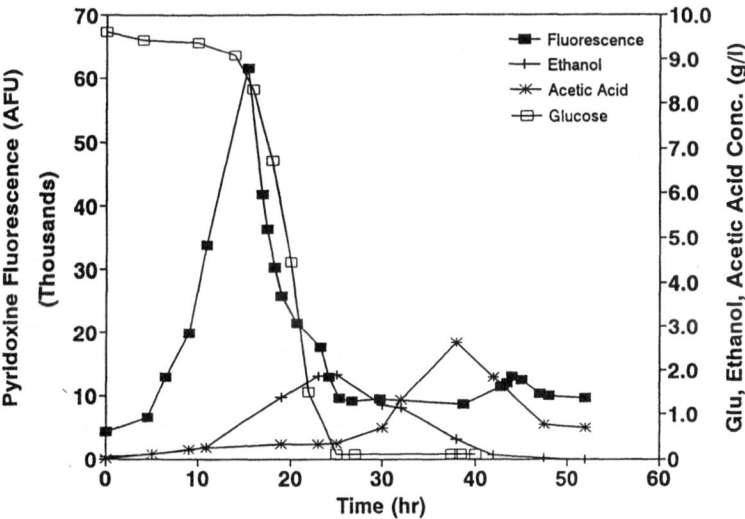

Fig. 8. Time course of Bakers' yeast fermentation in comparison to the pyridoxine fluorescence

2.2 Monitoring Metabolic Perturbations in Plant Cell Cultures

NAD^+ is an important electron carrier in transferring energy from catabolism and oxidative phosphorylation to anabolism. The ratio of $NADH/NAD^+$ regulates the TCA cycle flux. Since NADH fluoresces and NAD^+ does not, NADH fluorescence can be used as an indicator of the oxidation or reduction state of a culture system based on the fact that the sum of specific (per cell) NAD^+ and NADH concentration is constant regardless of the cellular metabolism. Through this phenomena, the behavior of cellular metabolism under a given set of environmental condition can be observed.

Growth of plant cells in suspension culture depends strongly on the oxygen supply [6, 7]. It is generally agreed that suboptimal gas exchange may reduce culture performance due to oxygen limitation. Oxygen limitation presumably affects the cells by reducing their ability to obtain metabolic energy. However, since cells are expected to response to dissolved oxygen concentrations, and not oxygen transfer rate, it is difficult to correlate O_2 transfer rates to cell physiology. This is especially true since cultivation at constant gas transfer rate will result in large changes in the dissolved oxygen concentrations if the culture's metabolic activity changes over time (e.g. batch culture).

In this study we used NADH fluorescence (measured by the commercial NADH probe) to specially examine the effects of dissolved oxygen on the oxidation/reduction state of *Catharanthus roseus* suspension culture and thus determine the level of dissolved oxygen at which cells switch from aerobic to anaerobic state. We also studied the effects of glucose perturbation on starved quasi-steady state cells in suspension culture. In this investigation changes in

NADH fluorescence were primarily attributed to mitochondrial and bound cytosolic NADH thus indicating catabolic functioning.

Catharanthus roseus was grown on B-5 medium [8] supplemented with 2% glucose, 1 mg/L 2,4-dichlorophenoxyacetic acid and 0.1 mg/L kinetin. The cultivation conditions were pH 5.5 and temperature 25°C . Dissolved Oxygen was controlled by manipulating the volumetric flow rates of O_2 and N_2. Total gas inlet flow was maintained between 0.3–0.4 L/min.

Figure 9 shows the culture response to aerobic-anaerobic-aerobic transition. It is seen that NADH fluorescence significantly increased when the cells switched to anaerobic condition. Accumulation of NADH in anaerobic condition is due to the lack of O_2 which serves as an electron acceptor in cellular respiration. Therefore by following NADH fluorescence, we can observe the critical DO level where the cells will switch from an oxidized to a reduced condition. This was done by conducting a step DO experiment. The results are

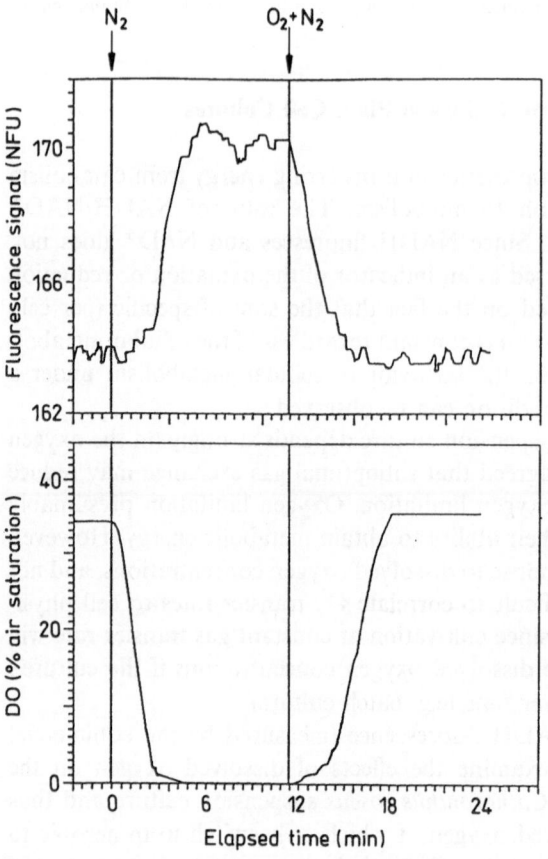

Fig. 9. NADH fluorescence and dissolved oxygen responses to aerobic–anaerobic–aerobic transition. (Aeration was turned off and on as indicated by vertical lines. Cell concentration in this run was 6.75 g/L)

shown in Fig. 10. These results suggest that this culture is still in oxidized state until the DO level felt below 11% air saturation. Above this level we believe there should be no oxygen limitation for *Catharanthus roseus* suspension culture.

NADH culture fluorescence can also be used to monitor oxidative glucose metabolism. Reduction of NAD^+ occurs due to several reactions in the glucose metabolic pathway. This includes such steps as glyceraldehyde-3-phosphate to

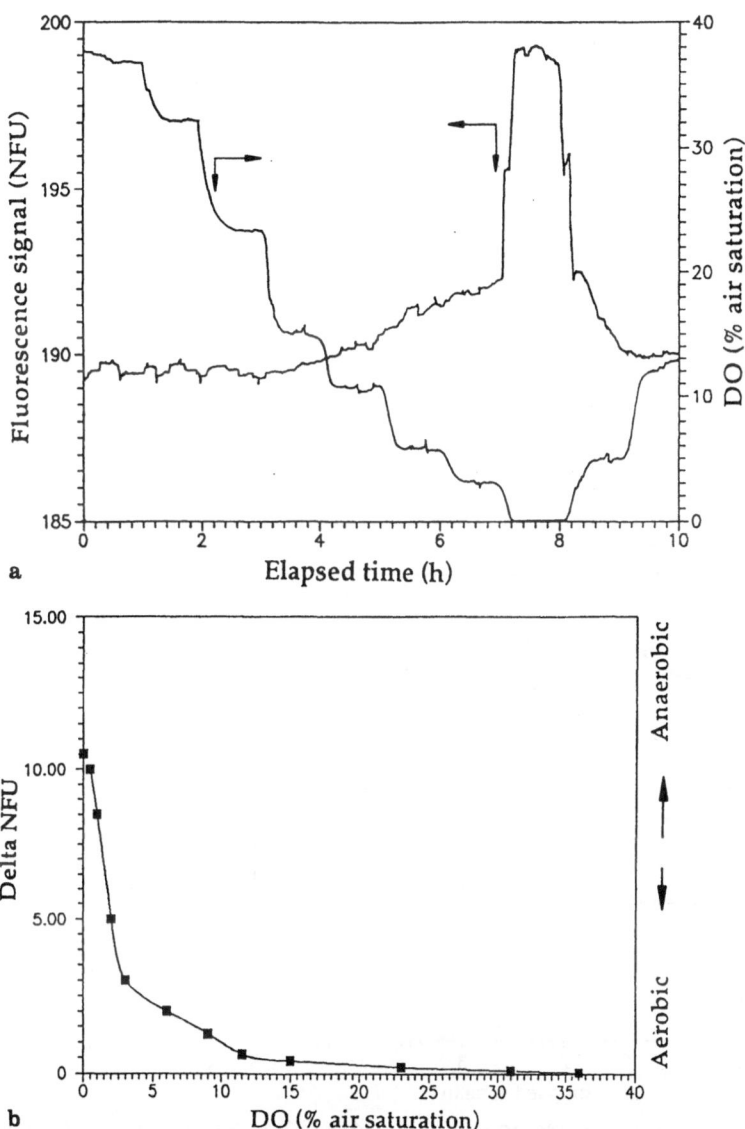

Fig. 10. (A) typical profile of various DO level vs NADH fluorescence signal. (B) the values of NADH fluorescence changes (ΔNFU) at different DO levels

3-phosphoglyceroyl-phosphate, pyruvate to acetyl-CoA, isocitrate to α-ketoglu-tarate, and malate to oxalacetate steps. A glucose perturbation of starved cells was performed. Characteristic NADH fluorescence for such an experiment is shown in Fig. 11. The NADH fluorescence profile qualitatively represents the NADH pool during glucose metabolism. Specific oxygen and glucose uptake rates appear to correlate with NADH oxidation and NADH generation rates, respectively. There are three different stages in the glucose uptake pattern. Stage one was characterized by the increase of NADH fluorescence which can be due

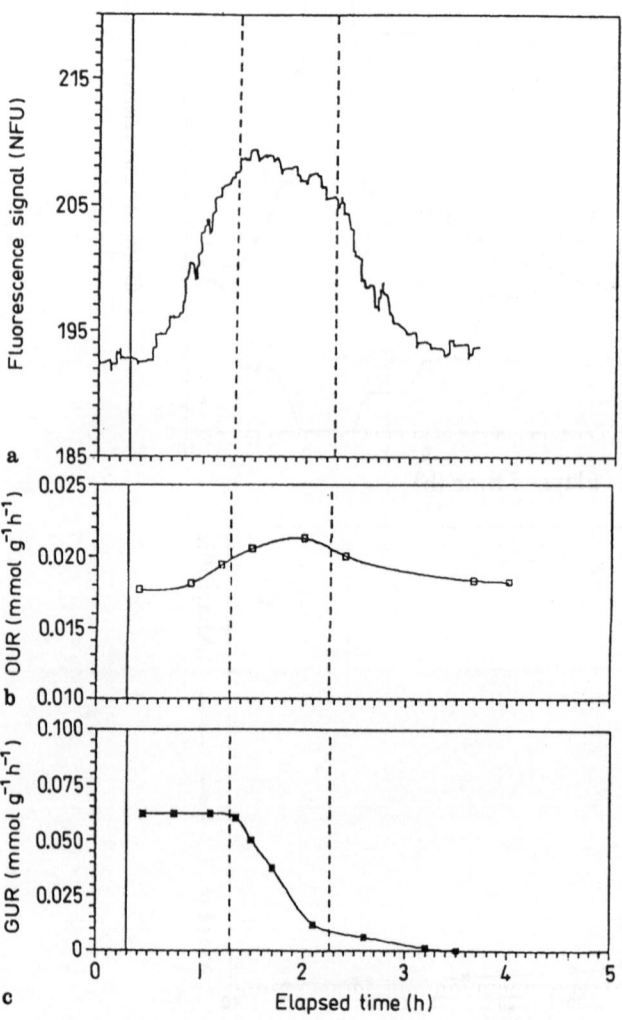

Fig. 11. Glucose perturbation at 15% DO saturation. (140 ppm of glucose was added as indicated by vertical lines. Cell concentration was 8.43 g/L) (**A**) NADH fluorescence profile (**B**) Specific oxygen uptake rate (OUR), (**C**) Specific glucose uptake rate (GUR)

to the high specific glucose uptake rate (0.062 mmol/g DW h). In the second stage the NADH fluorescence starts to decrease slowly which is concomitant with a decrease in the specific glucose uptake rate and a small increase in specific oxygen uptake rate. Stage three was characterized by a further decrease of NADH fluorescence until it reached its original level at which time glucose was completely depleted. These results suggested that NADH fluorescence monitoring system can be used on line as an indication of glucose depletion and for controlling a fed batch *Catharanthus roseus* plant cell culture system. A similar technique has been used by Ristroph et al. [5] to control a fed batch yeast fermentation.

2.3 Mixing Time and Gas Hold-up Measurements in Real Fermentation

The ability of fluorometry to monitor a specific compounds in a very complex mixture provides a unique opportunity to measure mixing times and hold-ups in a fermentations system [9]. The knowledge of the mixing time of a fermentation vessel is critical in the understanding of the mass transfer and yield characteristics of large fermentation vessels. The ability to measure mixing time or gas hold-up (liquid level) on-line during a fermentation is not presently available. If such information could be obtained, it would provide several benefits including the ability to optimize parameters such as optimal feeding port positions in a large fermenter, improved yield, and reproducible "setting" of a fermentation.

Choosing a proper inert fluorescence tracer is the first step to applying this technique. In this work, riboflavin was used as the inert tracer. Riboflavin is a very strong fluorophore. It provides many unique properties as well as being inexpensive ($0.13/g Sigma #R4500). Riboflavin is nontoxic to most fermentations. It is not naturally present in large quantities in most fermentations (minimizing the background fluorescence). It is very insensitive to changes in cell metabolic state and environmental conditions such as pH, DO, and temperature. It has great stability over time in response to varying physical conditions [1, 10].

The method of measuring mixing time involved the instantaneous injection of 10–20 mL of a 0.01 M solution of riboflavin into a 250 L fermenter (ABEC). Both the MEFS and NADH probes were used to monitor the fluorescence signal. For the MEFS, excitation was done at 365 nm and the integrated fluorescent emission between 490–530 nm was used for the determination of hold-up. A single emission wave length of 511 nm was monitored for mixing time determinations. Ports for the two probes were placed at 90° angle in the fermenter and a baffle introduced between them to reduce any interference. The mixing time was determined using the following differential equation.

$$\frac{dC}{dt} = (C_f - C)\frac{1}{t_{exp}},$$

where:

C = fluorescence signal,
C_f = final fluorescence signal,
t_{exp} = time constant which is the time when signal reaches 63.212% $(1 - e^{-1})$ of maximum response.

The results of the mixing time experiments indicated that there was a lag time in the signal response. To account for this, the parameter t_{lag}, which is defined as the time for first response, was added. Solving the above equation and linearizing it gives the following equation.

$$\ln\left(1 - \frac{(C - C_0)}{(C_f - C_0)}\right) = \left(\frac{t_{lag}}{t_{exp}}\right) - \left(\frac{t}{t_{exp}}\right),$$

where C_0 is the initial fluorescence signal. Mixing time is defined as $t_{exp} + t_{lag}$ or the time at which the signal is $(1 - e^{-1})$ of the full response. The advantage of using this definition is that the error in calculation a mixing time is reduced since the data is best fit over the middle of the observed range. Past mixing time studies utilized decay in oscillations in response to a pulse. This is useful only when there is close proximity of the detection probe to the tracer input. The present method allows the above equation to be fit over a wider range of data and does not require an arbitrary definition of when to conclude the oscillations have stopped.

A pH tracer method was used to compare the mixing time results obtained with the fluorescence probes. The results indicate that the NADH probe, which uses an integration of many wavelengths to get a signal, has a response time of

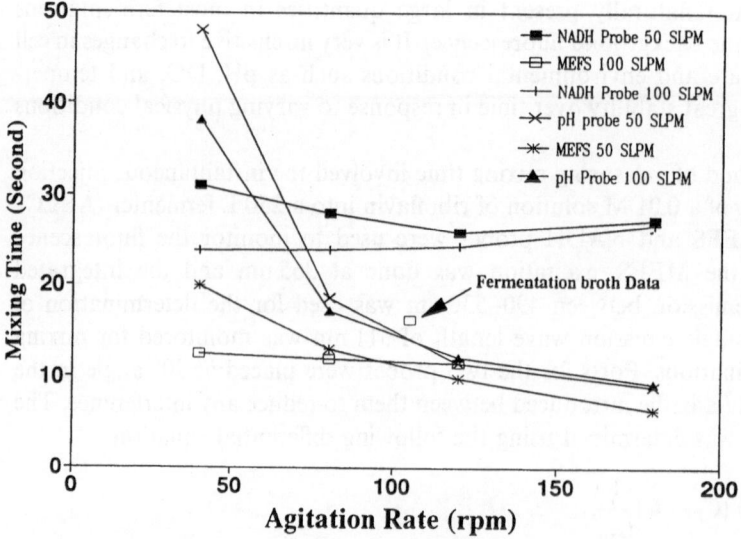

Fig. 12. Summarization of mixing time measurements of different agitation and aeration rates

about 15 s and was not, therefore, suited to be used in measuring mixing times of less than 30 s. The MEFS probe however has virtually instantaneous response and gave good agreement with the pH method. Typically it differed only by about 15% from mixing times obtained by standard pH methods (Fig. 12). The MEFS system does generation more noise than the pH method due to its greater sensitivity to bubbles. But by filtering the data noise, no problems were encountered in obtaining the mixing time values.

The MEFS probe was also used to measure the mixing time during the exponential growth phase of a *C. utilis* fermentation. The results, when compared with those for phosphate buffer medium, gave longer mixing times,

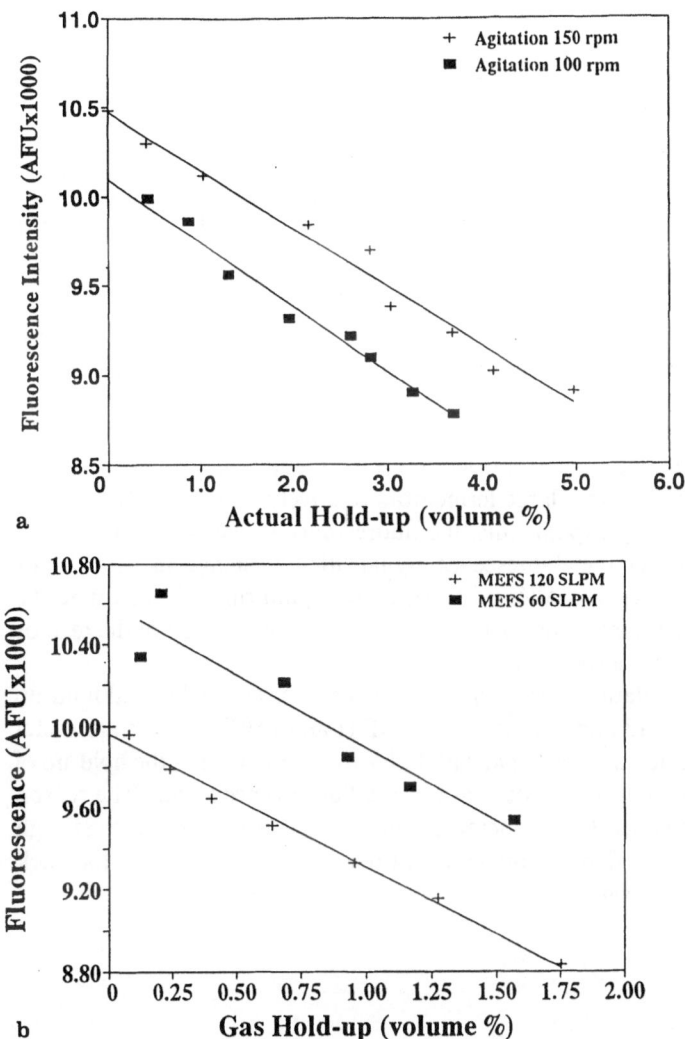

Fig. 13. Hold-up measurements in phosphate buffer by MEFS. (**A**) Fluorescence intensity vs actual hold-up at different agitation rates. (**B**) Fluorescence intensity vs actual hold-up at different aeration rate

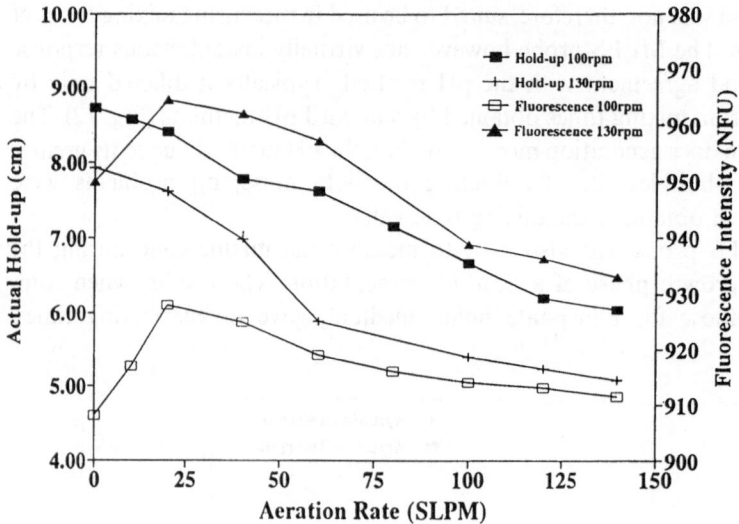

Fig. 14. Hold-up measurements in fermentation broth by NADH probe at different aerations. (Although hold-up is usually reported as the percent volume increase, the hold-up described here was utilized to enable the fluorometric data to be compared graphically with the hold-up measurements)

probably due to the fact that the fermentation broth had a higher viscosity plus higher bubble hold-ups. From the result we conclude that even for this simple fermentation system, the mixing properties of buffer and fermentation broths are very different. An apparently large error in predicting mixing characteristics of fermentation broth could occur if done in only aqueous solution. This system can be applied, even to very large fermentors relatively inexpensively.

For the gas hold-up experiments, the fluorescence signal was correlated to hold-up by measuring the fluorescence signal after a change in aeration or agitation and the vessel was allowed to come to equilibrium. At this time the actual hold-up was visually observed as the distance from the top of the reactor to the buffer or media level in the reactor.

Good linear correlations between fluorescence intensity and actual hold-up were observed (Figs. 13 and 14). Both the NADH and MEFS probes are useful in correlating the fluorescence signal with hold-up. An increase in the hold-up of the fermenter correlates with a decrease in the fluorescence signal. The reason for this can be explained by the increased air bubbles in the fermenter at high hold-ups, reducing the fluorophore concentration, and thus reducing the back scattered fluorescence signal.

3 Conclusions

From these studies it was concluded that the fluorescence intensity and behavior of various cellular fluorophores are very different for different fermentation systems. Monitoring a fermentation process using only NADH fluorescence can miss a lot of available data on fermentation behavior.

The most useful way of controlling a bioreactor using fluorescence as the monitoring signal is to monitor simultaneously the several fluorophores in the whole culture broth and relate these fluorescence signals to various biological parameters. The MEFS provides a way of doing this. The four biofluorophores —tryptophan, pyridoxine, NADH and riboflavin—can be used as model fluorophores since they possess a wide linear relationship between fluorescence intensities and concentrations over the range of interest in most fermentations.

Further, fluorescent probes can provide insight into the physical state of mixing characteristics and gas hold-up in real fermentations. Since the background riboflavin fluorescence signal is very weak in most fermentations and riboflavin gives the strongest fluorescence signal of the four biofluorophores at similar concentrations, it has the potential to be used as a nontoxic on-line fluorescence tracer in real fermentations.

Acknowledgement. The authors wish to express their appreciation to the National Institutes of Standards and Technology, Division of Chemical Process Metrology, for supporting this work.

4 References

1. Li J-K, Humphrey AE (1990) Biotechnol Bioeng 37: 1043.
2. Humphrey AE, Brown K, Horvath JJ, Semejian H (1989) In: Fiechter A, Okada H, Tanner RD (eds) Bioproducts and Bioprocesses. Springer, Berlin Heidelberg New York, p 309.
3. Armiger WB (1987) The fluromeasure system user's manual, BioChem Technology Inc.
4. Li J-K, Asali EC, Humphrey AE, Horvath JJ (1991) Biotechnol Progr 7: 21.
5. Ristroph DL, Watteew CM, Armiger WB, Humphrey AE (1977) J Ferment Technol 55: 559.
6. Pareilleux A, Vinas R (1983) J Ferment Technol 61: 429.
7. Scragg AH, Leckie F, Cliffe KC (1991) Biotechnol Bioeng 37: 364.
8. Dixon RA (1985) Plant cell culture, a practical approach, IRS Press, Oxford.
9. Li J-K, Gomez PL, Humphrey AE (1990) Biotechnol Tech 4: 293.
10. Li J-K (1991) Ph.D. Thesis, Lehigh University, Bethlehem, PA.

1.3 Computer Control of Glutamic Acid Production Based on Fuzzy Clustering of Culture Phases

Michimasa Kishimoto[1], Yoichi Kitta[1], Sougo Takeuchi[1], Mikio Nakajima[2], and Toshiomi Yoshida[2]

[1]Department of Biological Science and Technology, Science University of Tokyo, 2641 Yamazaki, Noda, Chiba, 278 Japan
[2]International Center of Cooperative Research in Biotechnology, Japan, Faculty of Engineering, Osaka University, 2-1 Yamadaoka, Suita, Osaka, 565 Japan

Contents

An algorithm for the clustering of culture status into several culture phases (lag, growth, transient, and production) was developed for the flexible control of glucose concentration in glutamic acid fermentation. Each phase has its own different type of control strategy. Phase recognition was carried out based on fuzzy sets theory.

For the recognition, three measurement variables (culture time, CO_2 evolution rate, and the amount of NH_3 feed) were selected as variables of the fuzzy antecedent, and several fuzzy levels for each variable were defined by the table of membership functions. The inference was carried out by the min–max fuzzy algorithm every three minutes, and the tubing pump for glucose feed was controlled to let the amount of feed follow the inferred amount. The control strategy was able to change smoothly from one to another during the cultivation and the glucose concentration was successfully kept constant.

1 Introduction

The reaction mechanism in a microorganism is much more complicated than that of an ordinary chemical reactor. The mechanism is fundamentally programmed by the DNA sequences of the organism involved, which can change from their original situation with time. It is difficult to keep all of the characteristics of a microorganism constant for a long period, and the improvement of strains by genetic engineering and/or screening techniques is often tried in order to improve industrial fermentation processes. As a result, models of a real fermentation process usually work only for a limited period, and it is virtually impossible to construct a comprehensive and robust model. Furthermore, only a few kinds of sensors are available for monitoring the culture states in fermentation processes, and we cannot measure the state inside cells directly. Therefore, a deterministic model for the application to the control or the simulation of fermentation processes cannot easily be constructed.

In almost all cases of fermentation processes, a computer is unable to decide a control strategy from deterministic models, but some operators can decide the appropriate operative conditions from their experience related with particular fermentation processes or from knowledge of the related fundamental information in the fields of microbiology, genetics, biochemistry, and etc. Process operators often try to use qualitative information to support or reinforce quantitative parameters when predicting the time course of a fermentation.

However, such human inference can consume considerable time, with success varying according to the individual, and its result cannot be expected to be precise. The human inference approach has generally been assumed to be quite different in nature from traditional process control techniques, in which deterministic models are used to simulate and/or control fermentation processes. Therefore, it has been desired that an inference system, which could act as a substitute for some part of the human inference procedure, would be developed for the optimum control of the fermentation process.

Fuzzy sets theory was developed for the effective use of qualitative information [1], and it has been applied to the simulation or control of fermentation processes such as alcohol production [2], glutamic acid production [3, 4], SCP production [5, 6], sake brewing [7–9], co-enzyme production [10], activated sludge process [11] and so on. In such cases, the control strategy, progress of the fermentation processes, or the physiological state of microorganisms has been estimated by using fuzzy production rules and membership functions [2,– 5]. In the present study, the fuzzy set theory was applied to the control of the glucose concentration in glutamic acid production. Several calculation methods has been proposed for fuzzy inference in addition to Mandani's min-max algorithm. We use Sugeno's method as a basic algorithm for fuzzy control. The concept of the clustering of culture phases resembles that of the physiological state adopted by Konstantin and Yoshida [5].

2 Materials and Methods

Brevibacterium sp., which was kindly supplied by the Ajinomoto Co., was used for the experiments of glutamic acid production in the present study. Table 1 shows the composition of the media used for the seed and main cultures. The seed culture was carried out for 20 h using 500 mL flasks, and the working volume was 100 mL.

The volume of the mini-jar fermenter (Mitsuwa Co.) used for the main culture was 5 L nominal, with an initial working volume of 2 L and maximum working volume of 3 L. A schematic view of the fermentation system is shown in Fig. 1. The aeration rate and temperature were kept constant at 2 L/min and 30°C, respectively. Agitation speed was automatically controlled so as to keep the concentration of dissolved oxygen above 40% air saturation. The pH was controlled at around 7.5 by using a glucose–NH_4OH solution. The computer control system was instrumented in order to monitor temperature, pH, DO, agitation speed, CO_2 concentration in the outlet gas, and the amounts of glucose solution and NH_3 solution feeds, as well as to display and to print the data. Furthermore the computer was able to control these items by sending signals to

Table 1. Media composition for seed and main cultures

Glucose	10.0–50.0 g/L
KH$_2$PO$_4$	1.0
MgSO$_4$·7H$_2$O	0.4
MnSO$_4$·4H$_2$O	2.0 ppm
FeSO$_4$	2.0 ppm
Growth factors	as required

Medium for feeding (only for main cultures).
30% NH$_3$ (pH control).
50% glucose (fuzzy control).

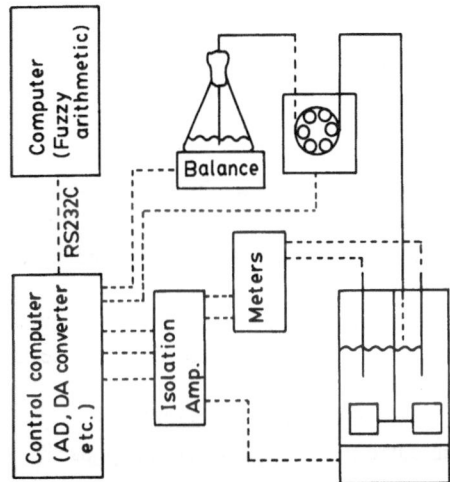

Fig. 1. Schematic diagram of glutamic acid fermentation with computer control

the controllers, and determine the amount of glucose feed based on fuzzy inference. In the case of preliminary experiments without computer control, glucose was mixed with NH$_3$ in aqueous solution to avoid exhaustion of glucose during glutamic acid production, with the mixing ratio of glucose to NH$_3$ being 5.0 (g/g).

Cell density was measured turbidometrically at 610 nm and converted into a dry mass concentration using a calibration curve. The concentration of glutamic acid was measured by the glucose oxidase method (Toyobo Co.). The concentration of glutamic acid was measured by the calorimetric method using glutamate dehydrogenase (Boehringer Mannheim Co.).

3 Structure of Software

The computer control system has two personal computers (NEC Co.), as shown in Fig. 2. One was instrumented for the monitoring and the control of the culture state. The other one inferred the current phase of the cultivation based on fuzzy

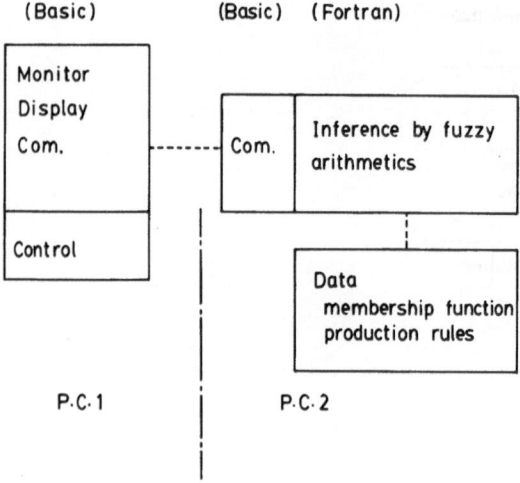

(Basic)   (Basic)   (Fortran)

Fig. 2. Software configuration

sets theory using a data file of production rules and membership functions, and determined the amount of glucose feed according to the control strategy which was set up a priori for each phase. The two computers were connected by a RS232C cable.

The tubing pump for glucose feed was regulated in an on-off manner following the signal from the first computer, so that the amount of glucose feed could follow the value calculated by the second computer.

An example of the production rules for fuzzy inference is as follows.

If the culture time is middle then the current state is production phase

CO_2 evol. rate is middle
NH_3 addition is middle

In the production phase
the amount of glucose feed $= a \times$ the amount of NH_3 addition $+ b$ ... (1)
a, b: constants which are determined by the least squares method using the experimental data.

A simple example of phase determination is as follows. First, the grades of the affiliation of the current measured data to the antecedents of each production rule are estimated by comparing the current measured data with the related data of membership functions, as shown in the left half of Fig. 3. In this case, there are three antecedents in each production rule. The minimum value of the possibility of the three antecedents was assigned to the adaptability of the production rule (the value of the membership function) to the current state of the fermentation process.

Each production rule has a consequent after "then" which describes a phase. The adaptability of the phase to the current state should not be higher than the

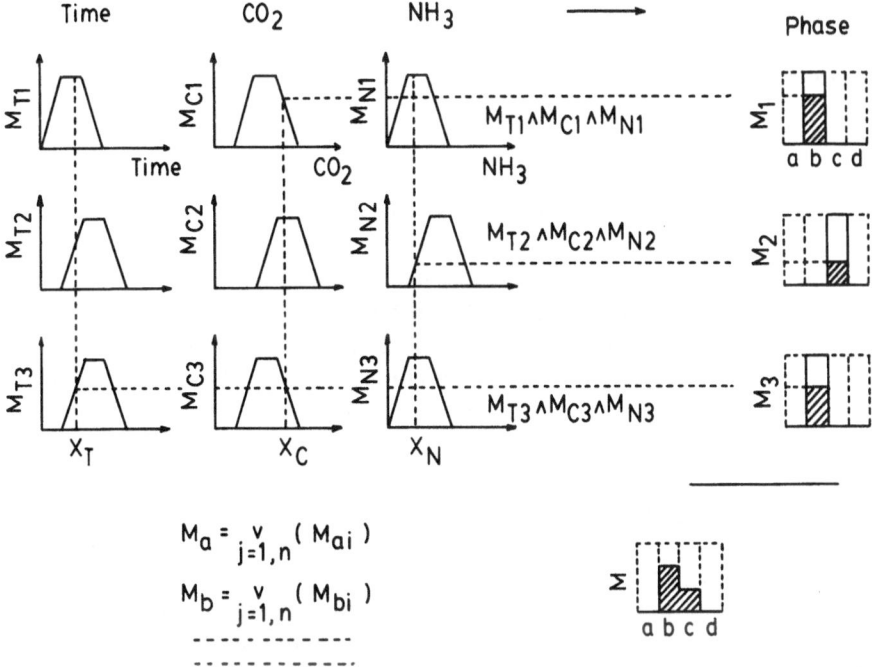

$$M_a = \mathop{V}_{j=1,n} (M_{ai})$$

$$M_b = \mathop{V}_{j=1,n} (M_{bi})$$

Fig. 3. Fuzzy inference for the determination of phase

adaptability of the production rules. The value of the adaptability of the production rules becomes an inferred possibility of the phase.

From each production rule, the phase possibility is inferred in a similar way as described above, and the final conclusion comes from a summarization of all of these possibilities by the max procedure, as shown in Fig. 3.

The inference procedures shown in Fig. 3 are displayed on the CRT during the cultivation. Operators can check each step of the current inference.

For each culture phase, a different type of calculation method for the estimation of the amount of glucose feed was assigned, and the amount of glucose feed (final inferred result) was determined by an arithmetical average of the estimated amount of glucose feed in each phase weighted by the possibility of the inferred phase. The data of the final inferred result was sent to the first computer, which controlled the tubing pump by comparing the inferred result and the monitoring result of the amount of glucose feed with the aid of the balance which measured the weight of glucose solution.

This computer control method can infer the culture phase easily by considering several culture conditions and several production rules related to the determination of the culture phase. Generally speaking, this method can avoid the sudden change of culture state which is often fatal in fermentation processes.

4 Results and Discussion

Figure 4 shows the relationship between the amount of glucose consumption and the amount of NH_3 addition for pH control in the production phase, for which experimental data were acquired by the preliminary experiments without computer control. The relationship was quite linear, as Nakamura reported (3), and suggested that in the production phase the amount of glucose consumption should be estimated by using the amount of NH_3 addition. The glucose concentration in the culture broth can be controlled based on the estimation. In the production phase, such a linear relationship is easily imagined because NH_3 is consumed for the production of glutamic acid, which is an acid, and NH_3 is automatically added into the culture broth for pH control. The values of a and b in Eq. (1) are determined by the least square method using the experimental data as shown in Fig. 4.

However, the relationship shown in Fig. 4 does not exist in the growth phase, because NH_3 and glucose are consumed not for the production of glutamic acid but for cell growth. A different control strategy should therefore be adopted for the growth phase, and for the other phases.

As shown in Fig. 5, we found a simple relationship between the amount of glucose consumption and the amount of CO_2 evolution in the growth phase. The amount of glucose feed was determined based on the amount of CO_2 evolution for the control of the glucose concentration.

In other phases, the amount of glucose feed was also estimated by using the relationships which were derived a priori experimentally, and a strategy for control of the glucose concentration was developed. An example of the computer control of glutamic acid production is shown in Fig. 6 and Fig. 7. Figure 6

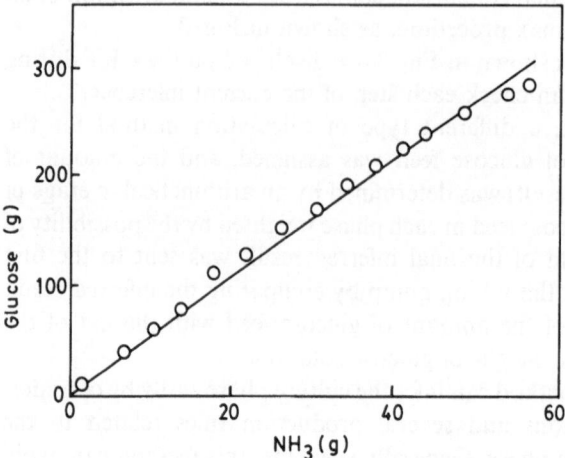

Fig. 4. Relationship between the amount of NH_3 addition and the amount of glucose consumption in the production phase

describes the time courses of OD_{610}, glucose concentration and glutamic acid concentration. The glucose concentration was successfully kept constant at around 25 g/L. Figure 7 shows the changes of phase during cultivation. The crosshatching in Fig. 7 indicates that the probability of the related phase is

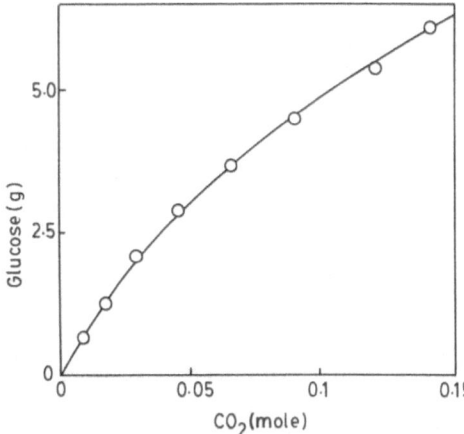

Fig. 5. Relationship between the amount of CO_2 evolution and the amount of glucose consumption in the production phase

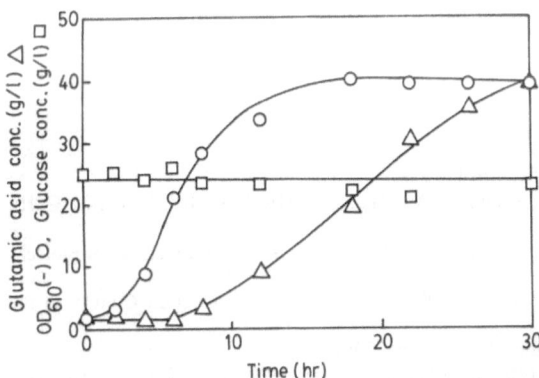

Fig. 6. Time course of glutamic acid production with computer control

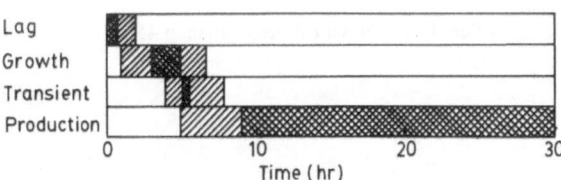

Fig. 7. Changes of inferred phase during the cultivation

100% in the period, and diagonal shading shows that there is some probability of the related phase in the period, but the phase cannot be ascertained for certain. In Fig. 7, the phases are seen to change smoothly, and the control strategy also moved from one to another smoothly. The results suggest that the computer control method based on fuzzy clustering is effective for the glutamic acid fermentation process.

In this study, we developed an algorithm for computer control, with fuzzy clustering of the culture status into several phases (lag, growth, transient, and production), and applied it to glutamic acid production. The control strategy could be changed smoothly from one to another. The glucose concentration was successfully kept constant. At first sight, the approach by fuzzy control seems to be effective in the computer control of fermentation processes. However, its effectiveness depends on the production control and membership function, which must be constructed manually based on experience and know-how related to the fermentation process. Thus, there can be variation to some degree, depending on the individuals concerned.

This kind of approach might be criticized for its lack of analysis of the process. However, it seems that in cases of problems with the process when it is difficult to use another more conventional approach, such an approach can play a role in improving both operation and control.

Acknowledgement. We are grateful to the Ajinomoto Co. for supplying us with the cultures of *Brevibacterium* sp. used in the study.

5 References

1. Zadeh LA (1965) Information and Control 8: 338.
2. Filev PD, Yoshida T, Sengupta S, Kishimoto M, Taguchi H (1984) Annual Report of ICME, Japan, 5: 379.
3. Nakamura T, Kuratani T, Morita Y (1985) Proc of IFAC Modelling and Control of Biotechnology Process, 211.
4. Kishimoto M, Yoshida T, Moo-Young M (1989) Proc. of IFAC Production Control in the Process Industry PS2.38-PS2.43.
5. Konstantin K, Yoshida T (1989) Biotechnol Bioeng 33: 1145.
6. Nakamura M, Nannba A (1989) Abstr Ann Meeting of Society of Ferment Bioeng Japan, p 65.
7. Tsuchiya Y, Koizumi J, Suenari K, Teshima Y, Nagai S (1990) Hakkou Kougaku Kaishi 68: 123.
8. Suenari K, Tsuchiya Y, Teshima Y, Koizumi J, Nagai S (1990) Hakkou Kougaku Kaishi 68: 131.
9. Oishi K, Tominaga M, Kawato A, Abe Y, Imayasu S (1989) Abstr Ann Meeting of Society of Ferment Bioeng Japan, p 66.
10. Yamada Y, Haneda K, Shiomi S, Murayama S (1989) Abstr 22nd Shuuki Taikai for Society of Chemical Engineering, Japan, p 201.
11. Suzuki K (1990) Abstr Ann Meeting of Society of Ferment Bioeng Japan, p 45.

1.4 National Science Foundation Engineering Centers Division – Biotechnology and Bioengineering Research Centers

Tapan Mukherjee, Ph.D., D.Sc., Program Director

Poster Abstract

The Engineering Centers Division supports university-based research centers aimed at enhancing industrial competitiveness by strengthening university/industry coupling in research and education. The division supports two major programs, Engineering Research Centers (ERC) and Industry/University Cooperative Research Centers (IUCRC).

The ERCs have special emphasis on developing strategic goals which will lead to new technological developments and an improved understanding of the integration and management of engineering systems. These centers are supported for five years and can be renewed for another five years.

1. *The Biotechnology Process Engineering Center (BPEC)* at the Massachusetts Institute of Technology, Director, Prof. Danny I. C. Wang. The primary research goal during the first six years of BPEC was to advance manufacturing concepts for the production of high-value therapeutic proteins. Within this overall theme, the three thrust areas were: Molecular Biology in Biotechnology, Engineering Principles in Biotechnology, and Downstream Processing in Biotechnology. Mammalian cells are targeted as the primary vehicle for the production of high-value proteins. During the next five years BPEC will expand its activities into products produced by prokaryotic organisms. Specifically, two new themes will be introduced which are antibodies and enzymes in biocatalysts, and new biomaterials through biotechnology.

2. *The Center for Interfacial Microbial Process Engineering (CIMPE)* at Montana State University, Director, Prof. William G. Characklis. The research goal for CIMPE is to develop a phenomenological model for cell accumulation and cell activity at interfaces relevant to industrial systems. The research thrusts are: Transport and Interfacial Transfer Phenomena, Physiological Ecology and Genetic Exchange, Extracellular Chemical and Electrochemical Phenomena, Sensors/Instrumentation/Monitoring, and Modelling and Information Management.

3. *The Center for Emerging Cardiovascular Technologies (CECT)* at Duke University and other North Carolina Institutions, Director, Prof. Theo Pilkington. By linking biomedical sciences and state-of-the-art engineering, the research program includes interventional stimulation and catheterization procedures aimed at developing new generation of devices to treat arrhythmias, atherosclerosis and other vascular diseases. Research on cardiovascular imaging

Bioproducts and Bioprocesses 2
Editors: Yoshida, Tanner
© Springer-Verlag Berlin Heidelberg 1993

systems is focusing on real-time 3D ultrasonic imaging, 3D magnetic resonance imaging, 3D radiopharmaceutical imaging and true 3D dynamic displays.

The IUCRCs encourage industry/university cooperation by developing research programs defined by the industries. The NSF role is catalytic and the support is provided for five years after which they must become self-supporting with a combination of industrial, State, and other Federal support.

1. *The IUC Research Center for Biological Surface Science* at SUNY-Buffalo, Director, Prof. Robert E. Baier. The research is focused around model substrata of polymers, metals and ceramics which are being used to study the deposits occurring during exposures to relevant interacting biosystems. Other studies are addressing biosystem compatibility with bacteria, mammalian cells, blood components, and micro algae.

2. *The IUC Research Center for Cell Regulation* at the University of Texas (San Antonio), Director, Dr. Barbara D. Boyan. The basic research in cell regulation includes five major areas: cell biology, immunology regulation, skeletal tissue, microbial pathogenesis, and mediators.

1.5 The Effect of Glutathione Depletion in the Glutamine Synthetase Gene Amplification System

Shue-Yuan Wang, Jane A. Reese, Robert Zivin, and Daniel R. Omstead

The R.W. Johnson Pharmaceutical Research Institute Raritan, New Jersey 08869, USA

The glutamine synthetase (GS) system is a novel gene amplification system in which vectors containing GS coding sequences are used as dominant markers amplifiable by selection with methionine sulfoximine (MSX). DNA encoding the desired protein is combined and co-amplified with the GS gene. Transfected cells are cultured in glutamine-free medium with 500 uM MSX to potentiate the amplification.

Analysis of cells cultured in the presence of MSX revealed that the selective agent reduced the growth potential of both the parental and the genetically engineered Chinese Hamster Ovary (CHO) cells. Since MSX is reported to inhibit enzyme activities of both GS and r-glutamylcysteine synthetase (GCS), the key enzyme in the glutathione synthetic pathway, glutathione depletion could be the cause of the observed loss in growth potential. As a major intracellular reductant, glutathione maintains proteins in reduced forms, facilitates the destruction of peroxides and facilitates amino acid transport.

To determine the specific effect of glutathione depletion, rescue experiments were carried out using glutathione or a glutathione monoester. The results showed that both forms protected CHO-K1 cells from the strong negative growth effect of buthionine sulfoximine (a potent specific inhibitor of the glutathione pathway). A limited effect was observed when MSX was added to the cells in the presence of glutamine. Preliminary data suggests that the glutathione depletion effect of MSX is reduced by the high concentration of glutamate in the medium, since glutamate can compete with MSX for the GCS binding site.

Bioproducts and Bioprocesses 2
Editors: Yoshida, Tanner
© Springer-Verlag Berlin Heidelberg 1993

1.6 Modeling Vapor Phase Water Droplet Extraction of Proteins from the Medium of an Air Fluidized Bioreactor

Robert D. Tanner,* Chever H. Kellogg and Prashant B. Kokitkar

Department of Chemical Engineering, Vanderbilt University, Nashville, TN 37235, USA

Contents

It has been observed that certain proteins are selectively removed from the media to the effluent gas stream of a humidified air fluidized bed bioreactor. Bakers' yeast growing on a potato substrate generated the proteins over a 24 to 36 h period.

This work explores the second of two mechanisms which appears to be essential in qualitatively describing the selective recovery and subsequent concentration of proteins from the media. The first mechanism, accounting for the precipitation and subsequent transportation of the produced proteins to the media-effluent air interface is attributed to bubble fractionation and its control by the underlying isoelectric point of the media. It was described at the Second US–Japan Biotechnology Conference in Lake Biwa, Japan [1]. Another way to describe this protein rejection (from the media) effect is to say its solubility in the water based solution is low (i.e. it is hydrophobic). An example of a hydrophobic protein is yeast invertase, which tends to favor attachment to the cellular membrane rather than dissolution in the intracellular cytoplasm (solution).

The second mechanism, discussed here, is postulated to be a partitioning step between these relatively insoluble proteins in the media phase and the micro-water droplet (in air) phase. This step is illustrated for the two proteins, invertase and α-amylase, by determining the partition coefficients for these enzymes over a range of solution concentrations at room temperature.

1 Introduction

It has been previously observed that in a semi-solid air fluidized bed bioreactor, with a potato providing the sugar substrate, and defined media providing the minerals and vitamins [2], certain proteins produced by the bakers' yeast growing on this aerated semi-solid are carried out of the bioreactor by the effluent air stream [3]. Since proteins can be separated from solution in traditional submerged fermentation processes at their isoelectric points by bubble fractionation [4], it seemed reasonable to propose that the isoelectric point may be an important variable in the air fluidized bed protein synthesis and separation process [5–7], as well. It turned out that for an initial 100 mg/L

---

* To whom all correspondence should be addressed.

Bioproducts and Bioprocesses 2
Editors: Yoshida, Tanner
© Springer-Verlag Berlin Heidelberg 1993

invertase concentration in an experimental model batch system (a shake flask), the separation between the air phase (or more precisely, micro-water droplets in the air phase) and the bulk liquid phase is not affected by changes in the initial pH concentration over the range $2 < (pH)_0 \leq 8$. This indicated that the separation between the media surface and the water droplet surface in the fluidized bed may not be sensitive to the bulk liquid pH level [8].

Since proteins have negligible vapor pressure, it was proposed [7] that the water droplets in the super-saturated humidified air in the air fluidized bed serve as the carriers for the entrained proteins (which leave the bed in the effluent air stream). Modeling the process as one controlled by an equilibrium step between the air micro-water droplets and the bulk media phase indicated that the partitioning between the phases achieved an efficiency of ca. 7–10% of the measured "equilibrium" value [8]. Direct coalescing of the micro-water droplets in the air phase (in the separatory funnel) by dry ice in acetone confirmed that invertase was concentrated in the micro-water droplets to about 50 times that of the original bulk phase [8].

The phenomena of transferring material from a bulk liquid water phase to small liquid water droplets just described seem to be similar to the mechanism by which material is transferred from the oceans to the earth's atmosphere. In an article discussing the problem of the "Red Tide" off the coast of Florida [9], Blanchard related the breakage of bubbles at the ocean surface to the transport of ocean debris to droplets emitted to the air as providing a possible mechanistic underpinning to the effect observed for proteins:

Many thousands of air bubbles are produced in the sea when whitecaps form. But what is the origin of that haze of droplets that drifts around in the air above where the bubbles are breaking? How does an air bubble, breaking at the surface of the water, produce drops?

In the laboratory, we watched a single small bubble rise to the surface of sea water. Then, so rapidly that the eye could detect nothing, the bubble simply disappeared. Simultaneously, in the air above the spot where the bubble had risen, four or five drops appeared.

We obtained a high-speed camera and with it succeeded in capturing on film what the eye had failed to distinguish. When the top of a bubble broke, the spherical air cavity in the drop began to collapse. From the bottom of the cavity, a pencil-like water jet moved rapidly upward. As it sped, the jet became unstable and shattered into four or five droplets, which continued to coast upward until their energy was expended. All this occurred in less than a few thousandths of a second. The speed at which the jet drops began their upward journey was exceedingly high, sometimes more than 110 miles per hour.

From that paper, partition coefficients for the "organic material in sea water" can be calculated from the typical sea water bulk concentration of 10 to 20 ppm and the concentration in rising droplets from the sea as high as 3500 ppm: $K = 175–350$, a concentration factor close to the 50-fold value observed for invertase [8].

2 Materials and Methods

The semi-solid air fluidized bed bioreactor used in the experiments previously alluded to is described in Ref. [2] and is compared with other bioreactors in Ref. [3]. Gel electrophoresis and the Bradford Coomassie Blue Protein determinations are described in Refs [1, 2, 4, 5]. Here the calibration curves for alpha amylase (From *B. subtilis*; Sigma I-1278) and invertase (From *S. cerevisiae*, Grade V: Practical, Sigma I-9253) concentration determinations from optical densities were developed in conjunction with the respective enzyme shake flask experiments. The shake flask used was a standard glass 600-mL separatory funnel. To minimize air dust (ca. 6 mg/600 mL in our laboratory, as reported in Ref. [8]) the air was drawn into the funnel through a cotton filter placed at the neck. Drawing the air into the flask (through the dry cotton filter) by draining an enzyme solution from the ca. 600 mL level to the 100 mL level, led to a dust content of ca. 0.5 mg/500 mL in the funnel air. The dust level was determined by measuring the protein concentration in the bulk phase following a shake flask experiment with deionized water, using the amylase calibration curve for protein concentration.

3 Results and Discussion

3.1 Mathematically Modeling Protein Partitioning in the Air Fluidized Bed

The modeling study used to simulate the protein operation in the air fluidized bed [7], is now summarized to give a sense of how the fluidized bed protein recovery process works. With the mathematical modeling study as background, we then discuss the results of our shake flask experiments in this paper.

The equilibrium constant, K, for extracellular yeast (*S. cerevisiae*) invertase with a molecular weight of ca. 270 kDa in the air-fluidized bed can be estimated from the gel electrophoresis data taken by Hong et al. [2] using the model developed previously [7]. The invertase concentrations in the process mixture in the bed and those in the overhead collector were obtained from the 1-D gel electropherogram [5], assuming that the band near 270 kDa was extracellular invertase. The actual concentration values were obtained by comparing on the gel the invertase band with those of the known α-amylase band since the amylase was added to the media along with (*S. cerevisiae*) and not generated to an appreciable extent during the fermentation process. The known concentration of α-amylase was 0.5 mg α-amylase per g solid substrate.

The mathematical model gives a reasonably good fit to the fluidized bed protein data, as seen in Fig. 1, for the invertase concentration in the bed, and in Fig. 2, for the presumed invertase concentration in the overhead collector. The value of K determined from the simulations was 62.85 g solid/liter of air, and the values of the parameters in the Luedeking–Piret equation used in the model, beta and alpha were $0.002\,\mathrm{h}^{-1}$ and 0.001, respectively, for the "best fit",

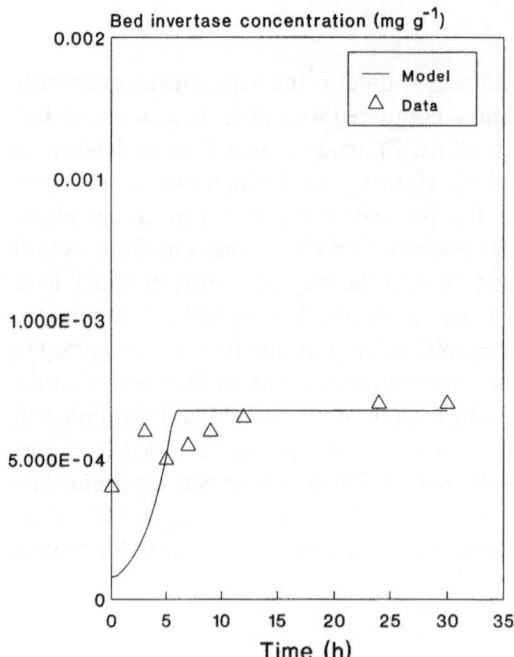

Fig. 1. Comparison of invertase concentration as a function of time in the fluidized bed, as found by experiment and calculated by the model (K_{eq} = 88.3 g L^{-1})

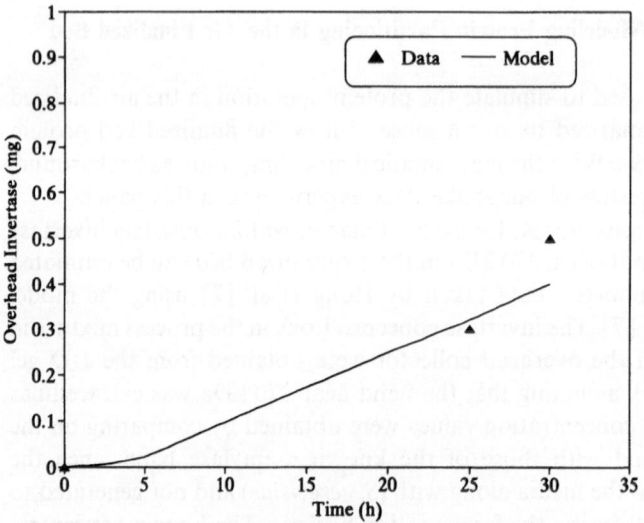

Fig. 2. Comparison of invertase concentration as a function of time in the overhead collector, as found by experiment and calculated by the model (K_{eq} = 88.3 g L^{-1})

following the procedure previously developed [7]. Here $\mathbf{K} \cdot \mathbf{E_T} = P_a/P_b$, where P_a = protein concentration in the air, mg/L air, P_b = bed protein concentration, mg/g solid substrate, and E_T is the efficiency of protein transfer (or the fraction of achieved equilibrium).

This K value was determined using an efficiency, (expressed as a per cent), (100) E_t, of 10.23%, estimated using the following postulated efficiency equation [7] for the air flow rate of 70 L min^{-1}:

$$E_T = \left[\frac{V_{max} - V}{V_{max} - V_0} \right]^2,$$

where

V = velocity of the air stream, m s^{-1},

V_0 = minimum fluidization velocity, m s^{-1},

V_{max} = elutriation (carry over) velocity, m s^{-1}.

Here, V_0 is obtained from the air flow rate of 15 L min^{-1} and V_{max} from 76.67 L min^{-1}. Since this K value of 62.85 g L^{-1} was low compared to the experimental value of 93 g L^{-1} [8], the two values were brought into closer agreement by recalibrating the constant V_0 in the above efficiency term. By assuming equilibrium at zero velocity ($V_0 = 0$), instead of at the above minimum fluidization velocity, the efficiency value was reduced to 7.28% and a new K value of 88.3 g L^{-1} was obtained (now within 5% of the experimental value). We note that an experimental value for K of ca. 90 g L^{-1} here is equivalent to the partition coefficient value of ca. 50 (mg protein/L water droplet)/(mg protein/L bulk water).

3.2 Direct Measurement of the Partition Coefficient For a Model Experimental System Using a Separatory Funnel

Examination of a typical depiction [7] of the semi-solid air fluidized bed fermentation process (Fig. 3) indicates that the bulk media phase (hatched lines) is in intimate contact with the effluent air phase (which contains the micro water droplets). This turbulent mixing in the air fluidized bed can be experimentally simulated by the inclusion of a known amount of enzyme (20–100 mg/L) in water, in place of the semi-solid bulk phase (of ca. 80% water) confined within a conventional 600 mL separatory funnel. Use of 100 mL water in vigorous contact with 500 mL of relatively dust-free air, as depicted in Fig. 4, provides the model system [8] for generating partitioned protein data. These data describe, by difference, the transfer of protein (here, either invertase or alpha-amylase) from the bulk water phase to the micro-water droplet phase.

The partition coefficient K is defined as the concentration of protein in the micro-water droplet phase to that in the bulk water phase. Based on previous work it is presumed that transferred protein lies on the surface of the water droplet [8,9]. The protein concentrations in the bulk phase are determined by the Bradford Coomassie Blue dye system. The protein concentrations are determined by a mass balance in this closed system, as enumerated in Tables 1 and 2. Prior work [8], by direct coalescence of the entrained micro-water

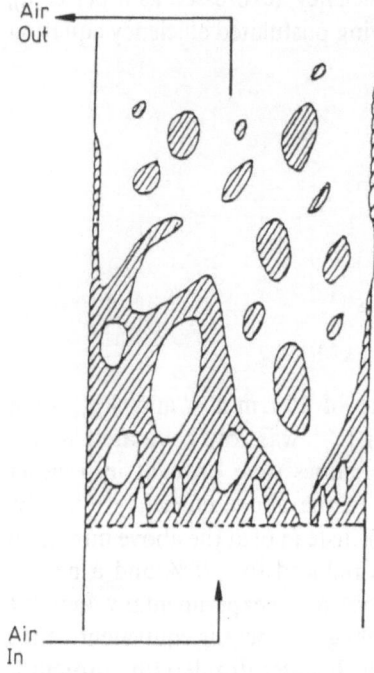

Fig. 3. Typical flow behavior of the semi-solid (potato) substrate in an air-fluidized bed fermentor. Taken from Fig. 2, Ref. [7]

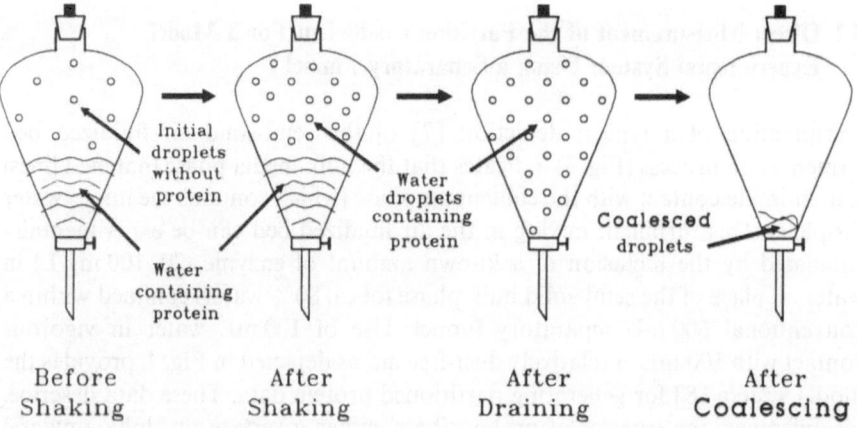

Fig. 4. Experimental simulation of the transfer of proteins from the bulk phase to the air–water droplet phase in an air-fluidized bed bioreactor. A capped separatory funnel was used for this experiment. Taken from Fig. 1, Ref. [8]

droplets, confirmed that the mass balance method (with a dust correction) is a reasonable first approximation to estimating the droplet protein concentration.

The resulting calculated partition coefficients for invertase and alpha amylase are graphed in Fig. 5. Fitting the data to a one parameter constant

Table 1. Yeast invertase concentration measurements in the bulk water phase and inferred (by mass balance) concentrations in the micro-water droplet phase. Bulk water concentration measurements taken after shaking in a 600 mL separatory funnel

Bulk phase initial protein concentration	Bulk phase protein concentration after shaking	Initial protein mass, m_0, in 100 mL liquid water bulk phase	Protein mass, m_B, in 99 mL liquid water bulk phase after shaking	Protein mass, m_D, in 1 mL (condensed) micro-water droplet phase after shaking, inferred from $m_D = m_0 - m_B$	Droplet concentration after shaking inferred from $m_D/(10^{-3})$ (L)
(mg/L)	(mg/L)	(mg)	(mg)	(mg)	(mg/L)
100	75.8 (or 7.5 mg in 99 mL water)	10	7.5	2.5	2500
80	55.6	8	5.5	2.5	2500
60	28.3	6	2.8	3.2	3200
40	15.2	4	1.5	2.5	2500
20	5.1	2	0.5	1.5	1500

Table 2. Bacterial alpha–amylase concentration measurements in the bulk water phase and inferred (by mass balance) concentrations in the micro-water droplet phase. Bulk water concentration measurements taken after shaking in a 600 mL separatory funnel

Bulk phase initial protein concentration	Bulk phase protein concentration after shaking	Initial protein mass, m_0, in 100 mL liquid water bulk phase	Protein mass, m_B, in 99 mL liquid water bulk phase after shaking	Protein mass, m_D, in 1 mL (condensed) micro-water droplet phase after shaking, inferred from $m_D = m_0 - m_B$	Droplet concentration after shaking inferred from $m_D/(10^{-3})$ (L)
(mg/L)	(mg/L)	(mg)	(mg)	(mg)	(mg/L)
100	85.9	10	8.5	1.5	1500
80	68.0	8	6.73	1.27	1270
60	40.7	6	4.03	1.97	1973
40	25.3	4	2.5	1.5	1500
20	4.65	2	0.46	1.44	1440

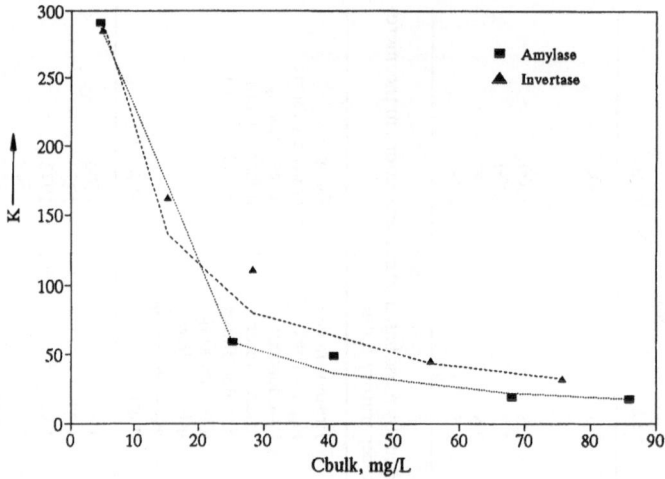

Fig. 5. Partition coefficient (relating protein concentration in the water droplet phase to that in the water bulk phase) as a function of the protein concentration in the water bulk phase

separation factor, α, model, used in the simplified binary Langmuir surface equilibrium model [10], of the form:

$$K = \frac{\alpha}{1 + (\alpha)\chi} = \frac{C_{\text{DROPLET}}}{C_{\text{BULK}}},$$

where χ = mole, x, or mass, X, fraction, indicated that the α model only fit the data approximately when χ was replaced by a concentration term. For example,

$$X = \text{mass fraction} = \frac{C/\rho_2}{1 + \dfrac{C}{\rho_2} - \dfrac{C}{\rho_1}},$$

where C = bulk mass concentration of enzyme, mg/L, ρ_1 = density of the protein (perhaps 2 or 3000 mg/L) and ρ_2 = density of water (1000 mg/L). Since C is ca. 100 mg protein/L, and

$$\left| C/\rho_2 - \frac{C}{\rho_1} \right| \ll 1,$$

it follows that:

$$K = \frac{\alpha}{1 + (\alpha)\, C/\rho_2}.$$

Qualitatively, the above relationship for K reasonably fits the data for invertase and amylase, as shown in Fig. 5. Fitting this hyperbolic equation to data on the reciprocal plot (shown in Fig. 6) gives an estimate within 40% [i.e. the fitted slope from the data for invertase is ca. 3.8×10^{-4} L/mg, as compared to the theoretical slope of $1/\rho_2$ of 10^{-3} L/mg ($\rho_2 = 1$ g/L].

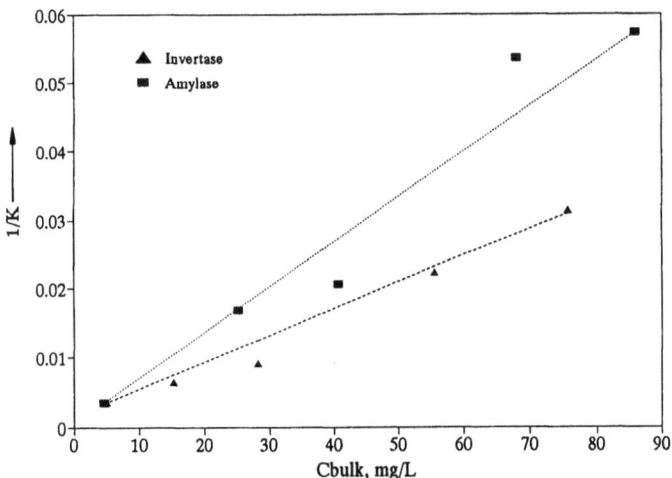

Fig. 6. Reciprocal partition coefficient (relating protein concentration in the water droplet phase to that in the water bulk phase) as a function of the protein concentration in the water bulk phase

A two parameter hyperbolic model of the more general form:

$$K = \frac{\alpha}{1 + (\beta)\,C/\rho_2}$$

leads to a better fit of the data, as given in the reciprocal plot on Fig. 6. Reasonable fits to the data (close to a least squares fit) can be made by forcing the lines on Fig. 6 through the end points to obtain the parameters from the slope, $(\beta/\alpha)\,1/\rho_2$, and the intercept, $1/\alpha$:

α alpha amylase $= 7143 = \alpha_A$

β alpha amylase $= 4751 = \beta_A$

and

α invertase $= 667 = \alpha_I$

β invertase $= 253.7 = \beta_I$

confirming that $\beta \neq \alpha$, as implied in the above ill-fitting one parameter model. Plotting

$$K = \frac{\alpha}{1 + (\beta)\,C/\rho_2},$$

with the above numerical values for α and β allows for a direct comparison with the data on the rectangular plot of Fig. 5. Physically, the inverse of the slope, $\alpha\rho_2/\beta$, represents the saturation value of C droplet as sketched in Fig. 7 (C droplet $\rightarrow \alpha\rho_2/\beta$, as C bulk $\rightarrow$ large value):

$$\left.\frac{\alpha\rho_2}{\beta}\right|_A = 1504 \text{ mg/L}$$

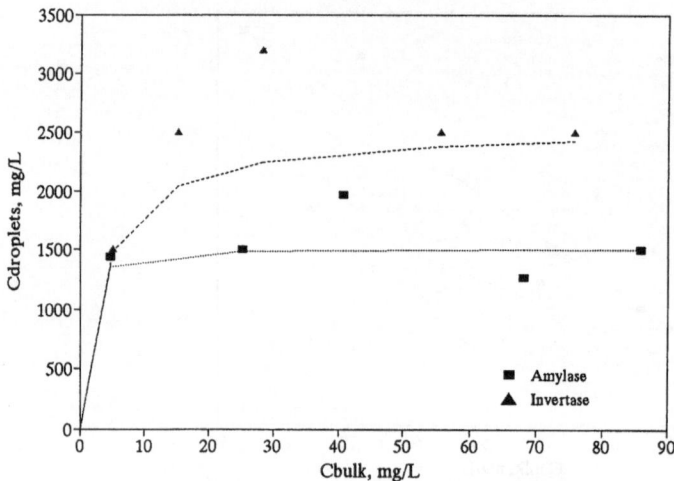

Fig. 7. Estimated (from a mass balance) protein concentration in the water droplet phase as a function of the protein concentration in the water bulk phase

and

$$\left.\frac{\alpha\rho_2}{\beta}\right|_{I} = 2628 \text{ mg/L.}$$

The parameter β is a ratio of the adsorption–desorption rate constants for the protein attaching as a monolayer to the droplet [10].

From the original relationship for K,

$$K = \frac{\alpha}{1 + \alpha X},$$

where X is the mass fraction of protein in the bulk solution, or the "simplified" relationship,

$$K \cong \frac{\alpha}{1 + (\alpha)C/\rho_2},$$

where C is the mass concentration of protein in the bulk solution, α represents the limit of K as either X or C passes to the limit of zero concentration:

$$\lim_{X \to 0} K = \alpha,$$

(or $C \to 0$).

It follows, therefore, that the partitioning of "organic material" in sea water, at very low concentrations cited by Blanchard [9] in the Introduction of this paper, of about:

$$175 \le K \le 350$$

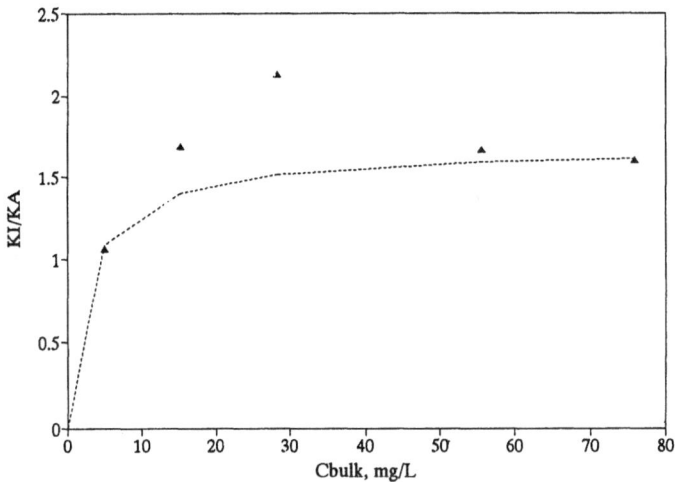

Fig. 8. Ratio of the individual yeast invertase to individual bacterial alpha-amylase water partition coefficients as a function of the respective protein concentration in the water bulk phase

is closer to the α of invertase than α-amylase. In other words, the "organic material" partitions in the droplet phase at very low concentrations (of organic material) about one half as much as invertase.

We observe in Fig. 7 and Tables 1 and 2 that the underlying mechanism seems to be that a constant amount of enzyme (protein) is transferred to the droplet phase (ca. 2.5 mg for invertase and 1.5 mg for amylase) for bulk enzyme concentrations above ca. 20 mg/L. This also helps explain how we may capture the effluent protein (in a fluidized bed process) in a large water quench tank without significant losses to air entrainment [5], since significant dilution shifts the bulk concentration to the negligible droplet concentration portion of the C_{droplet} curve in Fig. 7. Figure 8, a plot of the "water free" ratio of K_I/K_A, essentially expresses the fact that invertase is concentrated in the droplets in an amount about 50% greater than amylase for $C = C_{\text{bulk}} > 20$ mg/L, as also shown in Fig. 7.

4 Conclusions

The selective transfer of proteins from the medium of an air fluidized semi-solid fermentation process to the effluent micro-water droplet phase can be experimentally modeled by protein transfer in an agitated separatory funnel.

The two relatively insoluble enzymes alpha-amylase and invertase can be concentrated 15 to 50 times or more by transferring from a bulk liquid solution to micro-water droplets in air. For large enzyme concentrations in the bulk phase, about 50% more invertase than amylase transfers to the droplet phase.

Acknowledgements. We are most grateful to Lynda R. Phillips and Arun H.G. DeSouza for helping prepare this manuscript and the accompanying graphs.

5 References

1. Effler WT Jr, Tanner RD, Malaney GS (1989) Dynamic in-situ fractionation of extracellular proteins produced in a bakers' yeast cultivation process. In: Fiechter A, Okada H, Tanner RD (eds) Bioproducts and Bioprocesses, Springer, Berlin Heidelberg New York, p.235.
2. Hong K, Tanner RD, Crooke PS, Malaney GW (1988) Appl Biochem and Biotech 18: 3.
3. Kokitkar PB, Tanner RD (1990) Enzyme Microb Technol 12: 552.
4. DeSouza AHG, Tanner RD (1991) Appl Biochem and Biotechnol 28/29: 655.
5. Hong K, Tanner RD, Malaney GW, Danzo BJ (1989) Bioprocess Engin 4: 209.
6. Tanner RD, (1989) J Biomass Energy Society of China, 8: 73.
7. Kokitkar PB, Hong K, Tanner RD (1990) J Biotechnology, 15: 305.
8. Kokitkar PB, Tanner RD (1991) Appl Biochem and Biotechnol 28/29: 647.
9. Blanchard DC (January 1, 1972) The borderland of burning bubbles, Saturday Review p 60.
10. Adamson AW (1967) Physical chemistry of surfaces, 2nd edn, Interscience, New York, pp 570, 576.

2 Applied Genetic Engineering

2.1 Genetic Engineering and Protein Engineering on Chymosin and *Mucor* Rennin

Sueharu Horinouchi*, Jun-ichi Aikawa, and Teruhiko Beppu

Department of Agricultural Chemistry, The University of Tokyo, Bunkyo-ku, Tokyo 113, Japan

Contents

Chymosin and *Mucor pusillus* rennin are aspartic proteinases and important as milk-coagulants in the cheese industry. A system for production of chymosin in *Escherichia coli* cells and its refolding into the active form was established. A *Saccharomyces cerevisiae* system for production of *Mucor* rennin was also established. *Mucor* rennin was efficiently excreted by yeast as a heavily glycosylated form. Glycosylation affected both the secretion and the enzyme properties. By the use of the secretion-signal of *Mucor* rennin, pro-urokinase and human growth hormone (hGH) were excreted by yeast. Generation of a Lys–Arg linker (a *KEX2*-recognition sequence) between the prepro-sequence and the hGH-coding sequence led to extracellular production of mature hGH. Protein engineering on *Mucor* rennin for the purpose of its practical improvement as a milk-coagulant is also described.

1 Introduction

Chymosin (calf rennin), an enzyme obtained from the calf stomach has been used as a milk-coagulant in the cheese industry. Chymosin, a member of the aspartic proteinases, cleaves at a specific position (Phe105-Met106) of κ-casein, as a result of which, milk micelles are destabilized, leading to the clotting of milk. Chymosin is characterized by its high milk-clotting activity and very weak proteolytic activity. Recent success in X-ray crystallographic analysis has revealed its bilobal structure composed of two topologically similar domains

---

* To whom all correspondence should be sent.

Bioproducts and Bioprocesses 2
Editors: Yoshida, Tanner
© Springer-Verlag Berlin Heidelberg 1993

rich in β-structures [1]. At their junction is located the substrate-binding cleft and at the bottom two catalytic aspartyl residues, Asp32 and Asp215, are contained. Despite a wide variety of catalytic properties in many aspects, members of the aspartic proteinases possess well-conserved tertiary structure and well-conserved amino acid sequences covering the two catalytic aspartyl residues.

On the other hand, Arima *et al.* [2] succeeded in discovering a substituting enzyme from a microbial origin, a fungus *Mucor pusillus*. Their screening was motivated by a severe shortage of chymosin in 1950s. A similar enzyme was subsequently found from *Mucor miehei* by another group [3]. More than half the cheese now produced is done so with these microbial rennins called *Mucor* rennins. They also belong to the group of the aspartic proteinases and possess structures similar to that of chymosin.

These observations prompted us to produce chymosin in large amounts in microbial hosts from the practical point of view and to investigate its catalytic properties, for example, by site-directed mutagenesis. A better understanding of its catalytic features would also be useful in improving the practical properties of chymosin. The same holds true for *Mucor* rennin. In this paper, we describe systems for overproduction of chymosin in *Escherichia coli* and *Mucor* rennin in *Saccharomyces cerevisiae*. The two enzymes are produced in these hosts in such forms that they are ready as starting materials for examining the catalytic properties by site-directed mutagenesis.

2 Production of prochymosin in *E. coli* and Its Activation into Chymosin

2.1 Production and Activation of Prochymosin

Prochymosin cDNA was cloned and expressed under the control of the *trp* promoter in *E. coli* [4, 5]. Prochymosin was produced as a fusion protein with an additional peptide at its NH$_2$-terminus. Prochymosin with various amino acid sequences derived from β-galactosidase, TrpE, or Trp leader peptide were all produced in very large amounts, but as an inactive form of inclusion bodies (Fig. 1). The inclusion bodies were solubilized with 8 M urea and renatured to correctly refolded prochymosin by alkaline dialysis, which was then converted to chymosin by successive acid treatments (Fig. 2) [6]. This depends on the autocatalytic cleavage at the NH$_2$-terminal portion under acidic conditions, yielding 323 amino acid-chymosin from 365 amino acid-prochymosin [7]. Although the yields of correctly refolded prochymosins were still low (about 10%), they were purified easily by MonoQ anion exchange column chromatography for further analyses [8]. This procedure was applicable to almost all mutant chymosins generated by linker-insertion and site-directed mutagenesis.

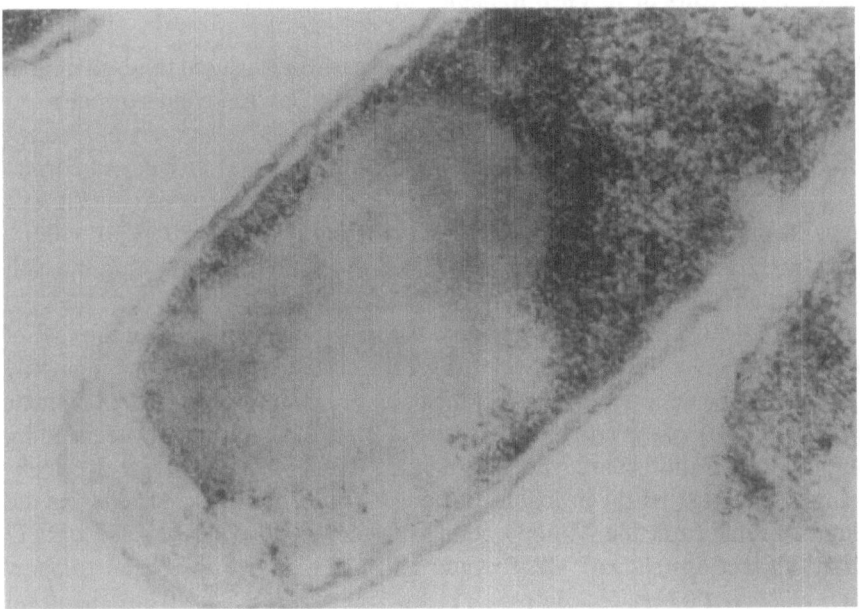

Fig. 1. Production of prochymosin as an inclusion body in *E. coli*

Preparation of prochymosin

cell pellet ← sonication
              5000g,10min
  ↓
 ppt
  ↓ ← urea(final 8M)
  ↓   40000g,1hr
 sup
  ↓ ← NaCl(final 0.5M)
  ↓ ← NaOH(pH10.5)
dialysis
 One step dialysis at pH10.5
              (Urea:8 → 0M)
          or
 Stepwise dialysis at pH10.5
              (Urea:8 → 4 → 2 → 1
                → 0.5 → 0.25 → 0M)
  ↓
dialysis at pH7.5 overnight
  ↓
sup " crude prochymosin"
  ↓
MonoQ anion exchange chromatography   **Fig. 2.** Scheme for preparation of prochymosin

2.2 Characteristics of In Vitro Refolding of Prochymosin

At the beginning we employed a one-step dialysis which caused dilution of urea from 8 to 0 M for allowing prochymosin molecules to refold correctly. As mentioned above, almost all prochymosins were refolded to some extent, except for CR601-prochymosin (Fig. 3). For improvement of the efficiency of correct refolding, we used a stepwise dialysis which gradually dilutes urea (Fig. 2). The stepwise dialysis greatly improved the efficiency of refolding of CR301, CR501 and CR712. As for the prochymosin directed by pCR601, its refolding was still very low, although improvement of it was seen (Fig. 4).

Since CR601-prochymosin was eluted in a fraction different from the case of the other prochymosins, as shown in Fig. 4, we supposed that it had a slightly different conformation from those of the other prochymosins. However, circular dichroism (CD) analysis to examine the extent of α-helix formation depending on the concentration of urea did not show any difference among these prochymosins, suggesting that the probable difference in conformation detected from the elution profile from the MonoQ column is too small to be detected by CD analysis (data not shown). CR501- and CR712-Prochymosins began to show their activity at a urea concentration below 2 M, whereas CR601-prochymosin showed its activity at a urea concentration of 0 M.

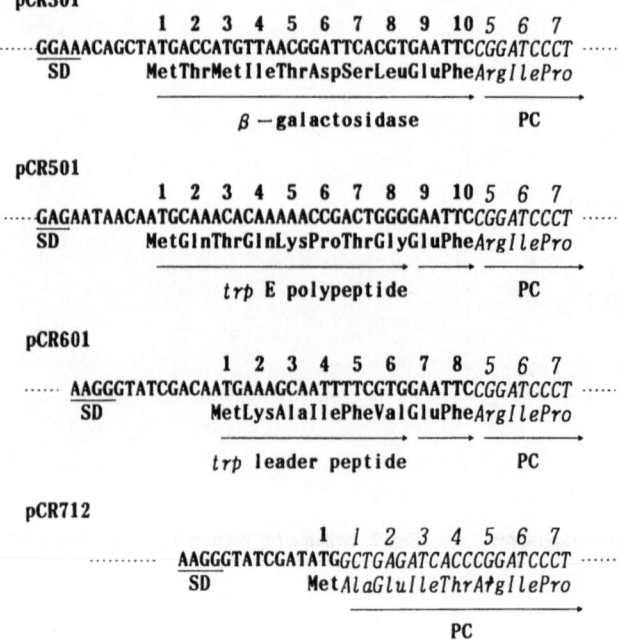

Fig. 3. NH$_2$-termini of prochymosins

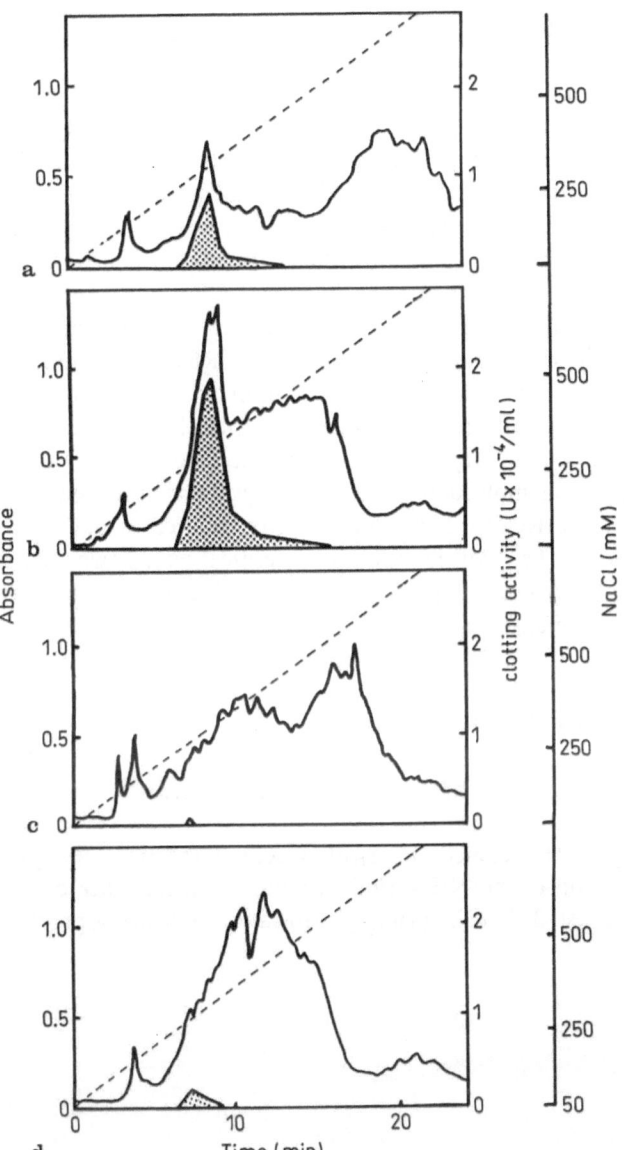

Fig. 4. Refolding of prochymosins. Profiles of elution by MonoQ column chromatography are shown. Dotted areas show the fractions containing autoactivable prochymosin molecules. **A.** CR712-prochymosin, one-step dialysis; **B.** CR712-prochymosin, stepwise dialysis; **C.** CR601-prochymosin, one-step dialysis; **D.** CR601-prochymosin, stepwise dialysis

2.3 Importance of Amino Acids at the NH$_2$-Terminal Portion for Efficient In Vitro Refolding

Our next question was what cause the less efficient refolding of CR601-prochymosin. The difference in the primary structure was the NH$_2$-terminal peptide fused to the prochymosin part (Fig. 3). We then focussed on the positively charged amino acid residue, Lys, at the second position. This Lys residue was changed into negatively charged amino acids, Asp and Glu, by site-directed mutagenesis. As expected, these two altered prochymosins were re-folded with a greater efficiency than was the unaltered CR601-prochymosin (data not shown). The stepwise dialysis was also useful in increasing the population of correctly refolded molecules. These altered CR601-prochymosins were eluted in the same fraction from the MonoQ column as those of the other prochymosins that were efficiently refolded. These data clearly indicated that only a single amino acid at the NH$_2$-terminus determined the efficiency of in vitro refolding of prochymosin molecules. Heterologous proteins produced in *E. coli* cells have been reported to form inclusion bodies in most cases, and these are very difficult to refold. Their correct refolding is critical as a downstream process in recombinant DNA technology. Our present findings give an important implication for the problem.

3 Secretion by Yeast of *Mucor* Rennin

3.1 Secretion by *S. cerevisiae* of the Mature Form of *Mucor* Rennin

Nucleotide sequencing of the cloned *M. pusillus* rennin (MPR) gene [9], together with determination of the NH$_2$-terminal amino acid sequence of the excreted MPR [10], suggested that the primary translation product, a prepro-

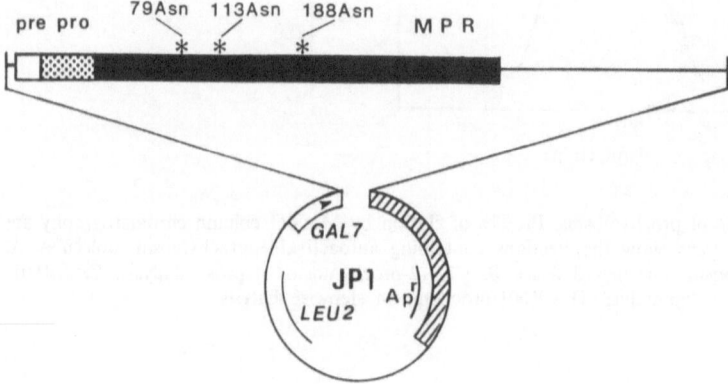

Fig. 5. Structure of a plasmid for expression of *Mucor* rennin. MPR contains three possible glycosylation sites indicated by *asterisks*

form, and the mature form were composed of 427 and 361 amino acids, respectively. When the gene was expressed under the control of the yeast *GAL7* promoter in *S. cerevisiae* (Fig. 5), the mature MPR was efficiently excreted into the medium with correct processing of its NH_2-terminal sequence [10, 11]. The secreted mature MPR amounted to more than 200 mg per liter.

During an early stage of culture, a slightly larger protein of 51 kDa was detected as a major extracellular protein, which was reactive with anti-MPR

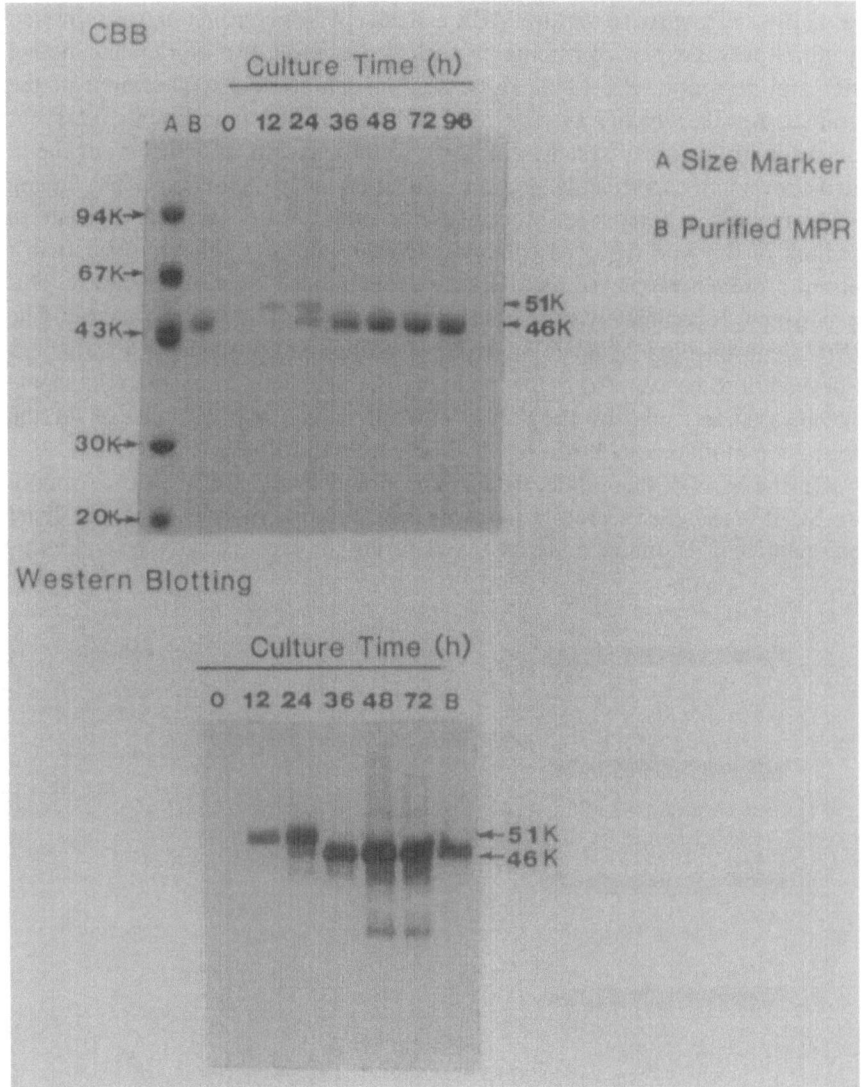

Fig. 6. Secretion of 51-kDa MPR by yeast and effect of acidic pH on the conversion of 51-kDa MPR

antibody (Fig. 6) [10]. Amino acid sequencing of the 51 kDa protein revealed that it contained the amino acid sequence from − 44 Arg to − 1 Phe at the NH$_2$-terminus of mature MPR. Incubation of the 51 kDa protein at pH 4.8 yielded mature MPR. This conversion occurred only under acidic conditions. We concluded that the 22 amino acids at the NH$_2$-terminus predicted by nucleotide sequencing were pre-sequence serving as a signal peptide to facilitate excretion of pro-MPR. This was the first observation on the secretion as a zymogen form of fungal aspartic proteinases.

The next question was how the pro-sequence is processed after being excreted in the medium. In addition to the above observations that pro-MPR was converted in vitro to mature MPR at acidic pH, its conversion was inhibited by inhibitors for aspartic proteinases, such as diazoacetyl-DL-norleucine methyl ester and pepstatin. Pro-MPR possessing an amino acid replacement at the catalytic Asp38, which was generated by site-directed mutagenesis, was not converted in this assay system. Contrary to these results obtained from the in vitro conversion experiments, however, the active-site mutant pro-MPR having no proteolytic activity was processed into the mature form in the culture medium of the wild-type yeast, but at a lower efficiency. This implied that a certain proteinase(s) played an alternative role in processing of pro-MPR. This processing was completely suppressed when a *pep4* mutant strain was used. The *PEP4* gene encodes a vacuolar aspartic proteinase, proteinase A, which is believed not to be excreted in the medium [12]. We therefore assume that some proteinase(s) activated by the action of proteinase A was responsible for the alternative in vivo conversion of pro-MPR to mature MPR.

On the basis of these data, we have proposed a model of the processing of pro-MPR in the yeast secretory pathway (Fig. 7). From prepro-MPR, the signal (pre-)peptide of 22 amino acids is cleaved during the export across the endoplas-

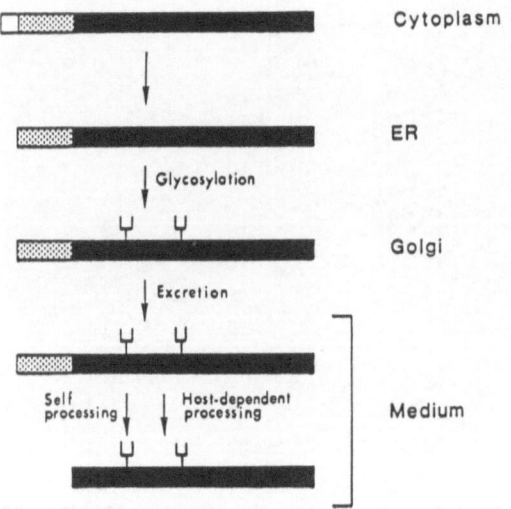

Fig. 7. Model for excretion of *Mucor* rennin from yeast

mic reticulum (ER) membrane. Two sugar chains are added, as will be described later, in the Golgi body during the transport from ER to the medium. The excreted pro-MPR is then processed mainly by autocatalytic proteolysis and also by a yeast proteinase activated by proteinase A.

3.2 Effects of Glycosylation on Secretion and Enzyme Activity of *Mucor* Rennin

Treatment of the excreted MPR with *endo-β-N*-acetylglucosaminidase H (*endo* H) reduced its size, which suggested the attachment of sugar chains. MPR contains three possible Asn-linked glycosylation sites, i.e, Asn79-Ile-Thr, Asn113-Val-Ser and Asn188-Asn-Thr. Generation of mutant MPR genes in which each of the three Asn residues was changed to Gln by site-directed mutagenesis and *endo* H-treatment of them revealed that Asn79 and Asn188 were glycosylated [13].

We observed that Gln79- and Gln188-MPR were not efficiently excreted into the medium. In addition, secretion of the double mutant Gln79/Gln188-MPR was reduced to about 1/20 that of the nonmutated MPR. Comparable amounts of proteins reactive with the anti-MPR antibody were accumulated in the cells. These data indicated that glycosylation at both the Asn residues was required for efficient secretion in the yeast secretory system.

In addition to the secretion, glycosylation of MPR also affected its enzyme activity. The commercial *M. pusillus* strain that had been improved for the productivity and enzymatic properties by mutagenesis produced a MPR with no glycosylation. Comparison of milk-clotting (*C*) activity and proteolytic (*P*) activity between the MPRs produced by yeast and the commercial *M. pusillus* strain showed that glycosylated MPR had a distinct lower *C* activity and a relatively higher *P* activity (Table 1). Removal of the sugar chains from the yeast MPR by an *endo* H-treatment improved the *C/P* ratio to a value very similar to that of the commercial MPR. When synthetic oligopeptides were used as the substrates, only small differences in K_m and K_{cat} were observed. The lengths of substrates may account for the differences in enzyme activities of glycosylated

Table 1. Clotting and proteolytic activities of MPR and its mutants

	No. of glycosylated residues	Clotting activity (U/μg)	Proteolytic activity (U/μg)	*C/P*
Yeast MPR non-mutated (− *endo* H)	2	3.04 (45.1%)	3.85 (198%)	0.790 (22.8%)
non-mutated (+ *endo* H)	0	5.61 (83.2%)	2.46 (127%)	2.28 (65.7%)
79Gln	1	4.39 (65.1%)	2.36 (122%)	1.86 (53.6%)
188Gln	1	4.08 (60.5%)	2.83 (146%)	1.44 (41.5%)
79Gln/188Gln	0	5.95 (88.3%)	2.18 (112%)	2.73 (78.7%)
Commercial MPR	0	6.74 (100%)	1.94 (100%)	3.47 (100%)

and nonglycosylated MPRs. Asn79, one of the glycosylated residues, which is located in the flexible flap region may perturb the flap structure, leading to the differences in enzyme activity. Asn188 locating at the junction of the NH_2- and COOH- terminal domain may influence the global tertiary structure of the enzyme, resulting in perturbation of enzyme activities.

3.3 Application of the Secretion-Leader of *Mucor* Rennin for Production of Human Growth Hormone

We used the secretion-leader of MPR for production of heterologous proteins in yeast. The galactose-inducible *GAL7* promoter was used, as shown in Fig. 5. Our first target was human growth hormone (hGH) of 191 amino acids [14]. Among several plasmids in which the hGH-coding sequence was connected at various positions downstream of the NH_2-terminal region of MPR, a plasmid with the structure containing the whole pre-sequence and part of the pro-sequence ($- 66$ Met to $- 40$ Lys) upstream of the hGH sequence directed the greatest secretion of the protein of 23 kDa in size, which contained the expected NH_2-terminal amino acids of additional eight amino acids derived from the pro-part of MPR. These data indicated the importance of part of the pro-region for efficient secretion of hGH. Further DNA manipulation including the addition of the *GAL10* transcriptional terminator downstream of the hGH gene improved the yield of secreted hGH up to 10 mg per liter of culture broth. Based on these findings, we chose human pro-urokinase as the second target [15]. Despite many attempt, however, efficient secretion was not observed in this case.

We also tried to produce hGH in its mature form [16]. For this purpose, an artificial Lys–Arg linker was inserted at the junction between the prepro-sequence of MPR and the hGH-coding sequence (Fig. 8). The Lys–Arg sequence is recognized by the *KEX2* protease, a Ca^{2+}-dependent serine protease, and is responsible for cleaving the precursor of yeast α-factor [17]. As shown in Fig. 8,

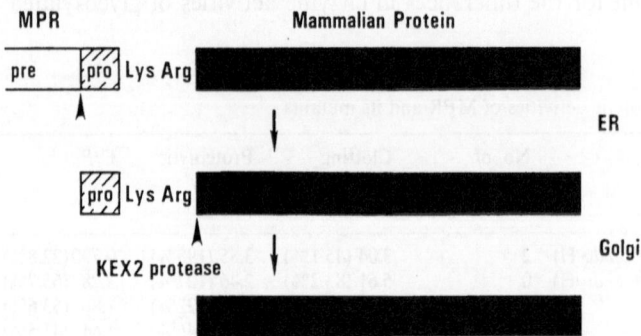

Fig. 8. Schematic representation of the structure of a plasmid directing the synthesis of mature human growth hormone

the hGH directed by the plasmid with this structure had the same NH_2-terminus as the native hGH, as determined by direct amino acid sequencing. The mature hGH recovered from the culture medium was indistinguishable from native hGH in biological activities.

4 Protein Engineering on Chymosin and *Mucor* Rennin

Chymosin and MPR are members of the aspartic proteinases. As mentioned above, all the aspartic proteinases possess a very similar bilobal structure, despite a deviation in the amino acid sequences. Insofar as we have established the systems for production of both the enzymes, they are good targets to investigate the structure–function relationships of aspartic proteinases and to improve the enzyme properties from the practical point of view. We have so far conducted protein engineering by changing several amino acids of both chymosin and MPR [8, 18]. In this section, however, we will focus on a few residues of MPR that are associated with heat-stability and in the C/P ratio (milk-clotting vs proteolytic activities). These two properties are important from the practical point of view. Low heat-stability and high C/P ratio are desirable properties in the milk-coagulant in cheese-making so as to avoid undesirable further reactions.

Residue Tyr75 of chymosin was found to be involved in not only substrate binding but also the catalytic function of the enzyme [8]. Mutants at this position Tyr75 Phe caused a distinct increase in the C/P ratio. On the basis of this observation, we changed Tyr75 of MPR to Phe and Asn and examined the properties of the mutant enzymes. Before the examination, the purified enzymes were treated with endo H to remove the sugar chains; this is a drawback in using the yeast-secretion system. As shown in Table 2, these mutations resulted in a decrease in milk-clotting activity, but their C/P ratios were much improved. In addition, the Tyr75 Asn mutation brought forth a dramatic decrease in heat-stability (Fig. 9). Thus, only a single mutation led to alterations of heat-stability and the C/P ratio. These findings imply that some other change(s) will greatly improve the enzyme properties of MPR and chymosin as milk-coagulants.

Table 2. Milk-clotting activity of yeast MPRs

Subsites	Enzymes	Activities
	Non-mutated	6.12
S_4	N219S	0.312
S_3	E12T	0.970
S_3, S_2	T218S	5.55
$S_2(,S_4)$	F220E	6.27
	F220L	5.25
S_2	1222T	6.73
S_1	Y75F	1.95
	Y75N	0.757

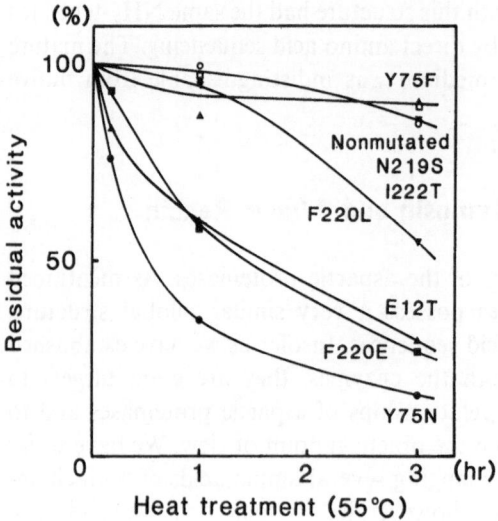

Fig. 9. Heat-stability of yeast MPRs. After heat-treatment at 55°C for indicated periods, the residual milk-clotting activity was measured with skim-milk as the substrate

The amino acids participating in forming the subsites are expected to be promising targets for improving the enzyme properties, because they are associated with substrate-recognition and catalytic properties. Table 2 and Fig. 9 also summarize the data on the mutations of the subsites. Among these, Phe220 → Glu retained its milk-clotting activity as high as the nonmutated enzyme but its heat-stability was greatly reduced. This mutant is a practically improved enzyme.

Acknowledgements. This work was supported in part by a Grant-in-Aid for Scientific Research on Priority Areas from the Ministry of Education, Science and Culture of Japan.

5 References

1. Kay J (1985) In: Kostka V (ed) Aspartic proteinases and their inhibitors. Walter de Gruyter, Berlin.
2. Arima K, Iwasaki S, Tamura G (1967) Agric Biol Chem 31: 540.
3. Ottensen M, Richert W (1970) CR Trav Lab Carlsberg 37: 301.
4. Nishimori K, Kawaguchi Y, Hidaka M, Uozumi T, Beppu T (1981) J Biochem 90: 901.
5. Kawaguchi Y, Kosugi S, Sasaki K, Uozumi T, Beppu T (1987) Agric Biol Chem 51: 1871.
6. Kawaguchi Y, Shimizu N, Nishimori K, Uozumi T, Beppu T (1984) J Biotechnol 1: 307.
7. Foltmann B (1966) CR Trav Lab Carlsberg 35: 14.
8. Suzuki J, Sasaki K, Sasao Y, Hamu A, Kawasaki H, Nishiyama M, Horinouchi S, Beppu T (1989) Protein Eng 2: 563.
9. Tonouchi N, Shoun H, Uozumi T, Beppu T (1989) Nucleic Acid Res 14: 7557.
10. Hiramatsu R, Aikawa J, Horinouchi S, Beppu T (1989) J Biol Chem 264: 16862.
11. Yamashita T, Tonouchi N, Uozumi T, Beppu T (1987) Mol Gen Genet 210: 462.
12. Hemmings BA, Zubenko GS, Hasilik A, Jones EW (1981) Proc Natl Acad Sci USA 78: 45.
13. Aikawa J, Yamashita T, Nishiyama M, Horinouchi S, Beppu T (1990) J Biol Chem 265: 13955.

14. Hiramatsu R, Yamashita T, Aikawa J, Horinouchi S, Beppu T (1990) Appl Environ Microbiol 56: 2125.
15. Hiramatsu R, Horinouchi S, Beppu T (1991) Gene 99: 235.
16. Hiramatsu R, Horinouchi S, Uchida E, Hayakawa T, Beppu T (1991) Appl Environ Microbiol 57: 2052.
17. Fuller RS, Brake A, Thorner J (1989) Proc Natl Acad Sci USA 86: 1434.
18. Suzuki J, Hamu A, Nishiyama M, Horinouchi S, Beppu T (1990) Protein Eng 4: 69.

2.2 Genetic Engineering of Carbon and Energy Metabolism

James E. Bailey*

Department of Chemical Engineering, California Institute of Technology, Pasadena, CA 91125, USA

Contents

Efforts to manipulate terminal synthesis pathways such as those leading to amino acids and secondary metabolites are ultimately bounded by the metabolic environment of these pathways. Genetic engineering to improve the bioprocessing characteristics of carbon conversion and energy metabolism can provide a foundation for further substantial improvements in process performance. The feasibility of effecting major improvements in central cell function under conditions of processing interest is illustrated through two examples involving cloned hemoglobin from the bacterium *Vitreoscilla*. *Escherichia coli* containing and expressing the gene for this hemoglobin synthesizes protein more efficiently under oxygen-limited conditions and attains substantially higher cell densities, while *Streptomyces coelicolor* expressing hemoglobin produces substantially more actinorhodin antibiotic under moderately aerated conditions than do control strains. Ethanol production by nongrowing yeast has been improved both by entrapment of *Saccharomyces cerevisiae* in calcium alginate and by introduction into this yeast of cloned phosphofructokinase. Sensitivity analysis of the flux to ethanol indicates that several steps should be manipulated simultaneously in order to obtain substantial increases in flux. Extension of this type of analysis to growing cells is complicated by couplings between the pathway of interest and the complex metabolic context within which this pathway operates. An approach to this problem based upon stimulus–response identification of a model for the pathway environment has been demonstrated in computational studies.

1 Introduction

Most previous research on genetic improvement of metabolic activities in industrial organisms has focused on improvement or modification or biosynthesis pathways. This strategy is reasonable because enhancement of key steps and alteration of feedback controls can increase the intracellular allocation of required intermediates and cofactors into production of the desired compound. Ultimately, however, any such manipulation is limited by the upstream kinetic and regulatory network which controls the concentration and the fluxes to biosynthetic intermediates, ATP, and redox cofactors. Although manipulation and improvement of the central pathways of carbon and energy metabolism is

---

* Present address: Institute of Biotechnology, ETH, CH-8093 Zürich, Switzerland.

Bioproducts and Bioprocesses 2
Editors: Yoshida, Tanner
© Springer-Verlag Berlin Heidelberg 1993

complicated by the extensive regulation which pervades these networks and by their interconnection with all other cellular activities, genetic restructuring of carbon and energy metabolism is an important arena for future research. Efforts in this area complement synthesis pathway engineering activities. Furthermore, successful improvements in carbon and energy metabolism may be beneficial over a broad spectrum of organisms and products.

2 Manipulation of Energy Metabolism via Cloned Bacterial Hemoglobin

The bacterium *Vitreoscilla*, an obligate gram-negative aerobe which lives in oxygen-poor environments, synthesizes large quantities of a novel hemoglobin when grown under oxygen-limited conditions. Based on the hypothesis that this might represent a novel genetic adaptation in nature to oxygen limitation and aware of the pervasive problem of oxygen supply in bioprocessing, Chaitan Khosla cloned the gene for this hemoglobin and studied the influences of its expression in *Escherichia coli* [1–3]. Several of these studies employ a construct in which a single copy of the *Vitreoscilla* hemoglobin gene and its promoter are integrated into a known locus in the *E. coli* chromosome (either in the *lac* or the *xyl* operon) and, in addition, a multicopy plasmid was inserted into the cell carrying a model cloned protein. Fed-batch fermentations conducted with these strains, compared to those conducted with wild-type *E. coli* containing the same plasmid, showed that the presence of hemoglobin enabled the recombinant strains to synthesize protein more rapidly, to accumulate substantially greater quantities of cloned protein product, and to achieve higher cell densities than wild-type plasmid-bearing strains under identical oxygen-limited growth conditions. Based upon measurements of medium glucose, acetate, and ammonium concentrations, it was argued that cloned hemoglobin improved the net ATP synthesis efficiency of *E. coli* under oxygen-limited conditions. These studies indicate that introduction of a heterologous active protein into this production organism for cloned proteins can substantially improve its process performance.

The hypothesis that a genetic improvement in energy metabolism might be broadly applicable is supported by subsequent studies involving antibiotic production by *Streptomyces* [4]. In these experiments, the hemoglobin gene was incorporated into an *E. coli–Streptomyces* shuttle vector and introduced into *Streptomyces coelicolor* which produces the antibiotic actinorhodin. Since introduction of plasmids into *Streptomyces* can alone alter secondary metabolism, another *Streptomyces coelicolor* construct bearing a plasmid identical to the first except for a deletion in the hemoglobin gene was employed as a control. In batch cultivations with moderate aeration, the hemoglobin-expressing strain accumulated 67 mg/L actinorhodin compared to only 7 mg/L actinorhodin for the control. Significantly, the cell density attained in both batch fermentations was identical. Therefore, the presence of hemoglobin affected the specific productivity of the cells, greatly enhancing secondary metabolite production. Based upon these and other investigations with this cloned hemoglobin, prospects for broad impact in aerobic biotechnology are excellent.

3 Analysis and Manipulation of Carbon Metabolism

Another approach to the design of beneficial genetic modifications is sensitivity analysis of the key processes involved. This approach has been developed and demonstrated in studies of the conversion of glucose to ethanol and other products by nongrowing yeast cells. Jorge Galazzo developed a detailed mathematical model for the pathways involved in this process, validating the in vivo rate expressions through a comprehensive combination of extracellular and intracellular measurements, the latter accomplished by NMR spectroscopy [5, 6]. These studies showed several important results in general terms. First, placing identical cells in a different environment can substantially shift the sensitivity of pathway flux to the activities of individual steps in the pathway. Such shifts in sensitivity of the glucose to ethanol bioconversion were observed accompanying changes in the reactor pH or changes in the yeast environment from liquid suspension to entrapment in alginate. Second, changes in the host-cell genotype, not obviously related to the pathway of interest, can again substantially change the operation and sensitivities of the production pathway. A mutated yeast used as a plasmid host, investigated without plasmid, exhibited greatly different pathway sensitivities than a wild-type yeast. In particular, the mutant yeast was much more sensitive to the activity of phosphofructokinase according to the analysis, and this result was confirmed experimentally by observing a substantial increase in ethanol production flux following expression of cloned phosphofructokinase in those cells.

4 Metabolic Sensitivity Analysis in Growing Cells

The modeling and sensitivity analysis of the yeast system discussed previously was greatly simplified by the nongrowing state of the cells during these experiments. However, in most situations of technological interest, cells grow simultaneously with operation of the desired pathway. In such situations and others, the pathway of interest operates embedded within a much larger and much more complex metabolic and transport network. Paul Schlosser has shown that, in such a situation, arriving at correct flux sensitivities for the desired pathway requires consideration of coupling between that pathway and the rest of the cell environment.

In order to address the objective of estimating sensitivities for pathways in growing cells, a new modeling structure has been devised in which the detailed pathway is embedded within an empirically identified metabolic context called the "black frame" [7]. A practically important attribute of this black frame concept is the possibility in principle of identifying the black frame model through a suitable series of stimulus–response experiments. A computational study illustrating this procedure and its results applied to the aromatic amino acid synthesis pathways in bacteria demonstrates the potential utility of the black frame approach.

5 Discussion

Significant opportunities exist for increasing cell productivity by genetic manipulation of carbon and energy metabolism. Much more needs to be learned about regulation and mechanism of both processes, and future efforts at manipulating carbon and energy metabolism will likely include not only genetic manipulation of certain enzyme activities but also manipulation of the regulation of expression and control of these activities. The successes demonstrated to date in improving aerobic process productivity using cloned hemoglobin should motivate greater efforts to identify novel carbon and energy metabolic motifs in nature and to evaluate their effects when introduced into commercially important organisms.

Many powerful tools for systematic improvement of metabolism are available. These include many methods and materials for gene cloning and manipulation, most notably improved recently through the introduction of PCR techniques. Nevertheless the available genetic systems for stable cloning and expression of cloned genes in many industrial organisms are still inadequate. Measurement methodologies such as NMR spectroscopy can be usefully applied, but more powerful, versatile, and sensitive approaches are needed. A core of mathematical tools exists for metabolic analysis, but their applicability to limited experimental data is difficult. Better mathematical strategies to utilize available data and to design and apply key experiments are also important.

Acknowledgements. Research on metabolic engineering in the author's library is supported by the National Science Foundation and by the Catalysis and Biocatalysis Program of the Advanced Industrial Concepts Division of the US Department of Energy.

6 References

1. Khosla C, Bailey JE (1988) Nature 331: 633.
2. Khosla C, Curtis J, DeModena J, Rinas U, Bailey JE (1990) Bio/Technology 8: 849.
3. Bailey JE, Birnbaum S, Galazzo J, Khosla C, Shanks JV (1990) Ann NY Acad Sci 589: 1.
4. Magnolo S, Leenutaphong D, DeModena J, Curtis J, Bailey J, Galazzo J, Hughes D (1991) Bio/Technology 9: 473.
5. Galazzo JL, Bailey JE (1990) Enzyme and Microb Technol 12: 162.
6. Galazzo JL, Bailey JE (1990) Biotechnol Bioeng 36: 417.
7. Schlosser PM, Bailey JE (1990) Math Biosci 100: 87.

2.3 Microbial Production of Useful Compounds with recDNA Technique

Akira Kimura

Kyoto University, Research Institute for Food Science, Uji, Kyoto 611, Japan

Contents

Recombinant DNA techniques have made a stunning and marvelous revolution in the field of microbiology. The techniques are quite helpful to analyze and strengthen the various functions of cells. Using these techniques, we have succeeded in breeding of various useful microorganisms such as a jumbo yeast, metal resistant yeasts, and other ones whose metabolic pathways have been modified. Some of these organisms could produce a large amount of ATP, glutathione, *S*-lactoylglutathione, CoA, tuna growth hormone, etc.

In this article, the author mentions the microbial production of tuna growth hormone, the analysis and modification of glycolytic methylglyoxal pathway, and production of glutathione and *S*-lactoylglutathione etc.

1 Brief Outline

As the result of the development of recDNA techniques, it has become possible to manipulate genes and breed useful microorganisms. Among various transformants bred in our laboratory to produce large amounts of useful compounds 2 topics will be described in this article. One is the breeding of a transformant of *E. coli* which accumulates tuna growth hormone. The other is the breeding of a jumbo yeast and production of *S*-lactoylglutathione.

Through the latter study we have proposed and proved the glycolytic methylglyoxal pathway. Its physiological role is still under study, but it seems to have a very important relationship to cell growth and cell division.

2 Microbial Production of Fish (Tuna) Growth Hormone

Interest in eating fish and fishery products is increasing, since Dyerberg et al. [1] suggested that a dietary intake of polyunsaturated fatty acids (especially eicosa-pentanoic acid) from fish may have some effect on morbidity due to coronary atherosclerosis.

Being surrounded by the sea, the Japanese have eaten large quantities of fish and fish products. In fact, about 60% of the intake of protein of the Japanese comes from seafood. However, the supply of marine products has been steadily declining, partly due to a decrease of natural resources caused by overfishing in the past, partly due to the establishment of 200 mile zone, and the prohibition of catching whales and some fish.

Under these circumstances, securing a sufficient quantity of marine products has become one of the most urgent and important problems not only in Japan, but also in the world. In order to solve' this problem, aquaculture and/or mariculture are proposed. However, there are several problems to be solved, for example, it takes too long for some kinds of fish to grow to commercial size. To make fish breeding feasible, the use of fish growth hormone (GH) has been proposed. Among the many fish consumed in Japan, tuna is one of the most popular and expensive fish, and moreover it is one of the most evolutionarily advanced ones. Since GHs of phylogenically higher organisms are generally effective on the growth of evolutionarily lower ones, we decided to clone the gene for growth hormone of tuna (*Thunnus thynnus*), and produce it in a microbial system.

The detailed review on growth hormone of fish, especially of tuna, was already published [2].

2.1 Fish Growth Hormones and Comparison with Mammalian Growth Hormones

Growth hormones (GHs) are polypeptides that are secreted from the pituitary glands of all vertebrate animals and involved in the regulation of postnatal somatic growth and maintenance of the metabolism of proteins, lipids, carbohy-drates, and minerals.

Many GHs from bony fish such as tilapia, carp, salmon, eel, yellowtail, flounder and bonito were isolated and purified. Comparisons of the physico-chemical properties and the primary structures of fish GHs with those of mammalian GHs [3] show that the overall molecular features are very similar (Fig. 1), although they are phylogenically distant. This means that the structure of GH has been conserved through evolution from fish to human beings [6]. The GHs are polypeptides with 2 S–S bonds. Their molecular weights are approximately 21 000–22 000. A genetically engineered porcine GH [7] and human GH [8, 9] have been crystallized and subjected to X-ray analysis. The former structure was predominantly helical and consisted mainly of four antiparallel alpha-helices.

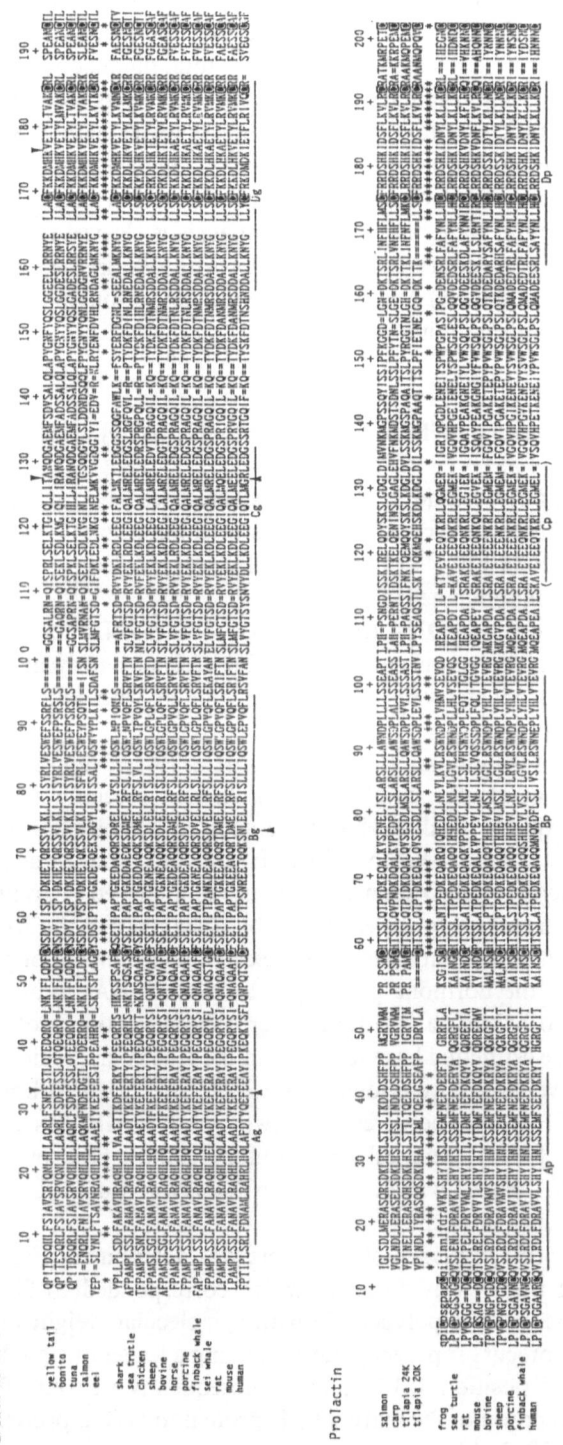

Fig. 1. Comparison of the primary structure of growth hormone and prolactin

We have isolated tuna GH (tGH), and later cloned its cDNA. The entire DNA sequence of this gene was determined and expressed in *Escherichia coli*, and also in the yeast, *Saccharomyces cerevisiae*. A large amount of tGH was produced by the recombinant strain of *E. coli*. The total number of amino acids was 187. The DNA complementary to mRNA of tGH was isolated from a cDNA library of the tuna pituitary gland with the use of synthetic DNA probes corresponding to the partial amino acid sequences. Sequence analysis of the cDNA revealed that the cDNA codes for a polypeptide of 204 amino acids [10]. A putative signal peptide consisting of 17 amino acids is cleaved from the pre-tGH of 204 residues, to produce the 187 residues with Gln in the *N*-terminus. The recombinant-tGH (r-tGH) was found effective for various phylogenically lower fish such as snappers, rainbow trout, etc.

2.2 Cloning and Sequencing of Tuna Growth Hormone (tGH) cDNA

The recombinant DNA technique is the most effective one for large scale preparation. The cDNAs for growth hormones of chum salmon [14], eel [5], yellowtail [15] and tuna [10, 11] have been cloned, and some of them were highly expressed in microbial cells [14, 16].

In order to clone tGH cDNA (tGHcDNA), total mRNA extracted from tuna pituitary glands was translated in a rabbit reticulocyte lysate and the labeled translation products were immunoprecipitated with affinity-purified tGH antibody.

An examination of the immunoprecipitate by SDS/PAGE showed that only one peptide with a molecular weight of 23 kDa was precipitable with the antibody, although the purified tGH peptide showed a molecular weight of 20–21 kDa. In order to clone the cDNA for tGH, a cDNA library was prepared and screened by colony hybridization with the use of a labeled oligonucleotide probe (No. 1) complementary to the estimated base sequence in the *N*-terminal region of tGH. About 40 recombinants were selected which had a longer insert sufficient to code for the hormone peptide of 23 kDa. cDNAs in all these 40 clones were shown to hybridize also to another oligonucleotide probe (No. 2) complementary to the estimated base sequence in the *C*-terminal region of tGH. Two kinds of positive clones were identified, which hybridized to both oligo-nucleotide probes [10]. The occurrence of two types of GH has also been reported in salmon [4, 14] and eel [5] pituitary glands. One of the most abundant tGH cDNA clones was designated pTTS339 and used for sequence analysis and determination of the restriction map.

The entire nucleotide sequence of the insert was determined (Fig. 2). One ORF (open reading frame) was detected, which corresponded to 204 amino acids. This ORF coded for a polypeptide with a molecular weight of 23 093, which represents the precursor polypeptide containing a leader or signal peptide sequence (amino acid position 1–17). This is subsequently cleaved off during processing and/or secretion of the mature tGH protein to yield a protein with a molecular weight of 20–21 kDa.

```
                     1                                                   50
5'------CTGCAGGGGGGGGGGAGATCGCACTGAAGAACTGAGCTCAGATCAGATTAACCAGAACCAGAACTGAACCCAGACCAGCC

                                                       100
ATG GAC AGA GTC TTT CTC TTG CTG TCA GTC CTG TCT CTG GGT GTC TCC TCT CAG CCA ATC ACA GAC AGC CAG CGT
Met Asp Arg Val Phe Leu Leu Leu Ser Val Leu Ser Leu Gly Val Ser Ser Gln Pro Ile Thr Asp Ser Gln Arg
                            -10                                 -1  1

            150                                                                 200
CTG TTC TCC ATC GCT GTC AGC AGA GTG CAA CAC CTC CAC CTG CTC GCC CAG AGA CTC TTC TCT GAC TTT GAG AGC
Leu Phe Ser Ile Ala Val Ser Arg Val Gln His Leu His Leu Leu Ala Gln Arg Leu Phe Ser Asp Phe Glu Ser
    10                                      20                                  30

                                            250                 *** *** *** *** ***
TCT CTG CAG ACA GAG GAG CAG CGT CAG CTC AAC AAA ATC TTC CTG CAG GAT TTC TGC AAC TCT GAT TAC ATC ATC
Ser Leu Gln Thr Glu Glu Gln Arg Gln Leu Asn Lys Ile Phe Leu Gln Asp Phe Cys Asn Ser Asp Tyr Ile Ile
                        40                                      50

            300                                                             350
AGC CCG ATC GAC AAG CAT GAG ACA CAA CGC AGC TCT GTT CTG AAG CTG CTG TCG ATC TCC TAT CGA TTG GTG GAA
Ser Pro Ile Asp Lys His Glu Thr Gln Arg Ser Ser Val Leu Lys Leu Leu Ser Ile Ser Tyr Arg Leu Val Glu
    60                                      70                                      80

                                            400
TCG TGG GAG TTC CCC AGC CGT TCT CTG TCC GGA GGT TCT GCT CCA AGG AAC CAA ATC TCA CCA AAA CTG TCT GAA
Ser Trp Glu Phe Pro Ser Arg Ser Leu Ser Gly Gly Ser Ala Pro Arg Asn Gln Ile Ser Pro Lys Leu Ser Glu
                        90                                      100

            450                                                                 500
CTG AAG ACA GGA ATC CAC CTG CTG ATC AGG GCC AAT CAG GAT GGA GCA GAG ATG TTC GCT GAC AGC TCT GCC CTC
Leu Lys Thr Gly Ile His Leu Leu Ile Arg Ala Asn Gln Asp Gly Ala Glu Met Phe. Ala Asp Ser Ser Ala Leu
    110                                     120                                         130

                                            550
CAG CTC GCT CCG TAT GGA AAC TAT TAT CAA AGT CTG GGA GCT GAC GAG TCA CTG AGA CGG AGC TAC GAG CTG CTT
Gln Leu Ala Pro Tyr Gly Asn Tyr Tyr Gln Ser Leu Gly Ala Asp Glu Ser Leu Arg Arg Ser Tyr Glu Leu Leu
                        140                                     150

            600 *** *** *** *** ***                                             650
GCC TGC TTC AAG AAG GAC ATG CAC AAG GTG GAG ACC TAC CTG ACG GTG GCT AAA TGT CGA CTC TCT CCA GAA GCT
Ala Cys Phe Lys Lys Asp Met His Lys Val Glu Thr Tyr Leu Thr Val Ala Lys Cys Arg Leu Ser Pro Glu Ala
    160                                     170                                         180

                                    700                                                 750
AAC TGC ACC CTG TAG CCCCGCCTCTCTGATGACGTCATCCTGTGTGTTCTGGAGCCCCGCCTCTCTGATGACGTAATCATGTGTGTTCTGTAGC
Asn Cys Thr Leu
                                    800                                             850
CCCGCCTCCATGTTCTCTTTGCTGGTTAGCATTAGCCTGTGATGGTTTTCTGATGTCATCATCAGATAAATAGTACAAGCTATGAACAGGAAGTGATGT
                                    900
CAGACTGTCGGCCTTTTCTCAGCATGTGA[AATAAA]CTGCGCTGAGTTGCATTCAAAAAAAAAAAAAAAAAAA----------3'
```

Fig. 2. Nucleotide sequence of tuna GH cDNA and amino acid sequence of precursor tuna GH. Precursor tuna GH codon sequence including leader peptide (amino acids: −17 to −1) and mature tuna GH peptide (amino acids: 1–187) starts at nucleotide position 66 and ends at 677. The sequences corresponding to those of probes 1 and 2 are indicated by asterisks (nucleotide position 264–278 for probe 1 and 603–617 for probe 2). Polyadenylation signal (nucleotide position 888–893) is boxed. Potential *N*-glycosylation site is marked by dashed line and three direct repeats are underlined

The analysis of amino acid composition of the tGH polypeptide indicated that the mature protein starts from glutamine residue at position 18 (Gln1).

The analyses of GHs, including tGH, revealed some unique structural features (Fig. 1). The GHs contain four cysteines and one tryptophan residue at nearly identical positions. The four cysteine residues may be essential for the stabilization of the peptide in a biologically active form through disulfide linkages.

In analogy with other GHs hitherto purified, tGH contained 4 cysteines (Cys52, Cys160, Cys177, Cys185) and 1 tryptophan (Trp85) residue.

As has been found for eukaryotic mRNAs [12, 13], codon usage in tGH was nonrandom, there being a tendency towards guanine and cytosine bases appearing preferentially in the third position of the codons.

The amino acid sequence of tGH was compared with those of other GHs. The tGH showed 67% amino acid and 66% nucleotide sequence homologies with salmon growth hormone, and 90% amino acid and 88% nucleotide sequence homologies with yellowtail growth hormone. The homologies of tGH with human growth hormone (hGH) were 32% and 46% for amino acid and nucleotide sequences, respectively. The analyses and comparison of GHs obtained from various sources will be useful for a better understanding of the interrelationships between structure, function and evolution of fish and mammalian hormones.

2.3 Construction of Plasmids and Their Expression in *E. coli*

We have cloned and sequenced tGHcDNA [10]. In order to express tGHcDNA in *E. coli* cells, plasmids were constructed in which the expression of the cDNA was under the control of the *trc* promoter [17]. Two kinds of plasmids were prepared, designated pTES8, and pTES8S, respectively; the former had a 5'-leader sequence. Since this leader sequence may repress expression [18, 19], we constructed the latter (pTES8S) which had no leader peptide region. The nucleotide sequences between the Shine–Dalgarno sequence (AGGA) and the initiation codon (ATG) in plasmid pTES8 and pTES8S were varied. The amino acid sequence in the amino terminal region was found to be arranged in the order of Met Ala Gln Pro etc. The *E. coli* cells of strain N1790 carrying pTES8 and pTES8S produced the new bands at the positions corresponding to precursor (molecular weight 23 000) and mature (molecular weight 21 000) tGHs, respectively. The protein precipitated from N1790 with pTES8 had a similar molecular weight (23 000) to that of precursor tGH, whereas the protein precipitated from N1790 with pTES8S showed a band of molecular weight 21 000, corresponding to mature tGH. These results showed that tGH cDNA was expressed in *E. coli* cells. The expression of our plasmids carrying tGH cDNA was slight, and only the Maxicell system allowed us to detect proteins specified by the plasmids.

2.4 Overproduction of Recombinant Tuna Growth Hormone

The poor expression of plasmids pTES8 and pTES8S in *E. coli* cells [20] suggested that sequence rearrangements of the functional regions were necessary to attain efficient expression of tGH cDNA in *E. coli* cells. Besides the distance between the SD sequence and the initiation codon (ATG) has been reported to affect the efficiency of translation of mRNA [21]. Therefore, plasmids having various distances were constructed. Numbers in the designation of the plasmids such as pUES10S, pUES12S, pUES13S, and pUES15 refer to the number of base pairs between the SD sequence and the initiation codon. The plasmids, thus constructed, were tested in various host strains to determine if they were expressed. Of the plasmids tested, pUES10S was found to be expressed in *E. coli* JM109, as was pUES13S in *E. coli* DH1 and JM109. The result indicated that an adjustment of the distance between the SD sequence and the initiation codon, and the screening of host strains was very important in achieving efficient expression of tGH cDNA.

A transformant *E. coli* JM109 carrying pUES13S synthesized a polypeptide of molecular weight 21 000, which was later proved to be the tGH. In order to improve the productivity of tGH, a hybrid plasmid (pUES13S-2) was constructed which contained two tandemly polymerized tGH cDNAs. The plasmid had the *tac* promoter and a transcription terminator (*rrn*B). The amount of tGH produced by the cells with pUES13S-2 (having 2 genes) was significantly lower (approximately 2–3% of total cellular protein) when compared with that of cells carrying pUES13S (having 1 gene). To increase the copy number of the plasmids, the cells were cultured at 40°C and induced for 5 h in the presence of IPTG. Under this culture condition, the amount of tGH synthesized increased and reached 20% of the total cellular protein. However, during incubation at 40°C, the cells lysed, aggregated, and the turbidity of the culture decreased to less than 30% of that of the culture of JM109 with pUES13S. The results indicated that the overproduction of tGH was deleterious to the growth of the host cells. The localization of tGH produced in cells of *E. coli* JM109 carrying pUES13S-2 was examined by electron microscopy, and tGH was found to be concentrated in inclusion bodies. Similar overproduction of fish GH in *E. coli* cells has been reported in the case of salmon GH [14].

2.5 Purification and Characterization of Recombinant Tuna Growth Hormone Produced in *E. coli*

To use tGH produced in *E. coli* as an efficient feed for fish, it is necessary to isolate it from inclusion bodies, and then solubilize it without destroying hormonal activity. Therefore, an attempt was made to isolate the inclusion bodies from *E. coli* cells and subsequently solubilize and renature the recombinant tGH (r-tGH) synthesized in *E. coli*. The r-tGH was produced in inclusion bodies on both ends of the cells. The insoluble materials in cell homogenates were subsequently sedimented by centrifugation and almost all the r-tGH was recovered in the pellet. The supernatant fraction contained no proteins with a molecular weight of 21 000 corresponding to tGH. The r-tGH was concentrated more than 90%.

Among several denaturants tested, guanidine-HCl was found to be effective on both solubilization and renaturation of r-tGH from inclusion bodies. The yield of r-tGH was about 30% of r-tGH initially contained in the inclusion bodies. The molecular weight of the purified r-tGH was 21 000 and 18 600 in the presence and absence of DTT (dithiothreitol), respectively. The r-tGH isolated from inclusion bodies through the solubilization and renaturation processes was homogeneous. Furthermore, the following evidence indicated that the r-tGH is chemically, structurally and physiologically the same as that of the native tGH.

1. Retention time on HPLC of the isolated r-tGH was identical to that of native tGH.
2. The amino acid composition and sequence of r-tGH was in good agreement with that of native tGH.
3. By Western blotting, the r-tGH bound to antibody against native tGH.
4. The behavior of r-tGH on SDS-PAGE in the presence or absence of reducing agents was the same as that of native tGH.
5. The circular dichroism spectra in the far ultraviolet region were similar for r-tGH and native tGH. This result indicated that the conformation of r-tGH is almost the same as that of native GH, and further that the secondary structure of r-tGH is correctly formed through the disulfide linkages between Cys52–Cys160 and Cys177–Cys185 [28].
6. The injection of r-tGH into intact juvenile rainbow trout gave appreciable increases in both length and weight, although clear dose–response relationship was not observed at the r-tGH concentrations tested. A dose–response relationship for both weight gain and length increase was however obtained when snapper (*Pagrus major*) was used instead of rainbow trout. Furthermore, the growth rate of r-tGH tested groups was very similar to those of native t-GH tested groups [28]. The r-tGH treated groups showed 1.5- to 1.7-fold higher food conversion than the control group.

2.6 Expression of Tuna Growth Hormone by *S. cerevisiae*

The feasibility of producing GH by microorganisms with recombinant DNA techniques has been extensively studied and several kinds of GH cDNA have been cloned from fish. Among them, salmon GH has been produced in *E. coli* [14] and yeast [22].

In order to develop a more efficient method for the isolation of active tGH, we investigated the expression of tGH cDNA in *S. cerevisiae* using the *PHO5* promoter region. Expression of a foreign DNA in a microorganism is affected by various factors, such as the copy number of the plasmid, strength of the promoter, the structures of the 5'- and/or 3'-flanking regions of the gene and so on. To express the tGHcDNA in yeast cells, a synthetic linker containing the *PHO5* 5'-flanking region was inserted between the *PHO5* promoter region and the initiation codon (ATG). The synthetic linker was designed to contain the conserved sequence, 5'-CACACA-3', which was found upstream of the ATG codon in several yeast genes [24], although such a consensus sequence was not found in the *PHO5* 5'-flanking region. It is thought that the consensus sequence and/or the distance between the *PHO5* promoter region and ATG are important factors for the efficient expression of tGHcDNA in yeast cells. The 3'-noncoding region seems very important, too, since it supposedly plays an important role in stability of mRNA in eukaryotic cells [23].

A plasmid containing tGHcDNA and the *PHO5* promoter was constructed and designated as pTGH82. It was introduced into cells of *S. cerevisiae* SHY2. The transformant cells were initially cultured in a minimal medium high in phosphate (Pi). The expression of tGHcDNA was induced by transferring the cells from high-Pi medium to low-Pi medium. The protein band corresponding to a molecular weight of 21 000 (native tGH lacking the signal peptide) was observed when the transformant cells of *S. cerevisiae* carrying pTGH82 were cultured in low-Pi medium. On the other hand, the 21 000 band was not observed in the homogenate prepared from cells grown in the high-Pi medium.

In order to identify the 21 000 protein as tGH, Western blotting was performed. The 21 000 protein produced by yeast cells carrying pTGH82 under inducible conditions strongly bound to an antibody against tGH. Under the conditions used, the amount of r-tGH approached 3% of the total cellular protein.

The r-tGH produced by yeast [25, 26] was detected in the cell pellet fraction. However, r-tGH appeared in the supernatant fraction when the cell pellet was sonicated for 30 min on ice in the presence of 0.1% Triton X-100, suggesting that r-tGH was localized in the membrane in the form of a membrane-bound protein. Similar results have been reported [27]. Thus, in yeast cells, r-tGH seemed to be produced without formation of insoluble inclusion bodies, while it was accumulated in inclusion bodies in *E. coli* cells [16, 25]. This finding is significant, since tedious and intricate denaturation/renaturation processes for r-tGH could be eliminated by producing the r-tGH in yeast cells. Furthermore, the recent finding [6] that the oral administration of the bovine serum albumin to rainbow trout was remarkably effective may be applicable to tGH. Yeast cells containing r-tGH could be directly used as a feed to promote the growth of fish.

Thus we have succeeded in the breeding of microorganisms (*E. coli* and *S. cerevisiae*), which can produce large amounts of tGH. This article was based on the doctoral thesis [29] of our colleague, Dr. N. Sato, who is continuing his study of the application of recombinant tuna growth hormone to the breeding of various fish on an industrial scale. Our final goal will be to breed the transgenic fish [30], which has recently attracted considerable attention.

3 Glycolytic Methylglyoxal Pathway and Production of Useful Compounds

3.1 Glycolytic Methylglyoxal Pathway

Glycolysis is a well-known metabolic pathway as an energy generating system and has been studied for many years.

In our course of study on metabolism of methylglyoxal, which is an endogenous toxic substance synthesized in living cells, we were able to reveal that the process constitutes the glycolytic methylglyoxal pathway (Fig. 3).

Methylglyoxal (MG) is a toxic metabolite usually found in all organisms at extremely low concentration. It is synthesized from one of the metabolites of glycolysis, dihydroxyacetonephosphate, by an enzyme MG synthase.

The question is why such a toxic substance is synthesized in cells?

MG is detoxified to lactate in two ways. One is through glyoxalase system consisting of glyoxalase I and II. The other through a red–ox system. In the former system MG is first detoxified to S-lactoylglutathione (S-LG) in the presence of glutathione. This reaction is catalyzed by glyoxalase I. S-LG synthesized is further converted to lactate to make glutathione free. MG and S-LG seem to be regulators of cell growth. In the latter system MG is reduced to lactaldehyde by MG reductase, then further oxidized to lactate. As a result a metabolic pathway is constructed from dihydroxyacetonephosphate to lactate

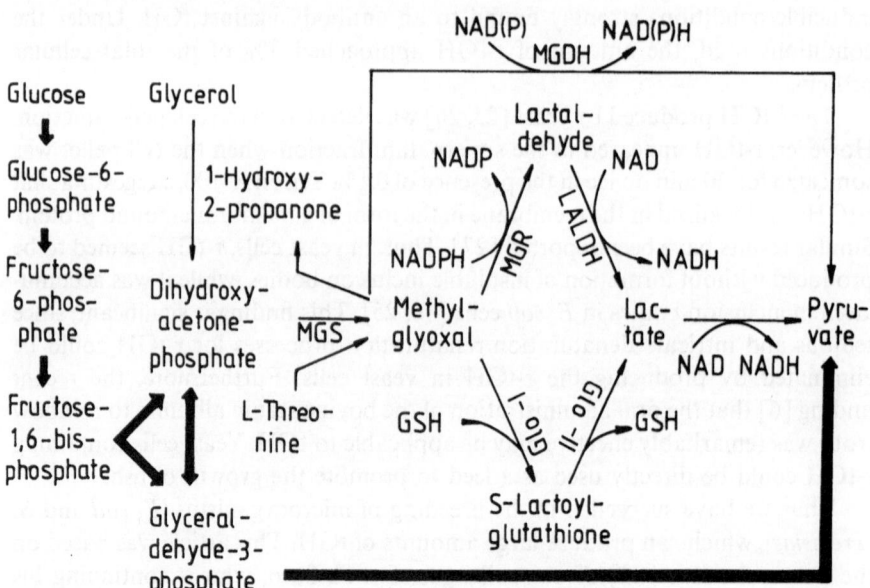

Fig. 3. Pathway for methylglyoxal metabolism in microorganisms. MGS, methylglyoxal synthase; Glo-I, glyoxalase I; Glo-II, glyoxalase II; MGR, methylglyoxal reductase; LALDH, lactaldehyde dehydrogenase; MGDH, methylglyoxal dehydrogenase; GSH, glutathione

through MG and S-lactoylglutathione or lactaldehyde. However, these pathways are not alternative to glycolysis, for they seem to have different functions. Glycolysis is an energy (ATP) generating system, and MG metabolism is not concerned with energy generation, but seems to be concerned with cell growth and division. Therefore, we designated it the glycolytic methylglyoxal pathway.

Concerning this metabolic pathway, we have cloned several genes. Some of their entire structures have been determined. Using one of these genes, we have been able to breed a jumbo yeast which is 2 times bigger in diameter, so 8 times bigger in cubic content. We also bred an *E. coli* which was resistant to the toxic methylglyoxal, keeping the normal size of the cell, while the control cells became longer in the presence of MG.

These microorganisms could produce a large amount of useful compounds such as glutathione, S-lactoylglutathione etc. These gene engineered organisms have already been described in several papers [31–34].

3.2 Production of Glutathione

Glutathione (GSH) is a detoxicating tripeptide consisting of L-glutamate, L-cysteine, and glycine. It is biosynthesized in two steps. The first step is catalyzed by the enzyme GSH-1 (gamma-glutamylcysteine synthetase), and the second by GSH-II (glutathione synthetase). Since both enzymes require ATP, this biosynthetic pathway should be coupled with the ATP regeneration process. The genes coding for GSH-I and II were cloned. By the introduction of them into *E. coli*, we were able to breed a transformant which could produce a large amount of glutathione [33]. The entire structures of both genes were determined, the tertiary structure of GSH-II was only recently determined. This is the first case in Japan, and 3rd or 4th case in the world in which the tertiary structure of the enzymes having an ATP binding site has been determined [35].

3.3 Relationship of Glyoxalase I Activity to Cell Growth and Division

Several enzymatic reactions are involved in the degradation of MG. Among them the glyoxalase system, consisting of glyoxalase I and II, seemed to have an important role not only in the degradation of MG, but also in cell growth and/or cell division for the following reasons. (1) Glyoxalase I is highly activated in cancer cells. (2) Glyoxalase I activity changes correspondingly quite closely to the cell growth and arrest. This is proved by the use of temperature sensitive CDC mutants of *S. cerevisiae*. (3) Phorbol ester (TPA: an activator for protein kinase C) activates glyoxalase I activity, while it inhibits glyoxalase II activity [36, 37]. This is the condition under which S-lactoylglutathione is accumulated. An increased level of S-lactoylglutathione is said to be related to the tumor-promoting properties of the compound [38]. The growth-inhibiting effect of MG and the growth-promoting effect of S-lactoylglutathione made us expect to find some anticancer chemicals by inhibiting glyoxalase I activity [39]. The activity of this glyoxalase system has been detected in a wide variety of organisms.

3.4 Breeding of a Jumbo Yeast and Production of S-Lactoylglutathione

Since MG scavenging reaction by glyoxalase I (Glo-I) is active in cancer cells, we anticipated being able to breed a yeast which can grow as rapidly as cancer cells, therefore we tried to clone the MG resistant gene. We were able to clone 2 kinds of genes; one increased glyoxalase I activity while the other MG reductase activity.

By the introduction of the DNA fragment enhancing MG scavenging activity, we bred a jumbo yeast, which was 2 times the diameter, i.e., 8 times the cubic content of normal yeast. This jumbo yeast produced about 5 g/L of S-lactoylglutathione (S-LG) [33, 40]. However, we also cloned the gene for glyoxalase I from *Pseudomonas putida*, which was expressed in *E. coli*. This transformant produced about 15 g/L of S-LG.

3.5 A Gene for Protein Controlling Glyoxalase I Activity

As the result of determination of base sequence of the MG resistant DNA fragment which increased glyoxalase (Glo-I) activity, the suitable ORF (open reading frame) was not found in the DNA fragment, since it coded a protein of 14.7 kDa which was much smaller than that of Glo-I (32 kDa). Therefore, we designated it as *GAC* (for Glyoxalase I Activation Conferring) gene. It has characteristically several cysteines near the *C*-terminal, which seems quite similar to the Zn finger structure of the C_6 type. Therefore, we think that this DNA fragment might be a kind of DNA binding protein, and regulate the Glo-I activity. By northern analysis we could prove the increase of mRNA of Glo-I gene in a transformant carrying the *GAC* gene.

We also tried to clone the real gene for Glo-I by immunoscreening. The base sequence of the DNA fragment harboring the real Glo-I gene is now being determined.

3.6 Increase of Glyoxalase I Activity through the Sexual Hormone Receptor and Phosphorylation of the Enzyme

Since MG was thought to play an important role in yeast cell growth and/or differentiation, we investigated the effects of yeast sexual response on the enzymes in the glyoxalase system. Then we found a phenomenon that the activity of Glo-I increased and that of Glo-II decreased in alpha-type yeast cells when they were exposed to a culture supernatant in which a-type cells had been grown [41]. This phenomenon was not observed when a mutated alpha-type yeast cells was treated, which were deficient in a-factor receptors. Therefore, we concluded this increase of Glo-I took place through the hormone receptor. We could also prove the phosphorylation of Glo-I protein when treated with the opposite sexual hormone [42], and that Glo-I activity was lost when it was dephosphorylated by phosphatase. Therefore, Glo-I activity seems to be regulated through phosphorylation and/or dephosphorylation of the enzyme. Since the products of the *STE7* and *STE11* genes which locate in receptors are reported to have some homology to protein kinases, they may play some important role in the information transfer system.

4 References

1. Dyerberg J, Bang HO, Hjorne N (1975) Am J Clin Nutr 28: 958.
2. Kimura A (1991) CRC review in Biotechnol 11: 113.
3. Farmer SW, Papkoff H, Hayashida T, Bewley TA, Bern HA, Li CH (1976) Gen Comp Endocrinol 30: 91.
4. Kawauchi H, Moriyama S, Yasuda A, Yamaguchi K, Shirahata K, Kubota J, Hirano T (1986) Arch Biochem Biophys 244: 542.
5. Kishida M, Hirano T, Kubota J, Hasegawa S, Kawauchi H, Yamaguchi K, Shirahata K (1987) Gen Comp Endocrinol 65: 478.
6. Kawauchi H (1988) Kagaku to Kogyo (Japanese) 41: 106.
7. Abdel-Meguid SS, Shien H-S, Smith WW, Dayringer HE, Violand BN, Bentle LA (1987) Proc Natl Acad Sci USA 84: 6434.
8. Clarkson J, Korber F, Christensen T, Junker F, Pedersen J, Hansen FB (1989) J Mol Biol 208: 719.
9. Skryabin KG, Rubtsov PM, Gorbulev VG, Schulga AA, Parsadanian AS, Kirpichnikov MP, Bayev AA, Pavlovskii AG, Borisova SN, Vainstein BK, Bulatov AA (1989) Protein hormones: Expression, structure and protein engineering. In: Jardetzky O, Holbrook R (eds), Protein Structure and Engineering. Plenum, New York.
10. Sato N, Watanabe K, Murata K, Sakaguchi M, Kariya Y, Kimura S, Nonaka M, Kimura A (1988) Biochem Biophys Acta 949: 35.
11. Sato N, Murata K, Watanabe K, Hayami T, Kariya Y, Kimura S, Sakaguchi M, Nonaka M, Kimura A (1988) Biotechnol Appl Biochem 10: 385.
12. Seeberg PH, Shine J, Martial JA, Baxter JD, Goodman HM (1977) Nature 270: 486.
13. Martial JA, Hallewell RA, Baxter JD, Goodman HM (1979) Science 205: 602.
14. Sekine S, Mizukami T, Nishi T, Kuwana Y, Saito A, Sato M, Itoh S, Kawauchi H (1985) Proc Natl Acad Sci USA 82: 4306.
15. Watahiki N, Tanaka M, Masuda N, Yamakawa M, Yoneda Y, Nakashima K (1988) Gen Comp Endocrinol 70: 401.
16. Sato N, Hayami T, Murata K, Watanabe K, Kariya Y, Sakaguchi M, Kimura S, Nonaka M, Kimura A (1989) Appl Microbial Biotechnol 30: 153.
17. Brosius J, Erfle M, Storella J (1985) J Biol Chem 260: 3539.
18. Rose JK, Shafferman A (1981) Proc Natl Acad Sci USA 78: 6670.
19. Nakahama K, Fujisawa Y, Ito Y, Ikeyama S, Kikuchi M (1986) Appl Microbiol Biotechnol 25: 262.
20. Sancar A, Hack AM, Rupp WD (1979) J Bacteriol 137: 692.
21. Gheysen D, Iserentant D, Derom C, Fiers W (1982) Gene 17: 55.
22. Hosoi N, Kuga T, Matsumoto T, Itoh S (1987) Int. cong. on Molecular Biology, Kyoto, p 125.
23. Zaret KS, Sherman F (1982) Cell 28: 563.
24. Dobson MJ, Tuite MF, Roberts NA, Kingsman AJ, Kingsman SM (1982) Nucleic Acids Res 10: 2625.
25. Sato N, Kawazoe I, Tamai T, Inoue Y, Murata K, Kimura S, Nonaka M, Kimura A (1989) J Ferment Bioeng 68: 79.

26. Hayami T, Sato N, Ichiryu T, Inoue Y, Murata K, Kimura S, Nonaka M, Kimura A (1989) Agric Biol Chem 53: 2917.
27. Sharma S, Godson GN (1985) Science 228: 879.
28. Kariya Y, Sato N, Kawazoe I, Kimura S, Miyazaki N, Nonaka M, Kawauchi H (1989) Agric Biol Chem 53: 1679.
29. Sato N, (1989) Studies on microbial production of tuna growth hormone, Doctoral Thesis.
30. Chen TT, Powers DA (1990) Trends Biotechnol 8: 209.
31. Kimura A (1986) Biotechnol Genet Eng Review 4: 39.
32. Kimura A (1986) In: Fiechter A (ed.) Adv Biochem Eng Biotechnol, Springer, Berlin Heidelberg New York Tokyo, Vol 33, p 29.
33. Kimura A (1989) In: Fiechter A, Okada H, Tanner RD (eds) Bioproduct & Bioprocess. Springer, Berlin Heidelberg New York, 173.
34. Murata K, Kimura A (1990) Biotech Adv 8: 59.
35. Yamaguchi H, Kato H, Hata Y, Nishioka T, Oda J, Katsube Y, Kimura A (1989) Protein Engineering '89, Second International Conference (abstract), p 142.
35a. Yamaguchi H, Kato H, Hata Y, Nishioka T, Kimura A, Oda J, Katsube Y (1993) J Mol Biol 229: 1083.
36. Gillespie E (1981) Biochem Biophys Res Commun 98: 463.
37. Murata K, Sato N, Inoue Y, Kimura A (1989) Agric Biol Chem 53: 1999.
38. Gillespie E (1979) Nature (London) 277: 135.
39. Vince R, Daluge S (1971) J Med Chem 14: 35.
40. Murata K, Inoue Y, Rhee H, Kimura A (1989) Can J Microbiol 35: 423.
41. Inoue Y, Choi B-Y, Murata K, Kimura A (1989) Biochem Biophys Res Commun 165: 1091.
42. Inoue Y, Choi B-Y, Murata K, Kimura A (1990) J Biochem 108: 4.

2.4 Amplification of Homologous Fermentative Genes in *Clostridium Acetobutylicum* ATCC 824

Lee D. Mermelstein[1], George N. Bennett[2] and Eleftherios T. Papoutsakis[1]*

[1]Northwestern University, Dept. Chemical Engineering, Evanston, IL 60208, USA
[2]Rice University, Dept. Biochemistry and Cell Biology, Houston, TX 77251, USA

Contents

Improvement in the fermentative characteristics of the solvent producing bacterium *Clostridium acetobutylicum* requires knowledge of the limiting enzyme activities and genetic regulation of product formation. The ability to amplify homologous genes in *C. acetobutylicum* is central to this endeavor. We have identified and characterized a type II restriction endonuclease in our host strain, *C. acetobutylicum* ATCC 824, and constructed a vector, pFNK1, that is minimally impeded by this enzyme during transformation. Clostridial genes cloned into pFNK1 in *Bacillus subtilis* may subsequently be transferred into 824 using an efficient electrotransformation protocol that we have developed. This has allowed us to amplify activities of the acid-formation phosphotransbutyrylase gene and the solvent-formation acetoacetate decarboxylase gene in *C. acetobutylicum*.

1 Introduction

Clostridium acetobutylicum, a natural producer of butanol, acetone and ethanol, is of interest as a potential commercial source of these solvents. The biochemistry by which *C. acetobutylicum* initially produces acetate and butyrate and subsequently the above solvents, in batch culture, is well understood. The genetics of this anaerobic, highly regulated, branched, primary metabolism is, however, poorly understood despite the fact that many of the genes responsible for the formation of the various acids and solvents have recently been cloned [1–5].

We are interested in using plasmid borne clostridial genes to amplify various acid or solvent formation enzyme activities. The effect of an amplified enzyme activity on the product profile should indicate, in a direct manner, whether that enzyme is limiting the metabolic flow through the corresponding pathway branch. This may suggest strategies for gene amplification or elimination that could lead to improved product yields and selectivities. Furthermore the

---

* To whom all correspondence should be addressed.

Bioproducts and Bioprocesses 2
Editors: Yoshida, Tanner
© Springer-Verlag Berlin Heidelberg 1993

amplification of homologous genes that have been altered in their coding or regulatory regions may help generate information on the biochemistry and genetics of clostridial metabolism that would be difficult, if not impossible, to attain using other hosts.

This paper summarizes the first amplifications of an acid- and a solvent-formation gene in *C. acetobutylicum*. This was accomplished by improvement of a transformation protocol, and the design of a vector to circumvent the restriction system that we identified and characterized in the host strain. Prior to this work, the *C. pasteurianum* leucine biosynthesis gene was the only cloned gene that had been expressed in *C. acetobutylicum* [6].

2 Results

2.1 Electrotransformation Studies

Plasmid DNA has been introduced into strains of *C. acetobutylicum* by conjugative transfer, protoplast transformation and electrotransformation (electroporation) [7]. *C. acetobutylicum* ATCC 824, the source strain for our cloned primary metabolic genes, has not previously been transformed, however. For assessment of limiting activities and regulatory behavior, we felt it was important to amplify clostridial genes in the strain from which they were cloned, because strains of *C. acetobutylicum* have been shown to vary widely in their physiological, biochemical and fermentative characteristics [7]. Since protoplast transformation of *C. acetobutylicum* has only been efficiently accomplished in an autolysin mutant and conjugative plasmid transfer is time consuming, inefficient and restrictive with respect to available vectors, electrotransformation was explored as a means of transforming ATCC 824.

An electrotransformation protocol reported for the transformation of *C. acetobutylicum* NCIMB 8052 [8] worked poorly for ATCC 824. Modifications in electroporation buffer, field strength, media, time constant and procedural parameters improved the transformation efficiency dramatically. Briefly, 60 mL of cells grown to late exponential phase (OD_{600} = 1.2) in Reinforced Clostridial Medium (RCM, Difco), pH 5.2, are harvested, washed and resuspended in 2.1 mL of cold ETB (5 mM NaH_2PO_4, 272 mM sucrose, pH 7.4). Aliquots of cells (0.7 mL) are chilled on ice for 5 min in 0.4 cm electrode gap × 1.0 cm cuvettes and plasmid DNA (10–500 ng), dissolved in H_2O, is added to the suspensions. After an additional 2.0 min incubation on ice, the suspensions are subjected to a 2.0 kV discharge from a 25 μF capacitor (τ = ca. 13.0 ms) using the Bio-Rad Gene Pulser. The cells are immediately transferred to 10 ml of prewarmed outgrowth medium (RCM, pH 5.2) and incubated for 4 h at 37°C prior to plating on RCM (pH 5.8, 40 µg/ml erythromycin). All manipulations are carried out anaerobically and centrifugations are performed at room temperature. Using the plasmid pIM13 [9] transformation efficiencies of 5×10^4 transformants (µg DNA)$^{-1}$ are routinely obtained.

A variety of plasmids containing gram-positive origins of replication and macrolide, lincosamide and streptogramin B antibiotic resistance (MLS^r) genes were obtained or constructed for evaluation as potential vectors for the amplification of primary metabolic genes. To facilitate cloning it was considered desirable to have a shuttle vector that could function in easily manipulated *E. coli* as well as *C. acetobutylicum*. However, it was found that vectors of streptococcal or Bacillus origin which transformed 824 to erythromycin resistance would no longer do so if fused to the *Escherichia coli* plasmids pUC19 or pBR322.

It was apparent that the inability to transform *C. acetobutylicum* with these chimeric plasmids was due to some aspect of the sequence or structure of the *E. coli* derived DNA. Toxicity of a protein was dismissed because transcription from these *E. coli* plasmid sequences in a gram-positive organism, such as *C. acetobutylicum*, was considered unlikely. Since the streptococcal and Bacillus DNA used in these constructs were, like the *C. acetobutylicum* genome, of high A + T DNA content, it was hypothesized that 824 might preferentially restrict non-A + T rich DNA as would be present in the *E. coli* portions of these plasmids. This hypothesis was confirmed upon identification and characterization of a type II restriction endonuclease within *C. acetobutylicum* ATCC 824.

2.2 Identification and Characterization of a Type II Restriction Endonuclease

Due to the high activity of non-sequence specific nucleases in 824, the restriction endonuclease had to be enriched considerably before it could be detected. By preparing extracts from washed protoplasts, the majority of secreted and cell-wall associated nucleases were eliminated. Cells were grown to $OD_{600} \sim 0.4$ in RCM (pH 5.2), supplemented with 0.4% glycine (to aid in protoplasts formation), and protoplasted at 37°C in 0.04 volumes of Clostridial Basal Medium containing 10% (wt/vol) sucrose, 25 mM $MgCl_2$, 25 mM $CaCl_2$ and 5 mg/mL lysozyme [1]. After approximately 85% protoplasts were obtained the cells were washed in cold protoplasting solution without lysozyme. Washed protoplasts were pelleted and lysed by resuspension in 5.0 mL of cold TEMK (40 mM Tris–Hcl pH 7.5, 10 mM EDTA, 6 mM 2-mercaptoethanol, 25 mM KCl) per g wet cell weight. Residual debris was removed by centrifugation to give the crude extract. Under proper assay conditions (see below) the resulting extracts possessed high restriction endonuclease activity, and contamination with other nucleases was sufficiently low to enable recognition sequence determination by restriction pattern analysis.

The restriction endonuclease was named Cac824I and was found to cleave the sequence 5'-GCNGC-3' where N can be any nucleotide. The recognition sequence determination was made by cleaving highly A + T rich DNAs of known sequence with Cac824I and analyzing the resultant digests by agarose gel electrophoresis along with DNA standards. DNAs used included Tn917 [10], pIM13 [9], the pAMβ1 MLS^r gene [11] and the *C. acetobutylicum* phosphotransbutyrylase [12], butyrate kinase [12] and alcohol dehydrogenase I [13]

[mM NaCl]

50   100 150 200   λ   Fnu4H1

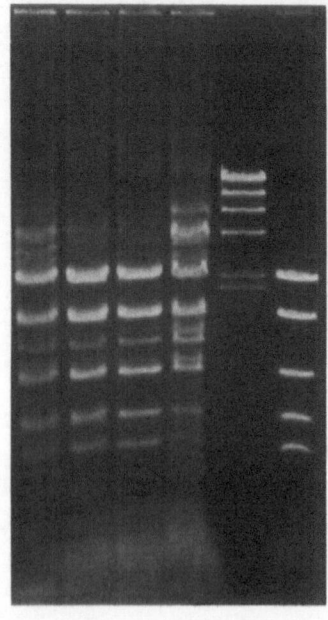

Fig. 1. NaCl dependency of Cac824I activity as analyzed by ethidium bromide stained digests of DNA
electrophoresed in 0.9% agarose. From left to right;
Cac824I digests of pMU1328 [14] (1 h at 37°C) in 10
mM Tris (pH 7.9), 10 mM MgCl$_2$, 2% crude extract
and NaCl at concentrations indicated above lanes,
HindIII digest of λ DNA, Fnu4HI digest of pMU1328

genes. Cac824I was found to have maximum activity under high salt (150 mM)
conditions (Fig. 1) and is an isoschizomer of Fnu4HI and FbrI. No methylase is
available commercially which can protect against Cac824I restriction.

2.3 Vector Construction

The plasmids pUC19 and pBR322 contain 19 and 42 Cac824I recognition sites,
respectively. It was apparent that fusions between gram positive functions and
such *E. coli* plasmids failed to transform 824 because they were restricted at one
of their many Cac824I sites before complete methylation at all such sites had
occurred. In contrast, the most efficiently transforming plasmid tested, the
naturally occurring *B. subtilis* MLS^r plasmid pIM13 (2246 bp), contains only
one such recognition site. This property of pIM13 is a consequence of its small
size and the A + T richness of its DNA. An *E. coli* plasmid with these properties
could not be located in the literature and therefore the possibility of constructing
a vector that transformed 824 and *E. coli* was not straightforward.

The plasmid pIM13 functions in relatively easily manipulated *Bacillus
subtilis* 168 and derivative strains, but contains few unique cloning sites that
would make it useful as a vector for subsequent transfer of genes to *C.*

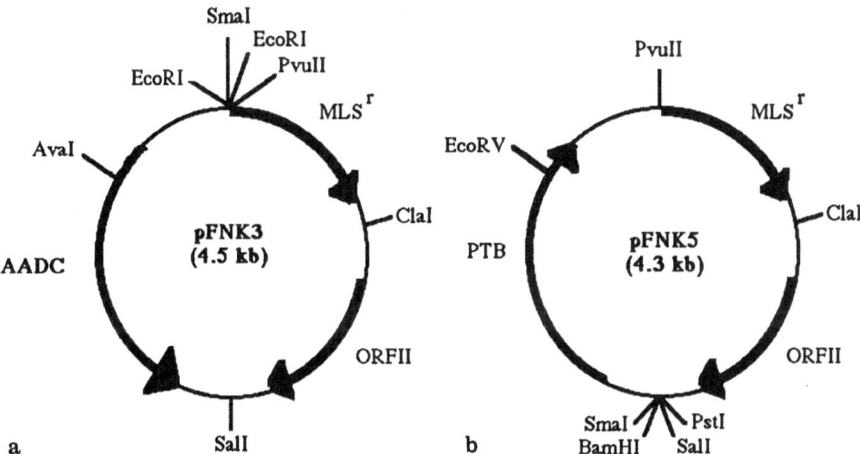

Fig. 2. Physical map of vector pFNK3 (A) and pFNK5 (B) containing the *C. acetobutylicum* ATCC 824 acetoacetate decarboxylase (AADC) and phosphotransbutyrylase (PTB) genes, respectively. The macrolide, lincosamide and streptogramin B antibiotic resistance gene (MLS[r]) and putative replication protein (ORFII) are derived from the *B. subtilis* plasmid pIM13 [15]. *Bold arrows* indicate structural genes and their transcriptional directions

acetobutylicum. Therefore the small, non-essential, HindIII fragment of pIM13 was replaced with the polylinker region of pUC9 to yield the vector pFNK1 (2.4 kb), which contains unique cloning sites for the following enzymes: BamHI, ClaI, EcoRI, PstI, PvuII, SalI and SmaI.

2.4 Gene Amplification

To evaluate the effectiveness of pFNK1 as a vector for delivery of homologous genes into 824, one acid and one solvent formation pathway gene containing their natural promoters were separately cloned into pFNK1 in *B. subtilis* ATCC 33677, repurified and electrotransformed into 824. The acetoacetate decarboxylase is the terminal enzyme in the pathway for acetone production, converting acetoacetate to acetone and CO_2. The acetoacetate decarboxylase gene and promoter were cloned into pFNK1 as an EcoRI/BglII fragment to yield the plasmid pFNK3 (Fig. 2A). The phosphotransbutyrylase is the branchpoint activity for butyrate production and transforms butyryl-CoA and inorganic phosphate into butyryl phosphate and reduced CoA. Butyryl-CoA can alternatively be converted into butanol in two enzymatic steps. The PTB gene was cloned as a SmaI/PvuII fragment into pFNK1 to yield pFNK5 (Fig. 2B). Both pFNK3 and pFNK5 were maintained intact in 824. M5 [16] a mutant of 824 lacking both of the acetone synthesis pathway genes was also transformed with pFNK3.

Transformants containing pFNK3 and pFNK5 exhibited enhanced enzyme activities corresponding to the resident cloned clostridial genes as indicated in

UNCONTROLLED PH BATCH FERMENTATION
WHOLE CELL EXTRACTS

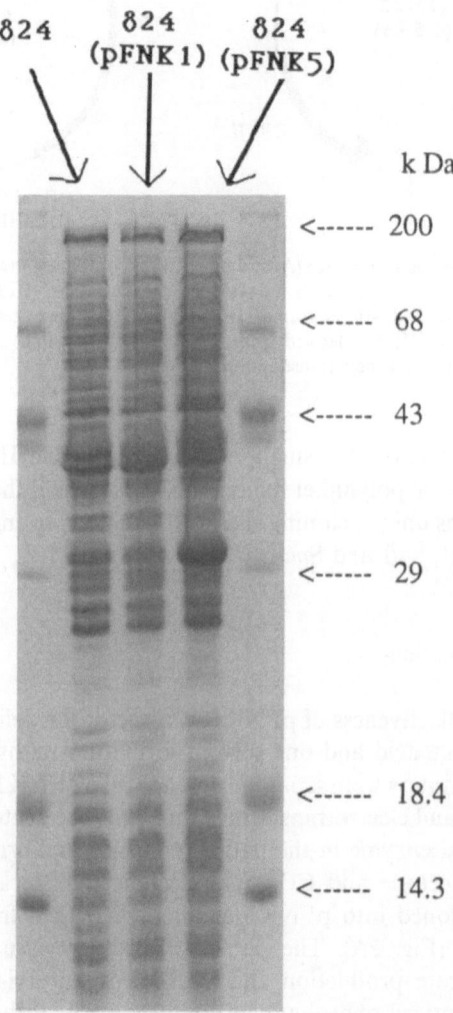

Fig. 3. Analysis of whole-cell extracts by Coomassie Blue stained 12.5% polyacrylamide–SDS gel electrophoresis. *C. acetobutylicum* ATCC 824 and the indicated plasmid containing derivatives were grown in CSM and harvested during late exponential growth. The outside lanes contain standards of molecular masses indicated to the right. A band of increased intensity at $\sim$ 31,000 Da that corresponds to the phosphotransbutyrylase subunit molecular mass can be detected in 824 (pFNK5)

Table 1. Acetoacetate decarboxylase (AADC) activities as determined by constant pressure differential respirometry of whole cell suspensions [3] and phosphotransbutyrylase (PTB) activities [17] of the indicated strains grown in CSM [18], pH 6.2, to exponential (E) or stationary (S) growth phases. One unit is defined as the amount of enzyme catalyzing the formation of one μmol of CO_2 (AADC) or 1 μmol of CoA/5,5'-dithiobis(2-nitrobenzoic acid) complex (PTB) per min, at standard temperature and pressure

| | U (mg protein)$^{-1}$ | | | |
| | AADC | | PTB | |
Strain	E	S	E	S
824	0.80	4.8	6.5	11
824 (pFNK1)	0.70	5.1	6.7	11
824 (pFNK5)	—	—	97	220
824 (pFNK3)	9.9	170	—	—
M5	0.0	0.0	—	—
M5 (pFNK1)	0.0	0.0	—	—
M5 (pFNK3)	7.8	220	—	—

Table 1. The acetoacetate decarboxylase was induced during the solventogenic late stage of growth like the wild type activity and was well expressed in the M5 mutant. Amplification was to a level clearly detectable by SDS–polyacrylamide gel electrophoresis (Fig. 3).

3 Discussion

The ATCC 824 strain of *C. acetobutylicum*, widely used in cloning, enzymatic and physiological studies, could not be transformed with conventionally constructed shuttle vectors that function in *E. coli*, even with a significantly improved electrotransformation protocol. We have shown that this is due to the presence of a type II restriction endonuclease, Cac824I, within 824 that frequently cleaves the *E. coli* portion of such vectors. Cac824I recognizes the sequence 5'-GCNGC-3', is the second restriction endonuclease identified in the species and is unique among the known clostridial restriction endonuclease identified in the species and is unique among the known clostridial restriction endonucleases. Considering the A + T DNA bias of the *C. acetobutylicum* genome, a Cac824I site would arise on average once every 3.0 kb. In the plasmids pUC19 or pBR322 however, such a sequence arises on average once every 110 bp, rendering shuttle vectors which contain such *E. coli* portions particularly susceptible to restriction in 824. Only one known restriction endonuclease, Cvb09, cleaves pUC19 or pBR322 DNA more frequently than Cac824I.

Using the *B. subtilis* plasmid, pIM13, which contains only one Cac824I recognition site, we were able to construct an efficiently transforming vector, pFNK1, possessing a variety of unique cloning sites and capable of replicating in both *B. subtilis* and *C. acetobutylicum*. Derivatives of pFNK1 were constructed in *B. subtilis* which contain the *C. acetobutylicum* 824 phosphotransbutyrylase or acetoacetate decarboxylase genes under the control of their natural promoters, and these were used to transform 824. The plasmid borne genes were expressed at high levels in the transformants. This was the first report of the amplification of fermentative genes in *C. acetobutylicum*. Given the number of genes that have recently been cloned from this strain such amplifications should make possible important determinations concerning limiting enzyme activities for product formation, the relative importance of isozymes and the characterization of regions of genetic control associated with genes.

Acknowledgements. This work was supported by Grant # BCS-8912209 from the National Science Foundation.

4 References

1. Cary JW, Petersen DJ, Papoutsakis ET, Bennett GN (1988) J Bacteriol 170: 4613.
2. Cary JW, Petersen DJ, Papoutsakis ET, Bennett GN (1990) Appl Environ Microbiol 56: 1576.
3. Petersen DJ, Bennett GN (1990) Appl Environ Microbiol 56: 3491.
4. Petersen DJ, Welch RW, Rudolph FB, Bennett GN (1991) J Bacteriol 173: 1831.
5. Youngleson JS, Santangelo JD, Jones DT, Woods DR (1988) Appl Environ Microbiol 54: 676.
6. Oultram JD, Peck H, Brehm JK, Thompson DE, Swinfield TJ, Minton NP (1988) Mol Gen Genet 214: 177.
7. Young M, Minton NP, Staudenbauer WL (1989) FEMS Microbiol Revs 63: 301.
8. Oultram JD, Loughlin M, Swinfield TJ, Brehm JK, Thompson DE, Minton NP (1988) FEMS Microbiol Lett 56: 83.
9. Mahler I, Halvorson HO (1980) J Gen Microbiol 120: 259.
10. Shaw JH, Clewell DB (1985) J Bacteriol 164: 782.
11. Brehm J, Salmond G, Minton N (1987) Nucl Acids Res 15: 3177.
12. Petersen DJ (1991) Characterization of the acid production pathway genes from *Clostridium acetobutylicum* ATCC 824. Thesis, Rice University, Houston.
13. Youngleson JS, Jones WA, Jones DT, Woods DR (1989) Gene 78: 355.
14. Achen MG, Davidson BE, Hillier AJ (1986) Gene 45: 45.
15. Monod M, Denoya C, Dubnau D (1986) J Bacteriol 168: 137.
16. Clark SW, Bennett GN, Rudolph FB (1989) Appl Environ Microbiol 55: 970.
17. Wiesenborn DP, Rudolph FB, Papoutsakis ET (1989) Appl Environ Microbiol 55: 317.
18. Roos JW, McLaughlin JK, Papoutsakis ET (1984) Biotechnol Bioeng 27: 681.

2.5 Cloning and Sequencing of the Alcohol Dehydrogenase Gene from *Bacillus stearothermophilus* and Alteration of the Optimum pH of the Enzyme by Protein Engineering

Tadayuki Imanaka and Hisao Sakoda

Department of Biotechnology, Faculty of Engineering, Osaka University, Yamadaoka, Suita, Osaka 565, Japan

Contents

Using *Bacillus subtilis* and pTB524 as a host and a vector plasmid, respectively, we cloned the thermostable alcohol dehydrogenase (ADH-T) gene (*adhT*) from *Bacillus stearothermophilus* NCA1503, and determined its nucleotide sequence. The deduced amino acid sequence (337 amino acids, molecular weight 36,098) was compared with those of alcohol dehydrogenases (ADH) from four different origins. The enzyme required both Zn ion and NAD as a coenzyme. We found that the enzyme was stable even after treatment of 65°C for 1 h. The amino acid residues responsible for the catalytic activity of horse liver ADH had been clarified based on the three-dimensional structure. Since those catalytic amino acid residues were fairly conserved in ADH-T and other ADHs, ADH-T was inferred to have basically the same proton release system as that of horse liver ADH. The putative proton release system of ADH-T was elucidated by introducing point mutations at the catalytic amino acid residues, Cys38 (cysteine at position 38), Thr40, and His43, with site-directed mutagenesis. The mutant enzyme, Thr40Ser (Thr40 was replaced by serine), showed a little lower activity than wild-type ADH-T. The result indicates that the OH-group of serine instead of threonine can be also used for the catalytic activity. To change the pK_a value of the putative system, His43 was replaced by the more basic amino acid arginine. As a result, the optimum pH of the mutant enzyme, His43Arg, was shifted from 7.8 (wild-type enzyme) to 9.0. His43Arg exhibited higher activity than wild-type enzyme at the optimum pH.

1 Introduction

Thermophilic bacteria produce various useful enzymes. Some of them have been purified and characterized, and their structural genes have already been cloned and sequenced [1–3]. *Bacillus stearothermophilus* NCA1503 was found to produce a thermostable alcohol dehydrogenase (ADH-T) amounting to 1–2%

Bioproducts and Bioprocesses 2
Editors: Yoshida, Tanner
© Springer-Verlag Berlin Heidelberg 1993

of soluble cell protein . By using this strain, ethanol was produced from sucrose or glucose as a carbon source under anaerobic condition at high temperatures [4, 5]. Alcohol dehydrogenases (ADH) were also isolated from both eucaryotes [6–9] and procaryotes [10, 11], and they exhibited various features. The ADH reaction mechanism was originally studied using horse liver ADH based on X-ray crystallographic analysis and kinetic studies [6, 12, 13]. The catalysis with horse liver ADH is performed by the proton release system consisting of a zinc atom, water molecule, and serine and histidine residues. By a series of intensive studies, the system including the 2′-hydroxyl group of the nicotinamide ribose was proposed for horse liver and human liver ADH [14–16]. Threonine and histidine of human liver ADH corresponded to serine and histidine of horse liver ADH, respectively, both of which functioned as catalytic sites [16, 17]. The amino acid residues responsible for substrate specificity in human liver ADH and yeast ADH were also proposed based on three-dimensional information on the horse liver ADH [18–20].

We describe here the following: (1) molecular cloning and nucleotide sequencing of the alcohol dehydrogenase gene, adhT, from B. stearothermophilus NCA1503, (2) comparison of the deduced amino acid sequence with different ADHs, and prediction of the catalytic system of ADH-T based on the crystallographically determined model of horse liver ADH, (3) construction of the modified enzyme carrying a different pH profile from the wild-type ADH-T.

2 Cloning of the Alcohol Dehydrogenase Gene from Bacillus stearothermophilus NCA1503

Bacterial strains, plasmids, and phage used are listed in Table 1. B. stearothermophilus NCA1503 was grown at 55°C in 2 L broth containing tryptone (20 g/L), yeast extract (10 g/L) and NaCl (5 g/L) and its pH being adjusted to 7.3 by HCl. Chromosomal DNA of B. stearothermophilus was prepared as described elsewhere [3, 26], and was partially digested with Sau3A I. Procedures for digestion of DNA with restriction endonucleases, ligation of DNA with T4 DNA ligase, agarose gel electrophoresis and SDS–PAGE are described elsewhere [2, 26, 27]. The fragments of around 6 kb were isolated by agarose gel electrophoresis and purified by Gene-Clean (Bio101 Inc., La Jolla, CA). These fragments were ligated into the BamH I site of a low copy number plasmid pTB524. Either the rapid alkaline extraction method or CsCl-ethidium bromide equilibrium density gradient centrifugation was used to prepare plasmid DNA [3, 26]. The ligation mixture was used to transform B. subtilis MI113.

For transformation of B. subtilis, competent cells were prepared as described previously (3). Transformants of B. subtilis with pTB524, pTB522 or their derivatives carrying a tetracycline resistance gene were selected on L agar containing tetracycline (25 μg/mL). Tc^r transformants were transferred on the modified aldehyde indicator plates and incubated at 37°C for 5 h. ADH producing colonies were selected on the aldehyde indicator plates as described

Table 1. Bacterial strains, plasmids, and phage.

Strain, Plasmid, or phage	Characteristics	Reference
Bacillus stearothermophilus NCA1503	DNA donor	[21]
Bacillus subtilis MI113	*arg-15 trpC2 hsrM hsmM*	[21]
Bacillus subtilis MI112	*leuA8 thr-5 arg-15 recE4 hsrM hsmM*	[22]
Escherichia coli TG1	*SupEΔ(lac-proAB) hsdΔ5* F′ *traD36 proAB*$^+$ *lacI*^{q} *lacZΔM*15	[23]
Plasmids		
pTB524	Tetracycline resistance, Tc^r	[24]
pTB522	Tc^r	[25]
Phage		
M13 mp18 or mp19		[23]

by Conway et al. [28], with slight modification. The plates were composed of antibiotic medium 3 (Difco Laboratories, Detroit, MI) (17.5 g/L), acting as a buffer at pH 7.0, ethanol (20 ml/L), pararosaniline (Sigma Chemical Corp., St. Louis, MO) (50 mg/L) and sodium hydrogen sulfite (250 mg/L). Ethanol diffuses into cells and can be converted by ADH to acetaldehyde which reacts with the reagents to form a Schiff-base intense red.

Out of around 3000 Tc^r transformants of *B. subtilis*, one alcohol dehydrogenase (ADH) positive clone was found on the modified aldehyde indicator plates. The transformant carried a recombinant plasmid in which a fragment of around 7 kb was inserted.

Lysate of the candidate cell showed a band of ADH activity staining at the same position as that of a DNA donor strain *B. stearothermophilus* NCA1503 (Fig. 1). Activity staining of the enzyme was performed according to the method described by Dowds et al. [29]. Crude enzymes were run on a 6% polyacrylamide gel with solutions and reagents from which SDS was omitted. The gel was stained for alcohol dehydrogenase activity by an alcohol-dependent nitro-blue-tetrazolium procedure. The gel was soaked in 500 mM Tris–HCl (pH 8.8) at 4°C for 15 min and then incubated at 37°C for 30 min in a staining solution containing 150 mM Tris–HCl (pH 8.8), NAD (0.132 mg/mL), nitro-blue-tetrazolium (0.163 mg/mL), phenazine methosulphate (0.03 mg/mL) and ethanol (10 mL/L).

Alcohol dehydrogenase was assayed by monitoring ethanol-dependent NAD reduction at 340 nm [10, 30]. Enzyme activity (unit) was expressed as μmoles of NADH produced per min, using a molar absorption coefficient of 6.22 mM^{-1} cm^{-1}. The standard alcohol dehydrogenase assay was performed at 55°C in a reaction mixture which contained 100 mM potassium phosphate buffer (pH 7.8), 1 mM NAD and 100 mM ethanol. The ADH activity of the recombinant plasmid carrier (1.48 u/mg dry cell) was about nine times higher than that of the DNA donor (0.17 u/mg dry cell), whereas the host cell showed a little activity (less than 0.002 u/mg dry cell). The candidate produced a 35-kDa protein which could be also found in the DNA donor but not in host cell, and moreover the protein was thermostable (Fig. 2). Therefore, the recombinant plasmid, which was designated as pTBAD70, was considered to carry the alcohol dehydrogenase (ADH-T) gene (*adhT*) from *B. stearothermophilus* NCA1503.

Fig. 1. Activity staining of alcohol dehydrogenase from crude cytoplasmic extracts. Each lane contains 5 μL of cell extracts or diluted sample. *Lane 1*, cell extract of *B. Subtilis* MI113; *lane 2*, cell extract of *B. stearothermophilus* NCA1503; *lane 3*. tenfold dilution of the cell extract from the ADH positive transformant

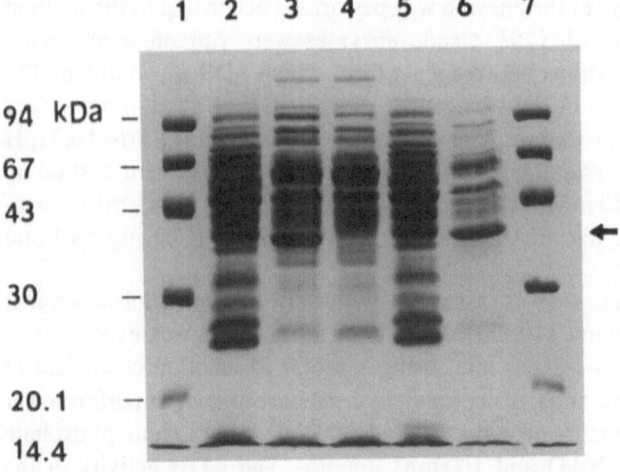

Fig. 2. SDS-PAGE analysis of cell extracts. Each lane contains 5μL of cell extract. Lane: *1,7*, molecular weight marker; *2, B. stearothermophilus* NCA1503; *3*, ADH positive transformant of *B. subtilis* MI113; *4, B. subtilis* MI113; *5*, cell extract of *B. stearothermophilus* NCA1503 heated (60°C, 10 min) and centrifuged (20 000 × g); *6*, cell extract of *B. subtilis* transformant heated (60°C, 10 min) and centrifuged (20 000 × g). An *arrow* indicates the position of ADH

3 Subcloning of the *adhT* Gene

Restriction maps of pTBAD70 and its derivatives are shown in Fig. 3. To localize the *adhT* gene, a *Bam*HI fragment (about 3.5 kb) and a *Hind* III fragment (about 4.0 kb) from pTBAD70 were subcloned in pTB524 and pTB522, and their recombinant plasmids were designated as pTBAD35 or pTBAD40, respectively. pTBAD70, pTBAD40 and pTBAD35-harboring cells showed ADH activity on the aldehyde indicator plate. In contrast, the strain carrying pTBAD35ΔSph I,which lacked a *Sph*I fragment (about 0.6 kb) from pTBAD35, had no ADH activity. Therefore, the *adhT* gene was considered to be located in the 2.2 kb *Hind* III-*Bam*HI fragment including the *Sph*I fragment (Fig. 3).

4 Nucleotide Sequence of the *adhT* Gene

DNA was sequenced by the dideoxy method of Sanger et al. [31] with a Sequenase sequencing kit (United States Biochem. Corp., Cleveland, Ohio). The cloned gene was digested with restriction enzymes and subcloned into M13 mp18 or M13 mp19. *E. coli* TG1 was used as a host cell.

Nucleotide sequence of the 2.2-kb *Hind* III-*Bam*HI fragment was determined. A large open reading frame was found in the fragment (Fig. 4). It was composed of 1011 bp corresponding to 337 amino acids. The molecular weight was estimated to be 36 098, which agreed with the result of SDS–PAGE (Fig. 2). The *N*-terminal amino acid sequence has been reported [32, 33], and it was nearly identical to the *N*-terminal sequence deduced from the nucleotide sequence. Amino acid composition which had been reported [34] was also agreement with the sequencing result in this work. It was, therefore, concluded that the open reading frame encoded the *adhT* gene. A Shine-Dalgarno (SD) sequence was found 10 bases upstream from the translation start site (ATG), and

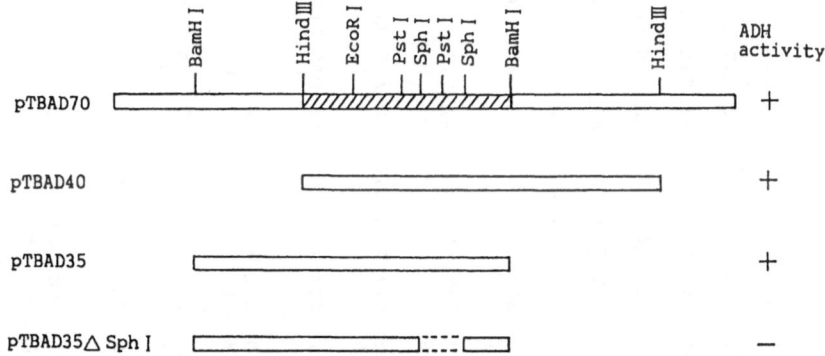

Fig. 3. Restriction endonuclease maps of the fragment carrying the *adhT* gene and its derivatives. The structures of derivative plasmids are shown below the physical map of pTBAD70. *Hatched box* indicates the region containing the *adhT* gene

```
                     EcoR I
-471  GAA TTC ATG GCA GCA TTG GTT ATA AAA CCC CGC AAG AGA TAG AAG ACA ACA TTC ATC GTA CAG CCT AAT  -403

-402  CAC TGT ATT AAG ATT GTG CAC CCG CTT TAC ATG TAC TAG CGG GCA TGA CCT TGT TTG GCG ATG CTG       -334

-333  ATG TTG TAA CAT CTG GCT TCT CGG CAG TAG GCT GCA GAG GCT CCA TAG GCG TCA TCG CTC               -265

-264  CCC ACT CTT AAA GGG GCA GGA ACT AGA ACA GAA ACC TCC AAC GTC GAT TAT ATT AAA TTT GTG TCC ATC   -196

-195  CAT TTA CAC TAT ACT AGA ACA GAA TTT TAT GAT GTC AAC TCC CGA ACC AAA TTT TTA ACT TTT           -127
                                                                          -35
-126  TAT CCA AAA ATA TTT TTC ATT TTT TTG AAC ATT TTA TTT GTG ATA TTT TTC ACA TAA ATG TTA ACT GCT  -58
                            -10                    -35              -10                      Pst I
-57   ACA CTA CAT ATG TAC AGA TCA AAA TCA AGT CCT TTT TTG CCC AGT AGG AGG ATT ATA ATC ATG AAA GCT GCA  12
                                                               -35         SD                 M   K   A   A    4

13    GTT GTG GAA CAA TTT AAA AAG CCG TTA CAA GTG AAA GAA GTG AAA GAA AAG CCT AAG ATC TCA TAC GGG GAA   81
5      V   V   E   Q   F   K   K   P   L   Q   V   K   E   V   K   E   K   P   K   I   S   Y   G   E    27

82    GTA TTA GTG AAA CGC ATC AAA GCG TGC CTT CCT ATC GAA GTA ATT GAA GAA GTG GAT GGC GTA ATC TCA   150
28     V   L   V   K   R   I   K   A   C   L   P   I   E   V   I   E   E   V   D   G   V   I   S    50
                                   Sph I
151   GTA AAG CCT AAA CTG CCT ATT CCT GGC GTC GGT GTA ATT GAA GAA GTG ATT CCT TGG CTT TAT TCG GGG   219
51     V   K   P   K   L   P   I   P   G   V   G   V   I   E   E   V   I   P   W   L   Y   S   G    73

220   GTA ACA CAT TTA AAA GTT GGA GAT CGC GTA GGT CGC GTG TGC CAT GGT TGC CTT TCG TAT CAT TGT GAC   288
74     V   T   H   L   K   V   G   D   R   V   G   R   V   C   H   G   C   L   S   Y   H   C   D    96

289   TAT TGC TTA AGC GGA CAA GAA ACA TTA TGC GAA CGT TAT GTC GAT TAT GGC GTC TCC GAT GGT GGT       357
97     Y   C   L   S   G   Q   E   T   L   C   E   R   Y   V   D   Y   G   V   S   D   G   G        119
                                                      Pst I
358   TAT GCT GAA TAT TGC CGT GCA GCC GAT AAC ATT CCT GAT AAC TTA TCG TTT GAA GAA                   426
120    Y   A   E   Y   C   R   A   A   D   N   I   P   D   N   L   S   F   E   E                    142

427   GCC GCT CCA CCA GCT GTA GGT GTA ACA ACA GCG CTC AAA GTA GCA AAA CCA GGT                       495
143    A   A   P   P   A   V   G   V   T   T   A   L   K   V   A   K   P   G                        165

496   GAA TGG GTA GTA GGC CTT GGG ATC TAC GCC ATT GGA CAT GTC GCA CAA TAC GCG ATG GGG .             564
166    E   W   V   V   G   L   G   I   Y   A   I   G   H   V   A   Q   Y   A   M   G               188
```

```
565  TTA AAC GTC GTT GCT GTC GAT TTA GGT GAT GAA AAA CTT GAG CTT GCT AAA CAA CTT GGT GCA GAT CTT  633
189   L   N   V   V   A   V   D   L   G   D   E   K   L   E   L   A   K   Q   L   G   A   D   L   211
                                                                              Sph

634  GTC AAT CCG AAA CAT GAT GAT GCA CAA GCA AAA ATA TGG AAA GAA AAA GTG GGC GGT GTG CAT GCG ACT  702
212   V   N   P   K   H   D   D   A   Q   A   K   I   W   K   E   K   V   G   G   V   H   A   T   234

703  GTC ACA GCT GTf TCA AAA GCC GCG TTC GAA TCA GCC TAC AAA TCC ATT CGT CGC GGT GGT GCT TGC      771
235   V   T   A   V   S   K   A   A   F   E   S   A   Y   K   S   I   R   R   G   G   A   C        257

772  GTA CTC GGA TTA CCG GAA GAA ATT CCT ATT CCA TAC AAA GTA ACA TTA AAT GGT GGA GTA              840
258   V   L   G   L   P   E   E   I   P   I   P   Y   K   V   T   L   N   G   G   V                280

841  ATT ATT GGT TCT ATC GTT GTT GGT ACG CGC AAA TTA GAC CAA CTT GCA CAA TTT GCA GCA GAA GGA AAA  909
281   I   I   G   S   I   V   V   G   T   R   K   L   D   Q   L   A   Q   F   A   A   E   G   K   303

910  GTA AAA ACA ATT GTC GAA GTC GAA GTG AAC ATT GAC CGT GTA TTC GAT CGT ATG TTA AAA GGG          978
304   V   K   T   I   V   E   V   E   V   N   I   D   R   V   F   D   R   M   L   K   G            326

979  CAA ATT AAC GGC CGC GTC GTG TTA AAA GTA GAT TAA AAA GAA GGC GTC TGA GGG CGC                  1047
327   Q   I   N   G   R   V   V   L   K   V   D   D   *                                           338

1048 CTT CTT ATT TTA CTT CAA CGG AAA ATA CTT GAT CAT GAA GCT CTT CCT TAT CGT CCC ACA AAA          1116

1117 CGT CCG ATA CGG TCG ATC AGA CGG AGG CTC AGG TAT ATA AGC TTA CCC GTG GTG CTA GAT CTC AAA      1185
                                                  BamH I

1186 CAA GCA TAA AAA TAG CCC TTG CAT CCC TTG CAT GAG GAT CC                                       1217
```

Fig. 4. Nucleotide sequence of the adhT gene and the deduced amino acid sequence of the encoded protein. The nucleotide sequence is counted from the putative initiation codon. The amino acid sequence is shown beneath the nucleotide sequence and the first amino acid of translation (Met) is counted as 1. A probable Shine-Dalgarno sequence and two promoter regions (− 35 and − 10 regions) are shown by solid lines below the nucleotide sequence. Terminator and inverted repeat at the 5'-flanking region are shown by arrows. The asterisk indicates a stop codon

two possible promoter sequences were found about 180 and 60 bases upstream of the SD sequence, respectively. Two tandem promoters of alcohol dehydrogenase gene were also found in the *Zymomonas mobilis adhB* gene [28]. The sequence resembling to typical prokaryotic terminators was found downstream of the open reading frame. The highly AT-rich sequence (about 200 bp) was found at the 5′-flanking region of the open reading frame (Fig. 4).

5 Comparison of the Deduced Amino Acid Sequence with Those of Four Different ADHs

Comparison of primary structures of the enzymes with the same function but of different origin is useful to determine the essential amino acids for activity, because active and substrate binding sites are highly conserved [35]. Homology of primary structure was analyzed according to the method described by Needleman et al. [36]. An NEC PC-9801RA computer (Nippon Electric Co., Tokyo, Japan) and the software DNASIS (Hitachi Software Engineering Co., Kanagawa, Japan) were used for the analysis.

We compared the amino acid sequences of five different ADHs (Fig. 5). The deduced amino acid sequence of thermostable ADH-T from *B. stearothermophilus* was fairly homologous (45%) with that of *Saccharomyces cerevisiae* [37], while the homologies of ADH-T with enzymes from maize [38], human [39] and horse liver [40] were about 35%. The amino acids which are reported to be indispensable for the catalytic activity of horse liver ADH [6] are highly conserved in all ADHs. The catalytic zinc atom is bound by three protein ligands, two sulfur atoms from Cys38 (cysteine at position 38 of ADH-T) and Cys148 and one nitrogen atom from His61. These amino acids are completely conserved. The ligands of second zinc atom, Cys92, Cys95, Cys98, and Cys106 are also conserved. Furthermore, one of the amino acids participating in the proton release system, serine residue of horse liver ADH corresponding to position 40 of ADH-T, is substituted to threonine in other ADHs including ADH-T. Serine or threonine (at position 40 of ADH-T) plays the same function through its hydroxyl group. Another amino acid, His43 is conserved except for maize ADH. Thr152, which has been reported to be important for proper positioning of NAD [41], is strictly conserved (Fig. 5). The nicotinamide ribose has been reported to be bound by hydrogen bond from 0–3′ to the carbonyl oxygen of isoleucine residue (at a position 239 in Fig. 5) of horse liver ADH [6, 14]. However, such a residue can not be found in ADH-T. Therefore, positioning of NAD could not be determined in the enzyme catalytic system of ADH-T.

6 A Putative Reaction Mechanism for ADH-T

According to the above-mentioned argument, it is believed that these ADHs have the same reaction mechanism (Fig. 6) as shown for horse liver ADH [6]. One equivalent of H^+ ion is released per equivalent of ethanol that is oxidized.

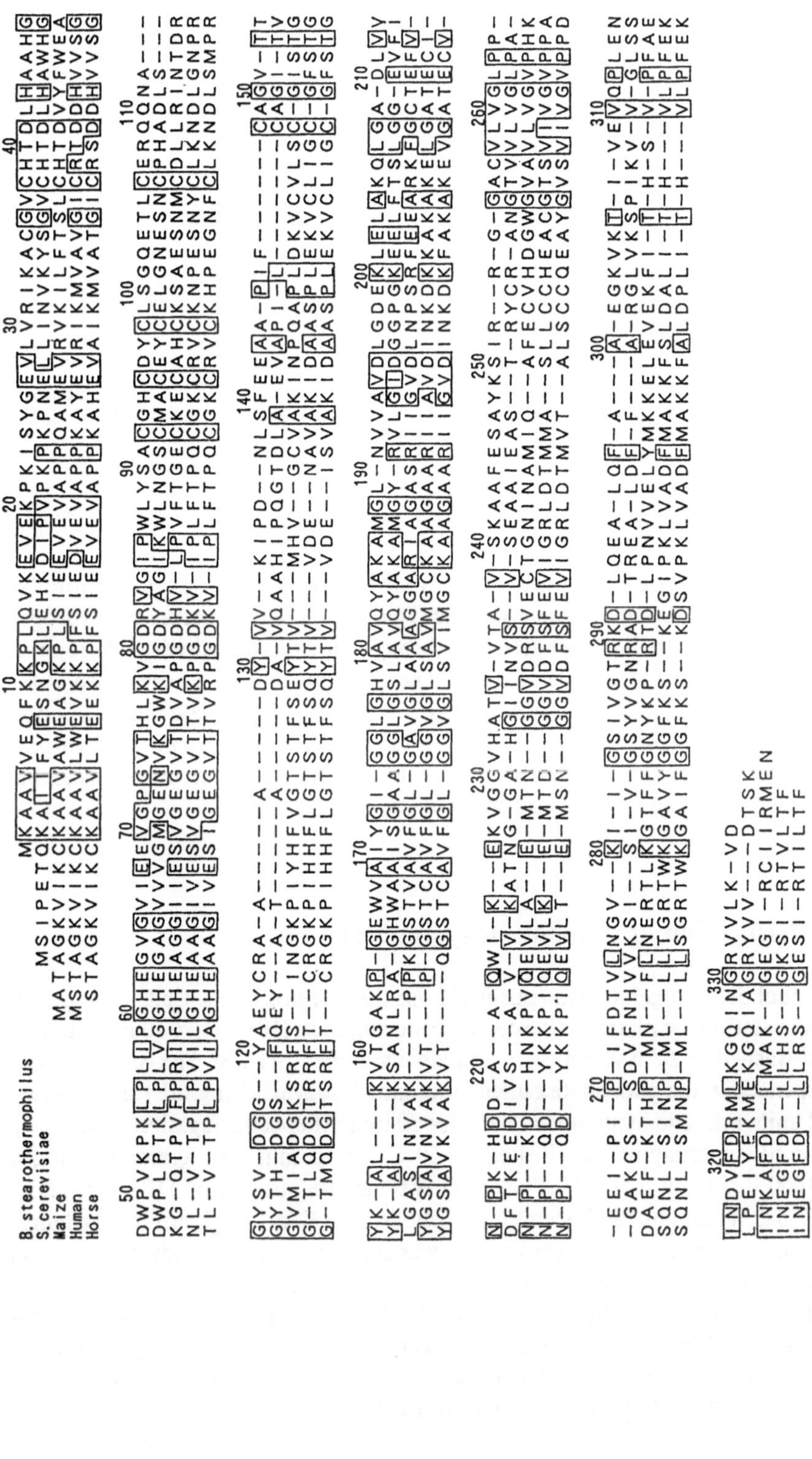

B. stearothermophilus
S. cerevisiae
Maize
Human
Horse

Fig. 5. Comparison of amino acid sequences of five different ADHs. Amino acids are numbered according to the amino acid sequence of *B. stearothermophilus*. Homologous sequences are surrounded by *rectangles*. Symbols; ————, gaps are introduced to obtain the maximal matching

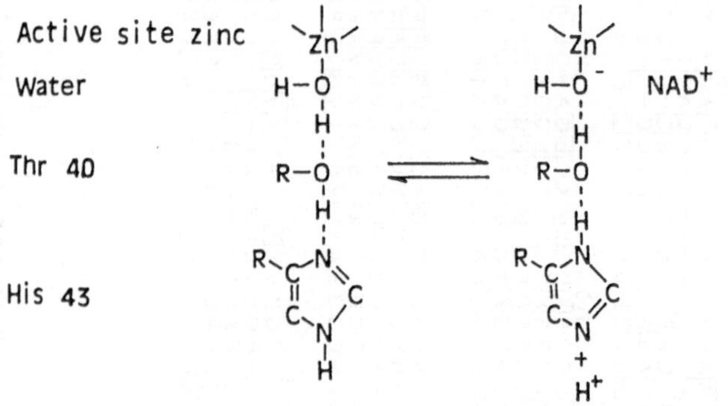

Active site zinc

Water

Thr 40

His 43

Fig. 6. A putative mechanism for the H^+ion release system for ADH-T. This system is composed of the zinc-bound water molecule, Thr40 and His43 and the H^+ion release is induced by NAD binding

This H^+ion release is associated with NAD binding and is dissociated from the water molecule bound to catalytic zinc. This H^+ion release occurs via hydrogen bond system from the water molecule through the side chain (hydroxyl group) of Thr40 to the imidazole ring of His43. The H^+ion would thus be released at the surface of the molecule and not into the interior of the substrate binding pocket (Fig. 6). Alcohol then binds to zinc as the alcoholate ion, displacing the hydroxyl ion. The zinc atom polarizes the alcoholate so that direct hydrogen transfer and subsequent rearrangement to aldehyde can occur [6, 12].

7 Purification of Alcohol Dehydrogenase

Alcohol dehydrogenase was purified from the transformants. Cells were grown to the stationary phase, harvested by centrifugation ($10\,000 \times g$, 10 min) at 4°C, and washed in 20 mM potassium phosphate buffer (pH 7.8). The cell pellet was suspended in the phosphate buffer containing lysozyme (1 mg/mL) and DNase I (10 units/mL) and incubated at 37°C for 30 min. After centrifugation ($55\,000 \times g$, 30 min), supernatant was heated at 60°C for 10 min and again centrifuged ($20\,000 \times g$, 10 min) at 4°C. The crude enzyme was purified by DEAE-cellulose (DE 52, Whatman BioSystems Ltd., Maidstone, Kent, England) ion-exchange column chromatography. The enzyme was eluted with linear gradient (0–1 M) of potassium chloride. Active fractions were dialyzed overnight at 4°C in 20 mM potassium phosphate buffer (pH 7.8). The final enzyme preparation was homogeneity by SDS–polyacrylamide gel electrophoresis (PAGE) analysis (Fig. 7). The protein concentration was measured by the method of Lowry et al. with bovine serum albumin as the standard [42].

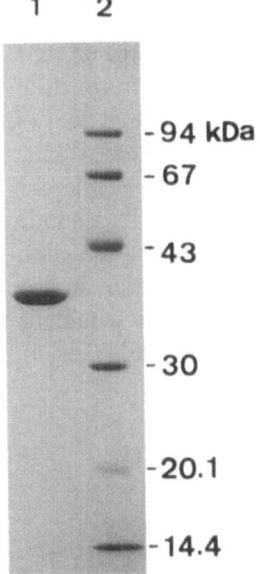

Fig. 7. SDS-PAGE analysis of purified enzyme. Lane: *1*, 5 µL of the purified wild-type ADH-T; *2*, molecular weight marker. The mutant enzymes, Thr40Ser and His43Arg, gave the same results (photograph not shown)

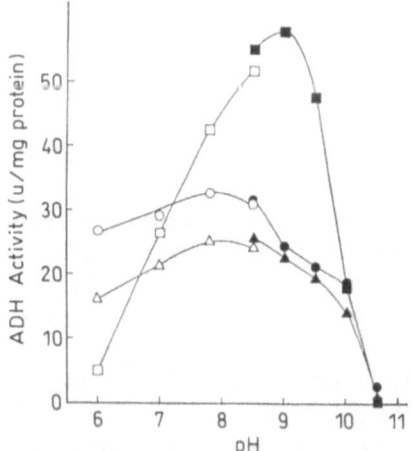

Fig. 8. pH profile for ethanol of wild-type ADH-T($\bigcirc$, $\bullet$), Thr40Ser mutant enzyme ($\triangle$, $\blacktriangle$) and His43Arg ($\square$, $\blacksquare$). *Open symbols* indicate enzyme assay in 100 mM potassium phosphate buffer and *closed symbols* in 100 mM glycine–KOH buffer. Enzyme activity was assayed under the standard condition written in the text except for buffer pH. When the NAD concentration in reaction mixtures was reduced to 0.2 mM, nearly the same results were obtained. Therefore, 1.0 mM NAD was actually in excess under different pH conditions

8 Analysis of the H⁺-Ion Release System by Amino Acid Substitution

To make sure the reaction mechanism, these amino acids were substituted by site-directed mutagenesis. Point mutations were introduced into a gene with oligonucleotide-directed in vitro mutagenesis system (Amersham, Buckinghamshire, United Kingdom), using chemically synthesized oligo-nucleotides. The mutant enzyme Cys38Ser (Cys38 as a catalytic zinc ligand was replaced by serine) had no ADH activity. Other mutants, Thr40Ala and His43Ala, also completely lost the enzyme activities. In contrast, Thr40Ser showed the enzyme activity. The wild-type ADH-T and the mutant enzyme were purified to homogeneity as shown in Fig. 7. Thr40Ser had the same pH profile as that of wild-type enzyme, although the enzyme activity was lower (Fig. 8). These results indicate that His43 and the hydroxyl group of Thr40 or Ser40 are essential for enzyme activity and lower activity of Thr40Ser might be explained by a subtle change of steric conformation. These data agree with the possible reaction mechanism shown in Fig. 6. Cys38 would function as a ligand for a catalytic zinc atom which is indispensable for catalytic activity and the proton would be transported through hydroxyl group of Thr40 and released from His43 (Fig. 6).

9 Alteration of the pH Profile of ADH-T by Site-Directed Mutagenesis

His43 was substituted by arginine to alter the pK_a of the side chain, i.e., the pK_a imidazole ring of histidine is 6.0 and that of guanidino group of arginine is 12.5. We contemplated that this mutation might disturb the pK_a of the proton release group and result in a different pH dependence from that of wild-type enzyme. The mutant enzyme, His43Arg, exhibited ADH activity and was purified to homogeneity. Wild-type ADH-T showed its maximum activity at around pH 7.8 corresponding to the pK_a 7.6 of the H^+ ion release group of horse ADH in the presence of NAD. His43Arg exhibited lower activity under acidic conditions, but higher activity under alkaline conditions than the wild type. It might be explained by substitution of His43 by arginine slowing down the H^+-ion release reaction under acidic conditions. The maximum activity was observed at pH 9.0. Furthermore, its maximum activity was about two times higher than that of wild type (Fig. 8). Thus the optimum pH of the enzyme was shifted from neutral to alkaline by replacing a catalytic amino acid histidine by arginine without any loss of ADH activity.

Generally speaking, the pK_a value of the active center of an enzyme can influence the pH profile. In other words, the pH profile of enzyme could be altered by changing the pK_a value of a catalytic amino acid. For example, the active site histidine residue of serine protease acts as a general base in enzyme catalysis and its pK_a regulates enzyme activity. Increasing the overall negative charge on the enzyme should raise the pK_a of the active site histidine by stabilizing the protonated form of the histidine, whereas increasing the positive charge should lower the pK_a by destabilizing the protonated form of the histidine. Its activity under acidic condition increased, when the number of lysine residues of the enzyme surface were increased by site-directed mutagenesis [43].

Since enzymes are proteins containing many ionizable groups, they exist in a whole series of different states of ionization. However only one of the ionic form of active center is catalytically active [44, 45]. Our experiment shows that pK_a value of an active site is responsible for pH profile of the enzyme and the optimum pH would be altered by substituting a catalytic amino acid.

10 Conclusion

Using *Bacillus subtilis* and pTB524 as a host and vector plasmid, respectively, we cloned the thermostable alcohol dehydrogenase (ADH-T) gene (*adhT*) from *Bacillus stearothermophilus* NCA1503, and determined its nucleotide sequence. The deduced amino acid sequence (337 amino acids) was compared with those of alcohol dehydrogenases (ADH) from four different origins. Since the amino acid residues responsible for the catalytic activity of horse liver ADH, whose catalytic mechanism had been analyzed based on the three-dimensional structure, were fairly conserved in all ADHs, ADH-T was inferred to have the same proton release system as that of horse liver ADH. According to the system, threonine at position 40 was replaced by serine. The mutant enzyme, Thr40Ser, showed a little lower activity than the wild-type enzyme. The result indicates that the OH-group of serine instead of threonine can be also used for the catalytic activity. To change the pK_a value of the putative system, a catalytic histidine residue was replaced by the more basic amino acid arginine. As a result, the optimum pH of the mutant enzyme, His43Arg, was shifted from 7.8 (wild-type enzyme) to 9.0 without any loss of ADH activity.

1 References

1. Aiba S, Kitai K, Imanaka.T (1983) Appl Environ Microbiol 46: 1059.
2. Fujii M, Takagi M, Imanaka T, Aiba S (1983) J Bacteriol 154: 831.
3. Kuriki T, Okada S, Imanaka T (1988) J Bacteriol 170: 1554.
4. Atkinson A, Ellwood DC, Evans CGT, Yeo RG (1975) Biotech Bioeng 17: 1375.
5. Runswick MJ, Harris JI (1978) FEBS Lett. 92: 365.
6. Brändén C-I, Jörnvall H, Eklund H, Furugren B (1975) In: Boyer PD (ed), The enzymes, vol 11. Oxidation-reduction part A. Academic, New York p.103.
7. Edenberg HJ, Zhang K, Fong K, Bosron WF, Li TK (1985) Proc Natl Acad Sci USA 82: 2262.
8. Estonius M, Karlsson C, Fox EA, Höög J-O, Holmquist B, Vallee BL, Davidson WS, Jörnvall (1990) Eur J Biochem 194: 593.
9. Höög JO, von Bahr-Lindström H, Hedén L-O, Holmquist B, Larsson K, Hempel J, Vallee BL, Jörnvall H (1987) Biochemistry 26: 1926.
10. Lamed RJ, Zeikus JG (1981) Biochem J 195: 183–190.
11. Nakayama T (1961) J Biochem 49: 240-251.
12. Eklund H, Nordström B, Zeppezauer E, Söderlund G, Ohlsson I, Boiwe T, Brändén C-I (1974) FEBS Lett 44: 200.
13. Taniguchi S, Theorell H, Akeson A (1967) Acta Chem Scand 21: 1903.
14. Eklund H, Plapp BV, Samama J-P, and Brändén CI (1982) J Biol Chem 257: 14349.
15. Eklund H, Horjales E, Vallee BL, Jörnvall H (1987) Eur J Biochem 167: 185.
16. Eklund H, Müller-Wille P, Horjales E, Futer O, Holmquist B, Vallee BL, Höög J-O, Kaiser R, Jörnvall H (1990) Eur J Biochem 193: 303.
17. Ehrig T, Hurley TD, Edenberg HJ, Bosron WF (1991) Biochemistry 30: 1062.
18. Creaser EH, Murali C, Britt KA (1990) Prot Engng 3: 523.

19. Ganzhorn AJ, Green DW, Hershey AD, Gould RM, Plapp BV (1987) J Biol Chem 262: 3754.
20. Stone CL, Li TK, Bosron WF (1989) J Biol Chem 264: 11112.
21. Takagi M, Takada H, Imanaka T (1990) J Bacteriol 172: 411.
22. Tanaka T (1979) J Bacteriol J 139: 775.
23. Sambrook J, Fritsch EF, Maniatis T (1989) In: Molecular cloning, 2nd edn. Cold Spring Harbor Laboratory Press.
24. Nakamura, K, Imanaka T (1989) Appl Environ Microbiol 55: 3208.
25. Imanaka T, Himeno T, Aiba S (1985) J Gen Microbiol 131: 1753.
26. Imanaka T, Fujii M, Aramori I, Aiba S (1982) J Bacteriol 149: 824.
27. Imanaka T, Fujii M, Aiba S (1981) J Bacteriol 146: 1091.
28. Conway T, Sewell GW, Osman YA, Ingram LO (1987) J Bacteriol 169: 2591.
29. Dowds BCA, Sheehan MC, Bailey CJ, McConnell DJ (1988) Gene 68: 11.
30. Wills C (1976) Nature 261: 26.
31. Sanger F, Nicklen S, Coulson AR (1977) Proc Natl Acad Sci USA 74: 5463.
32. Bridgen J, Kolb E, Harris JI (1973) FEBS Lett 33: 1.
33. Jeck R, Woenckhaus C, Harris JI, Runswick MJ (1979) Eur J Biochem 93: 57.
34. Sheehan, MC, Bailey CJ, Dowds BCA, McConnell DJ (1988) Biochem J 252: 661.
35. Imanaka T, Shibazaki M, Takagi M (1986) Nature 324: 695.
36. Needleman, SB, and Wunsch CD (1970) J Mol Biol 48: 443.
37. Russell DW, Smith M, Williamson VM and Young ET (1983) J Biol Chem 258: 2674.
38. Dennis ES, Gerlach WL, Pryor AJ, Bennetzen JL, Inglis A, Llewellyn D, Sachs MM, Ferl RJ, Peacock WJ (1984) Nucleic Acids Res 12: 3983.
39. Ikuta T, Fujiyoshi T, Kurachi K, Yoshida A (1985) Proc Natl Acad Sci USA 82: 2703.
40. Jörnvall, H (1970) Eur J Biochem 16: 25.
41. Jörnvall, H, Eklund H, Brändén CI (1978) J Biol Chem 253: 8414.
42. Lowry, OH, Rosebrough NJ, Farr AL, Randall RJ (1951) J Biol Chem 193: 265.
43. Russell, AJ, Fersht AR (1987) Nature 328: 496.
44. Dixon M, Webb EC, Thorne CJR, Tipton KF (1979) In Enzymes. Longman Group Ltd., London.
45. Michaelis L, Davidsohn H (1911) Biochem Z 35: 386.

2.6 Bioprocess Kinetics and Optimization of Recombinant Fermentation: Genetic and Engineering Approaches

Dewey D.Y. Ryu and J.Y. Kim

Department of Chemical Engineering, University of California, Davis, CA 95616, USA

Contents

As part of our continuing endeavor to improve and/or optimize the recombinant fermentation process, the genetic and engineering approaches are considered and some highlights of our recent research results are presented in this paper. A few selected examples of bioprocess strategies which enable incremental improvements of recombinant fermentation processes include: (1) use of *par* sequence to improve the growth rate of the recombinant, (2) increasing translation efficiency of the recombinant by increasing the growth rate and productive cell concentration, and (3) use of optimized temperature and dilution rate control strategies.

List of Symbols

q_p	the specific production rate of the gene product
k_o	the overall biosynthetic rate
ε	the gene expression efficiency
G_p	the gene or plasmid DNA concentration
μ^+ and μ^-	the specific growth rates of plasmid-containing and plasmid-free cells, respectively
A and B	the empirical constants given in Eq. (2)
ρ	the concentration of a gene product
X	the cell concentration
t	time
b	the parameter related to intrinsic cellular biosynthetic rate as given in Eqs. (1) and (3)

Discussion of Results

When a recombinant fermentation process is to be optimized and/or improved on a rational basis, many important parameters closely related to the cellular physiology and metabolism, genetic characteristics of the recombinant having

Bioproducts and Bioprocesses 2
Editors: Yoshida, Tanner
© Springer-Verlag Berlin Heidelberg 1993

foreign DNA, bioreactor environmental and operational conditions and all of their effects on productivity must be well understood and appropriate bio-process strategy must be developed. In recombinant fermentation systems, there is a heterogeneous cell population consisting of productive plasmid-harboring cells and non-productive plasmid-free cells. Their cell population dynamics and fraction of productive cell population must be understood. Determination of key genetic parameters and assessment of their effects on the heterogeneous cell population and the productivity of genetically engineered recombinant organisms are very important to the design, control, and optimization of large-scale recombinant fermentation processes [1].

A kinetic model for the heterogeneous cell population dynamics of recombinant fermentation system has been developed [2]. The key genetic parameters can be estimated from the dynamic cell population model and applied to the development of kinetic model for gene product formation. The specific production rate (q_p) and its functional relationship with the overall biosynthetic rate (k_o), gene expression efficiency (ε), the gene or plasmid DNA concentration (G_p), the specific growth rate of plasmid-containing and plasmid-free cells $(\mu^+$ and μ^- respectively) and other parameters can be formulated and the productivity kinetic model developed [2].

The product formation kinetics may be expressed as

$$q_p = k_o \varepsilon G_p (\mu^+ + b). \tag{1}$$

When we compare the recombinant protein production kinetics with the Leudeking–Piret empirical equation,

$$\frac{dp}{dt} = A \frac{dX^+}{dt} + BX^+, \tag{2}$$

the empirical coefficients A and B now show some biological significance and meaning, namely,

$$A = k_o \varepsilon G_p \text{ and } B = Ab. \tag{3}$$

Based on experimental data, the specific production rate, q_p, and specific growth rate, μ^+, can be correlated and the level of gene expression (or gene expression efficiency, ε) can be estimated from the plasmid DNA concentration, G_p, and the protein biosynthetic rate of the host cell under the conditions of no metabolic stress, k_o. (Fig. 1). This kind of bioprocess kinetics is a powerful tool enabling us to estimate the performance of recombinant organisms for practical purpose and large scale applications.

The gene expression efficiency (ε) may have several different values depending on the reference values of the gene expression. For the purpose of evaluating the performance of recombinants, the possible reference expression values are:

1. the protein biosynthetic rate of the non-recombinant host cell,
2. the biosynthetic rate of other recombinants having different vectors, promotors, and/or design and construction of the cloned gene,

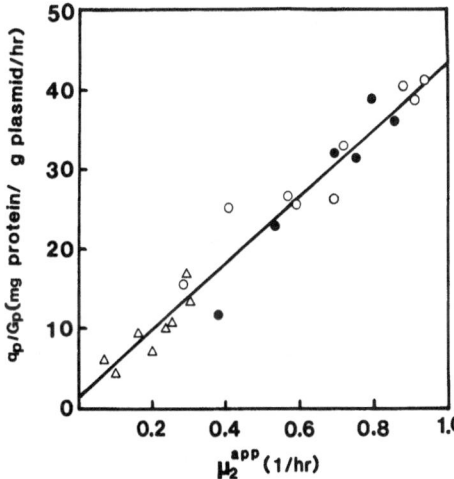

Fig. 1. Relationship between q_p/G_p and μ_2^{app}

3. the biosynthetic rate of other recombinants having different genetic parameters such as gene copy number and stability related parameters,
4. the biosynthetic rate of the same recombinant under different chemical and physical conditions of cellular environment, and others.

The recombinant fermentation process can then be optimized in terms of the operational and environmental bioreactor conditions by making use of the productivity kinetic model developed. For example, the optimal temperature and dilution rate control strategies can be established for a given set of host cell and vector systems and other conditions of recombinant fermentation.

As an example, the optimal temperature control strategy can be developed for a two-stage recombinant fermentation system in which the gene expression is induced by a temperature-sensitive gene switching system [3].

The kinetic model developed and given in Eq. (1) can be used to quantify the specific gene expression rate for the development of optimal temperature control policy. A constant temperature control policy and temperature profiling control policy including temperature cycling were studied and compared [3]. The results of this study showed that maximum average production rate was obtained from a temperature control policy in which the second stage was operated initially at a higher temperature and followed by a gradually but only slightly decreasing temperature profile (Fig. 2). The loss of productivity at the higher temperature level where the specific gene expression rate is relatively high is due to the more rapid loss of plasmids at this temperature. This analysis also showed that temperature cycling between high and low temperature levels does not lead to the maximum production rate.

As a second example of recombinant bio-process optimization, the effect of dilution rate on the productivity of a two-stage recombinant fermentation system can be considered [4]. The dilution rate determines the averaged or

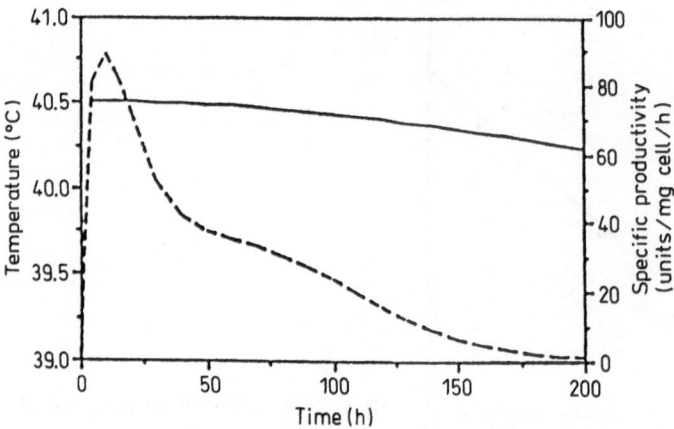

Fig. 2. Optimal temperature control policy: (———) temperature; (- - -) specific production rate

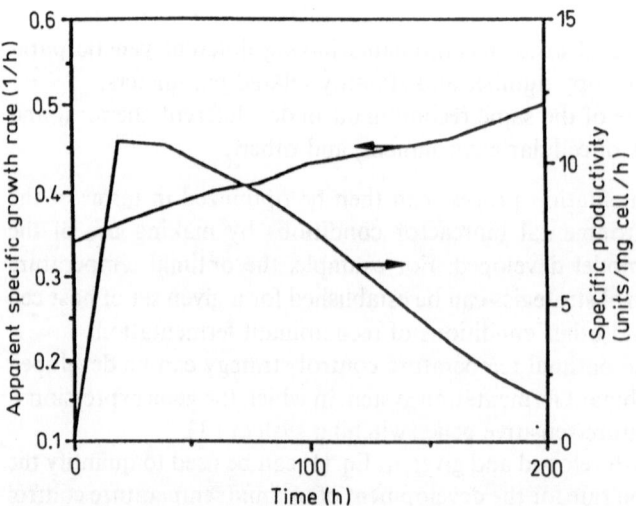

Fig. 3. Optimal control profiles of apparent specific growth rate and/or dilution rate and specific productivity profiles

apparent specific growth rate of a heterogeneous population of plasmid-containing and plasmid-free cells. The specific growth rate affects the stability of the plasmid-harboring cells, the plasmid concentration and the specific gene expression rate.

The simulation of plasmid stability in the first stage showed that for longer fermentation periods the plasmid concentration remained higher at higher dilution rates. The optimal apparent specific growth rate for maximum productivity in the second stage was found at a certain dilution rate. At this dilution rate, the growth rate coincides with the maximum plasmid content in the second stage. A

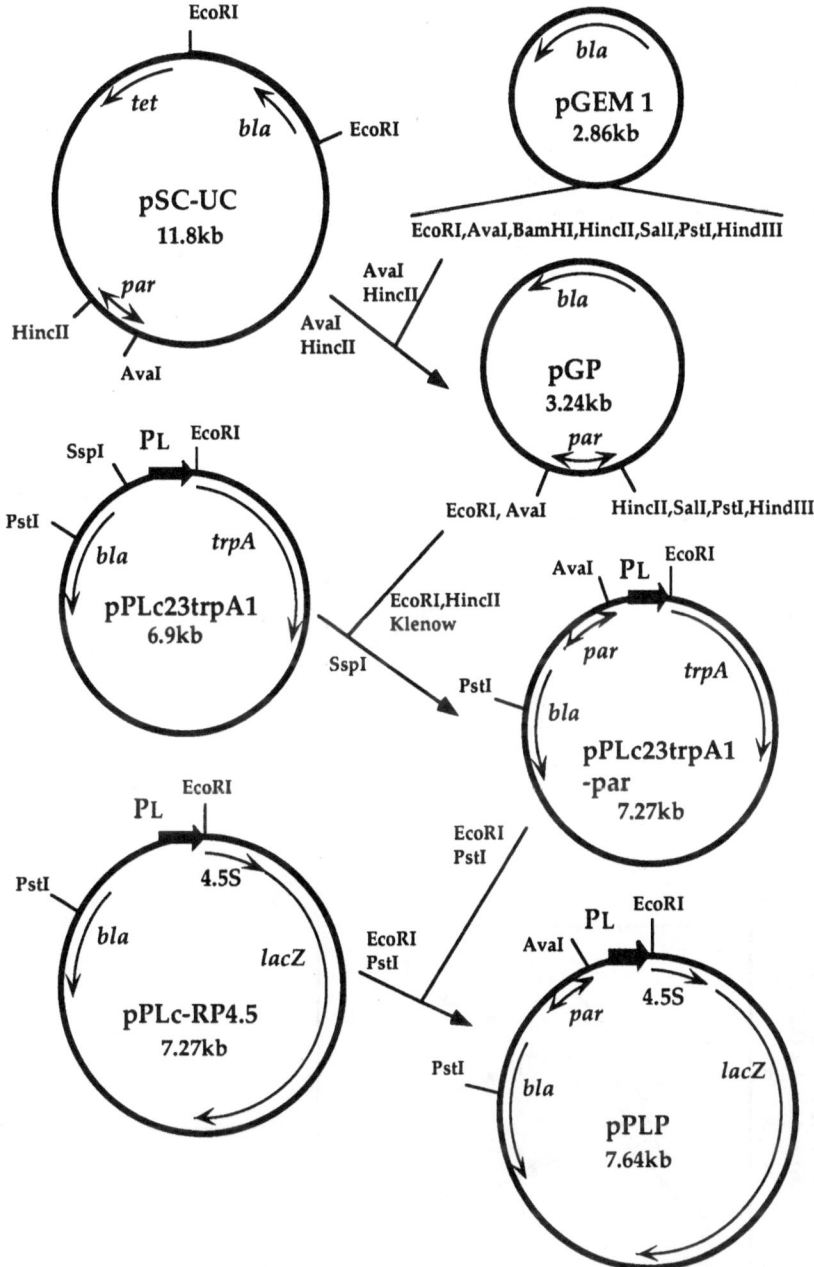

Fig. 4. Plasmids constructed and used in this research, pPLP. The *par* sequence is incorporated into the control plasmids

significant increase in the overall productivity can be achieved by a linear time dependent specific growth rate control in the course of fermentation time (Fig. 3).

The recombinant fermentation process can also be optimized by improvement of genetic/plasmid stability [5]. From our recent studies on the effect of the *par* sequence on the growth rate of recombinant *E. coli*, it was found that the *par* sequence showed a positive effect on the growth rate of recombinant cells which

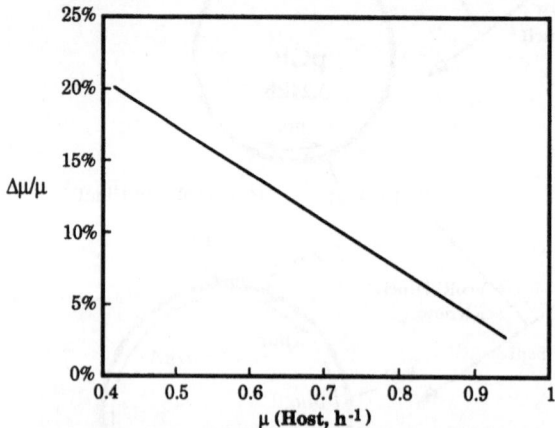

Fig. 5. The ratio of growth rate differential to specific growth rate of the host cell vs specific growth rate of the host cell

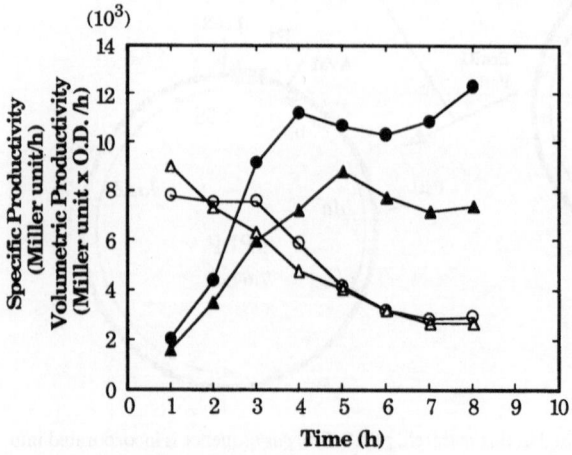

Fig. 6. Specific and volumetric productivity of β-galactosidase as a gene product

●: pPLP/K12ΔH1Δ*trp*EA (Specific productivity)
▲: pPLc-RP4.5/K12ΔH1Δ*trp*EA (Specific productivity)
●: pPLP/K12ΔH1Δ*trp*EA (Volumetric productivity)
▲: pPLc-RP4.5/K12ΔH1Δ*trp*EA (Volumetric productivity)

usually grow at a slower growth rate than the host cell. A decreased growth rate of recombinant cells is usually observed when a cloned gene protein encoded in a multicopy plasmid is induced from a strong promoter. This negative effect of multicopy plasmids and cloned genes on the cell growth rate may be the consequence of alternate use of reallocation of energy, precursor metabolites, and protein synthesing machinery.

Several plasmids including newly constructed vectors were introduced into the *E. coli* host cell. These plasmids, pPLcRP4.5, pPLP, and pGP, are different with respect to the presence of *par* sequence from pSC101 (Fig. 4). The difference in growth rates between the host cells with and without plasmids were evaluated [6]. Results obtained with the pPLc-RP4.5 bearing strain showed that the

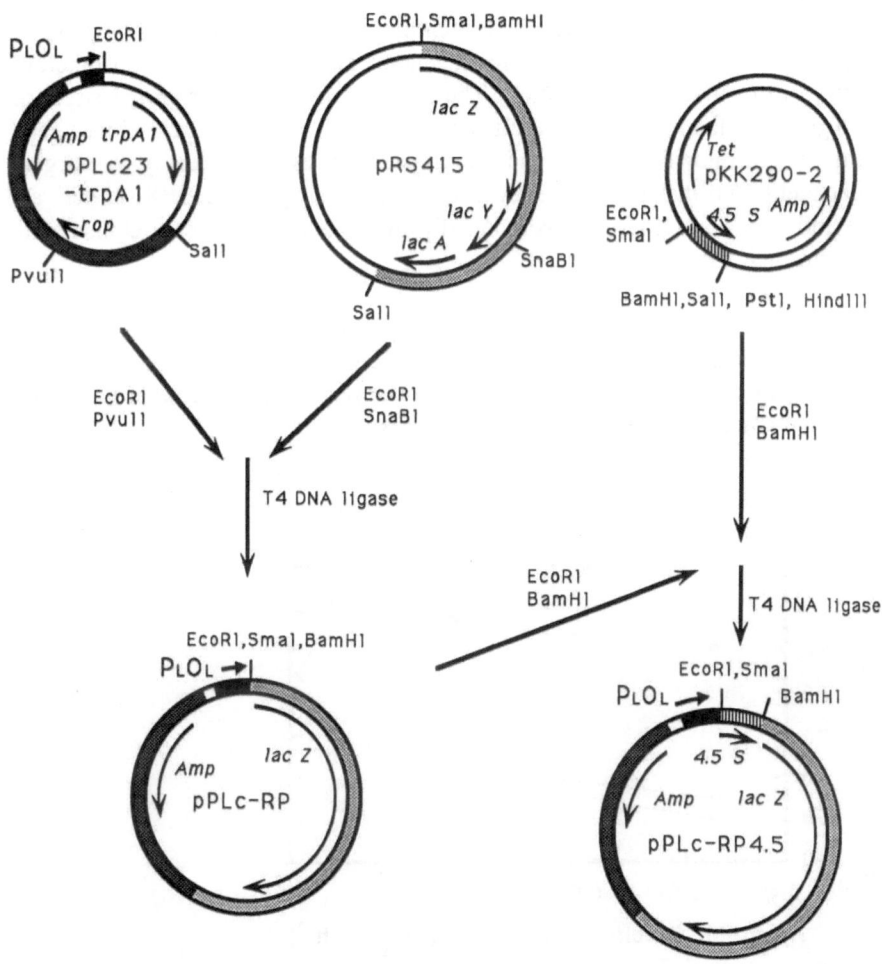

Fig. 7. Construction of plasmid, pPL$_c$-RP.4.5

relative growth rate differential, as the ratio of the growth rate differential to the specific growth rate, decreased linearly as the specific growth rate increased (Fig. 5). In batch cultur, the strain carrying pPLP plasmid which has the *par* sequence and *lacZ* gene, showed a significant increase in cell density and volumetric productivity of the gene product as compared to the strain with pPLc-RP4.5, which does not have the *par* locus (Fig. 6).

The new vector (pPLc-RP4.5) was used in order to investigate how plasmid content, transcription efficiency and translation efficiency affect the productivity of cloned gene protein [7]. The vector has P_L promoter, *lacZ* as structural gene and 4.5S RNA gene between P_L promoter and *lacZ* gene. We took advantage of the characteristic that 4.5S RNA is accumulated inside cell and can be quantitatively measured (Fig. 7).

A two-stage continuous culture system in combination with a temperature sensitive gene switching system was used to study the performance of the recombinant DNA (pPLc-RP4.5) fermentation. It was found that the plasmid content as varied by the dilution rate in production stage showed different pattern from that in growth stage. The results showed that promoter strength had greater influence on the overall gene expression efficiency of the cloned gene than the plasmid content, and the overall gene expression efficiency was largely dependent on translation efficiency when a multicopy plasmid (pBR322 derivative and *rop⁻*) and a strong promoter (P_L) were used to express a cloned gene protein in *E. coli* (Fig. 8).

Based on the results discussed in this paper, the productivity of a cloned gene product can be increased by increasing the specific growth rate of recombinants which, in turn, enhances the translation efficiency. One could also maximize and/or make incremental improvement of the recombinant productivity using both the engineering and genetic approaches.

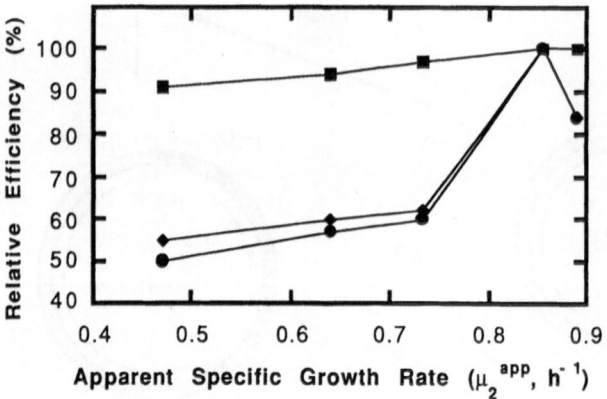

Fig. 8. The effects of apparent specific growth rate on the overall gene expression efficiency (●), transcription efficiency (■) and translation efficiency (◆)

2 References

1. Ryu Dewey DY, Kim JY, Lee SB (1991) Bioprocess kinetics and modeling of recombinant fermentation: In Rehm HJ, Reed G (Eds.), *Biotechnology* (pp. 485–505), Weinheim, VCH.
2. Lee SB, Ryu Dewey DY, Siegel R, Park SH (1987) Biotech Bioeng 31: 805.
3. Hortascu A, Ryu Dewey DY (1990) Biotech Prog 6: 403.
4. Hortascu A, Ryu Dewey DY (1991) Biotech Bioeng 38: 831.
5. Kim JY, Ryu Dewey DY (1992) Biotech Bioeng (in press).
6. Kim JY (1992) Ph.D. Dissertation, Univ of Calif Davis, CA.
7. Kim JY, Ryu Dewey DY (1991) Biotech Bioeng 38: 1271.

3 Biocatalysis

3. Blockpläne

3.1 Comparison of Primary, Secondary and Tertiary Structures of Xylanase of *Bacillus pumilus* and Cellulase of *Aspergillus acleatus*

Hirosuke Okada

Department of Applied Microbial Technology, Kumamoto Institute of Technology, 422 Ikeda, Kumamoto-shi, Kumamoto, 860 Japan

Contents

The tertiary structure of xylanase produced by *Bacillus pumilus* was determined by X-ray crystallographic analysis at the resolution level of 2.2 Å. It consisted of three β-sheets and one α-helix, and has a new topology. Glu[93] and Glu[182] were estimated to be the catalytic residues, from their location in the cleft, conservation among xylanases of the same evolutional origin, chemical modification, and site directed mutagenesis. The tertiary structure of cellulase (Fl-CM-cellulase) of *Aspergillus acleatus* was also studied by X-ray crystallographic analysis and found to be very similar to that of xylanase of *B. pumilus*.

1 Introduction

Cellulose and xylan are two main components of plant cell mass and are polysaccharides polymerized mainly by β-1,4-glycosylic linkage. The major difference between the two is that the sugar unit of cellulose is glucose while that of xylan is xylose. The chemical difference of these two sugar moieties is that one of the hydrogen atom of xylose at carbon 5 is replaced by a hydroxymethyl group in glucose. From the similarity of chemical structure of xylan and cellulose, it is interesting to compare the enzymes acting on these two polysaccharides. We have cloned and sequenced the genes coding xylanase of *Bacillus pumilus* IPO and cellulase of *Aspergillus acleatus* and have succeeded in solving the tertiary structure of these enzymes by X-ray crystallographic analysis. In this paper, I will discuss the homology of the tertiary structures though no significant homology in the amino acid sequence is observed.

2 Xylanase of *B. pumilus*

More than 50 kinds of xylanase of microbial origin including *Aspergillus*, *Bacillus*, *Streptomyces*, *Clostridium*, and *Trichoderma* have been purified and characterized. From the results, xylanase can be classified into two groups, of large size and small size from their molecular mass with some exceptions [1].

Bioproducts and Bioprocesses 2
Editors: Yoshida, Tanner
© Springer-Verlag Berlin Heidelberg 1993

Xylanase of smaller size has a molecular mass of 16 000–25 000 Da and that of xylanase of larger size is 43 000–55 000 Da. Among xylanases of smaller size, in addition to the molecular size, homology in amino acid sequence is observed so far as sequenced (Fig. 1). They are xylanase of *Bacillus pumilus* IPO, *B. subtilis*, *B. circulans*, *Trichoderma harzianum* and *Schyzophilum commune*. These xylanases contain 185–201 amino acid residues. Xylanase of *B. pumilus* IPO which we have studied in our laboratory is a typical xylanase of smaller size.

We have isolated a bacterial strain which produces high level of xylanase from a soil sample from Thailand. The xylanase gene was cloned to *Escherichia coli* [2] and the nucleotide sequence of 1070 bp covering the entire xylanase gene and its flanking regions was determined [3]. The amino acid sequence deduced from it is shown in Fig. 1. From the result, xylanase is produced as a prexylanase with a signal peptide of 27 amino acid residues at the amino terminal of the mature enzyme.

Crystals of xylanase were obtained by the micro-dialysis method [5]: 100 μL of the enzyme solution (4 mg/mL) containing 5% (w/v) polyethylene glycol 6000 and 50 mM phosphate buffer solution (pH 6.5) was dialyzed against 20 mL of a solution containing 15% polyethylene glycol 6000 and 50 mM phosphate buffer of the same pH at 20°C. Crystals with dimensions up to 0.5 mm × 0.3 mm × 0.3 mm grew within a month. These crystals belong to a space group of $P2_1$, having unit cell parameters of $a = 4.062$ nm, $b = 6.664$ nm, $c = 3.469$ nm, $\beta = 103.10°$.

Two heavy-atom derivatives, $PtCl_4$-derivative and UO_2-derivative, were prepared by soaking the native xylanase crystals in a 20% polyethylene glycol 6000 solution containing the corresponding heavy-atom compound. An electron density map was calculated at 2.2 Å resolution by the multiple isomorphous replacement method, supplemented by anomolous dispersion effects due to the Pt and U atoms; the figure of merit was 0.70.

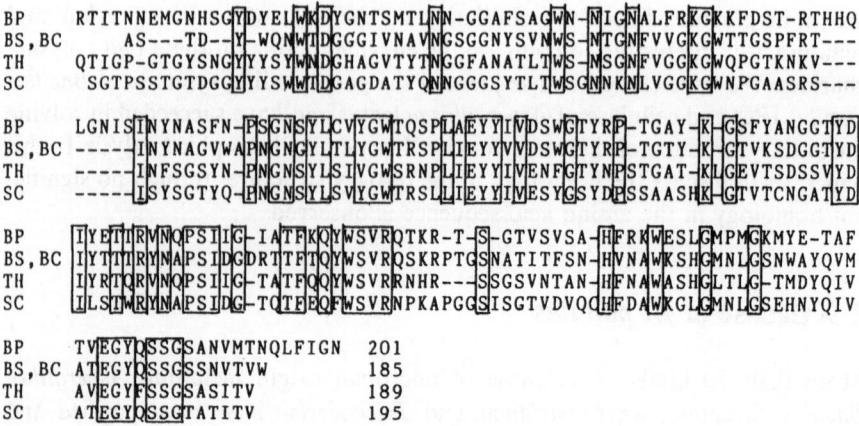

Fig. 1. Alignment of amino acid sequences of xylanases of smaller molecule size. *BP*; *B. pumilus* [3], *BS*; *B. subtilis* [4], *BC*; *B. circulans* [4], *TH*; *Trichoderma harzianum* [personal communication], *SC*; *Shizohilum commune* [personal communication]

The electron density map was of good quality and showed no main chain break. The starting point for the chain tracing was the identification of the aminoterminal end near the surface of the enzyme molecule. Fortunately, the $PtCl_4^{-2}$ ion in the $PtCl_4$ derivative was located on the N-terminus on the electron density map. Side chain densities were also well defined.

A structural model of the enzyme was built by fitting the ideal chemical-bond models into the electron density contours with a computer graphics display system, and the structure was refined by the Hendrickson–Konnart least-squares method. The final model has an R-value of 26%. Bond lengths and angles are restrained to root mean-squared deviations of 0.03 Å and 2° from their ideal values.

The drawing of the C-atoms and the schematic representation of the backbone structure are presented in Figs. 2A and 2B, respectively. The enzyme molecule has an ellipsoidal-shape of approximately 40 Å × 35 Å × 35 Å. The enzyme has a well defined deep cleft running down one side of the ellipsoidal

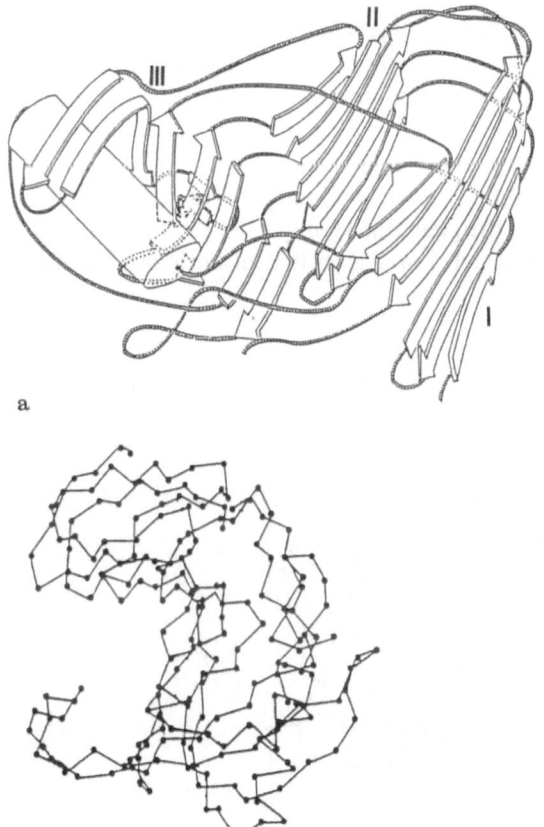

a

b

Fig. 2. A drawing of the C-atoms of xylanase (**B**) and a schematic representation of the backbone structure (**A**)

molecule to bind the substrate. As shown in Fig. 2B, the structure is predominatly characterized by three large-sheets; sheet I, sheet II and sheet III. The structure of these three sheets involves about 57% of the total residues in the polypeptide chain. The structure in the cleft is dominated by polypeptide chains of sheets II and III which spread apart, forming a "V"-shaped framework.

Sequence comparisons among xylanases of smaller molecule size shown in Fig. 1 indicate that many of the residues are conserved. It is generally said that important residues for the catalytic activity are conserved among series of enzymes of the same evolutional origin. In the case of hen egg-white lysozyme, within the active site cleft are two acidic groups, Glu[35] and Asp[52]; which are in close proximity to the bond to be cleaved when a substrate binds to the cleft. Assuming that the catalytic mechanism of xylanase may be somewhat similar to that of the lysozyme. Two acidic residues in the cleft and conserved among xylanases of the same evolutional origin may be the catalytic residues. Three candidates are Asp[21], Glu[93] and Glu[182].

As mentioned above, we prepared the UO_2-derivative for the crystal structure determination. The UO_2^{2+} ions bind to two sites in the enzyme, causing complete loss of the activity. This inactivation was protected by the addition of tetra-xyloside. The structure determination has permitted us to prove that one of these site is located at Glu93 which is conserved among the species, whereas the other site is at Glu[176], but this residue is not conserved. The specific chemical modification of Glu[93] with ^{14}C-glycineamide in the presence of water soluble carbodiimide resulted in complete inactivation. These results suggest that Glu[93] in the cleft plays an important role in the catalytic reaction.

The structure of the vicinity of Glu[93] is shown in Fig. 3. The carboxyl group of Glu[93] forms a salt bridge with the guanidino group of Arg[127]; thus the pK_a value of Glu[93] is expected to be much lower than that of normal Glu. In the

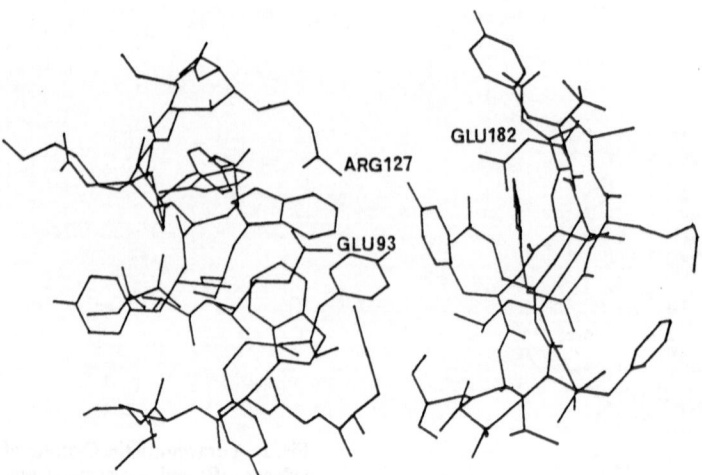

Fig. 3. Structure around the active site of xylanase produced by *B. pumilus*

Table 1. Catalytic parameters of the wild and mutant xylanases

Relative activity		V_m	K_m
Wild type	100		
-1M	103	383.7 U/mg	6.3 mg/ml
D21E	7	86.0	22.7
D21S	24	175	18.6
E93D	0.07	0.4	16.7
E93S	< 0.01		
E182D	< 0.01		
E182S	< 0.01		

vicinity of Glu^{93}, Glu^{182} is the only acidic residue conserved among the species, and the distance between their carboxyl groups is 7 Å. Glu^{182} is in a somewhat more hydrophobic environment than Glu^{93} and probably has a substantially elevated pK_a value. This suggests that the ionized carboxylate of Glu^{93} and the protonated carboxyl of Glu^{182} work as base and acid, respectively, in the general-acid–base catalysis. This is comparable to the situation around the catalytic residues, Asp^{52} and Glu^{35}, in the hen egg-white lysozyme.

In order to confirm the active residues proposed here, mutant enzymes were prepared using site-directed mutagenesis. The part of the xylanase gene which codes the excretion signal was replaced by ATG, so the resultant gene codes mature xylanase added methionine at its *N-terminal* (– 1 M). This gene was subjected to site directed mutagenesis to change Asp^{21} to serine (D21S) or glutamic acid (D21E), Glu^{93} to serine (E93S) or aspartic acid (E93D), and Glu^{182} to serine (E182S) or aspartic acid (E182D). These mutant enzymes were produced in *E. coli*, purified and characterized, and the results are shown in Table 1.

The mutant D21S and D21E still have considerable activity, suggesting that Asp^{21} is not the catalytic residue. The mutant E93S produced by replacing Glu93 by serine completely lost activity, as did the mutants E182S and E182D. On the other hand, the mutant E93D had a very slight activity. If the structural change by a single amino acid replacement is very localized, these results would strongly suggest that Glu^{93} and Glu^{182} represent the catalytic residues.

3 Cellulase of *Aspergillus acleatus*

Aspergillus acleatus produces 9 kinds of cellulose hydrolyzing enzymes [6]—3 endo-β-glucanases, 1 cellobiohydrolase, 2 exo-β-glucanases and 3 β-glucosidases. One of the endo-β-glucanases, named F1-CM-cellulase, has a molecular mass of 25 000 Da and not glycosylated. Similar cellulase of the same size and not glycosylated has been isolated from *Trichoderma viride* [7]. So, Fl-CM-cellulase is not very specific cellulase but has no homology to other cellulases whose amino acid sequences have been determined. The cDNA and the genomic

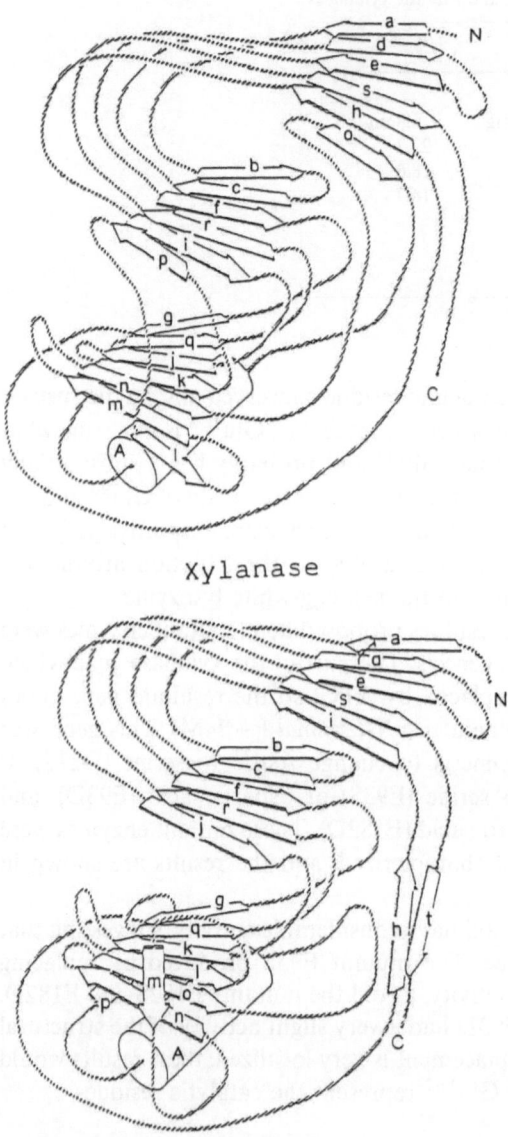

Xylanase

Cellulase

Fig. 4. Comparison of the backbone structure of xylanase of *B. pumilus* and cellulase of *Asp. acleatus*

DNA of this cellulase were cloned and sequenced [8]. The amino acid sequence deduced is 19% homologous to that of xylanase by the best fit alignment method.

Monocrystals suitable for X-ray crystallographic analysis were prepared and the tertiary structure was solved at 2.3 Å resolution level, though refinement experiments are still progressing.

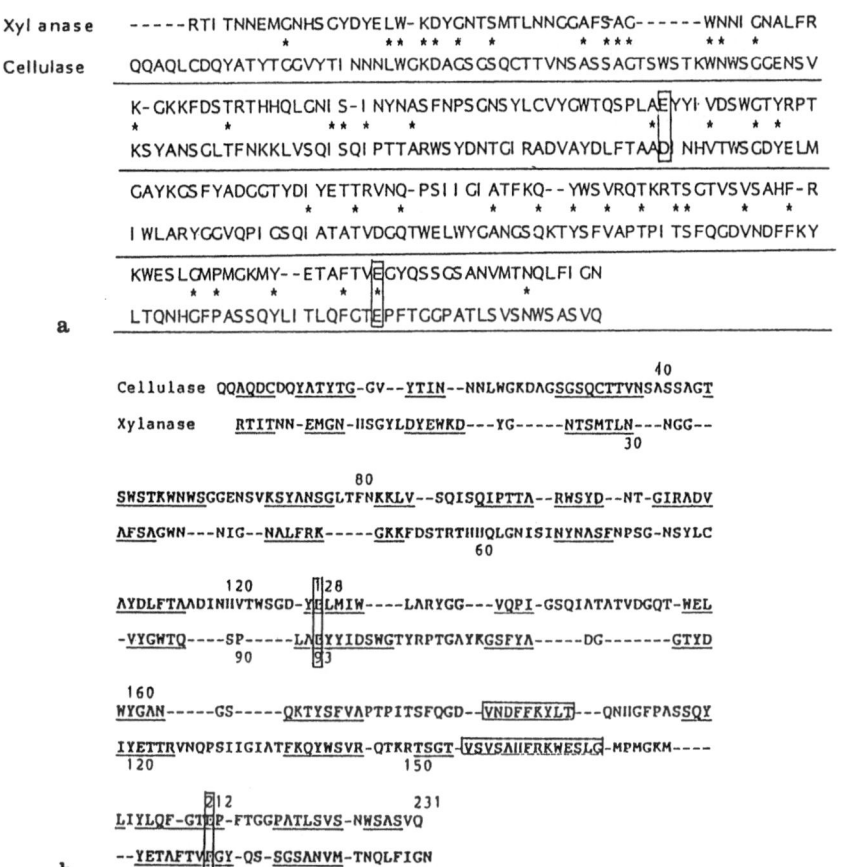

Fig. 5. Alignment of the amino acid sequences between xylanase of *B. pumilus* and cellulase of *Asp. acleatus* from the standpoint of the best fit (**A**) and tertiary structure (**B**)

Similar to xylanase of *B. pumilus* xylanase, cellulase of *Asp. acleatus* consisted of three β-sheets and one small α-helix as shown in Fig. 4. Both enzymes have basically the same arrangement of secondary structures. The topology of xylanase is unique among proteins so far determined, but the fact that cellulase of *Asp. acleatus* also has the same topology indicates that it might be widely distributed among hydrolases of small molecular mass.

The best fit alignment of the amino acid sequences between xylanase of *B. pumilus* and cellulase of *Asp. acleatus* is shown in Fig. 5A, which is completely different from that based on the tertiary structure as seen in 5B, in which comparable secondary structures are aligned to be in the same position.

4 References

1. Wong KY, Tan LUL, Saddler JN (1988) Microbiol Rev 52: 305.
2. Panbangred W, Kondo T, Negoro S, Shinmyo A, Okada H (1983) Mol Gen Genet 192: 335.
3. Fukusaki E, Panbangred W, Shinmyo A, Okada H (1984) FEBS Lett 171: 197.

4. Paice MG, Bourbonnais R, Jurasek L, Yaguchi M (1986) Arch Microbiol 144: 201.
5. Moriyama H, Hata Y, Yamaguchi H, Sato M, Shinmyo A, Tanaka N, Okada H, Katsube Y (1987) J Mol Biol 193: 237.
6. Murao S, Sakamoto R, Arai M (1988) Methods in Enzymol 160A: 274.
7. Voragen AGJ, Beldman G, Rombouts FM (1988) Methods in Enzymol 160: 243.
8. Ooi T, Shinmyo A, Okada H, Murao S, Kawaguchi T, Arai M (1990) *Nucleic* Acid Res 18: 5884.

3.2 Synthesis of Chiral Intermediates for D-Pantothenate Production by Microbial Enzymes

Hideaki Yamada and Sakayu Shimizu

Department of Agricultural Chemistry, Kyoto University Kitashirakawa, Sakyo-ku, Kyoto 606, Japan

Contents

D-Pantoyl lactone and D-pantoic acid are important chiral building blocks for the chemical synthesis of D-pantothenic acid. We found that *Candida parapsilosis* produced a carbonyl reductase which converts ketopantoyl lactone to D-pantoyl lactone. Similarly, conversion of ketopantoic acid to D-pantoic acid can be carried out in a better yield with the ketopantoic acid reductase of *Agrobacterium radiobacter*. Optical resolution of unmodified DL-pantoyl lactone can be carried out with a novel fungal enzyme which specifically hydrolyzes D-pantoyl lactone to D-pantoic acid. The enzyme catalyzing this stereospecific hydrolysis has been shown to be a new lactonohydrolase. These enzymes may be applicable to the industrial production of D-pantothenic acid.

1 Introduction

In recent years, the most significant development in the field of synthetic chemistry has been the application of biological systems to chemical reactions. Reactions catalyzed by enzymes or enzyme systems display far greater specificities than more conventional forms of organic reactions; some have been shown to be useful for synthetic or biotechnological applications.

We have recently been carrying out studies on the synthesis of various biologically, pharmacologically or chemically useful coenzymes, amino acids and other chemicals, using microbial enzymes as catalysts [1-6]. Here, we summarize the results of our recent work on the production of D-pantothenate, including the evaluation of useful enzymatic reactions for the synthesis, screening of potent microorganisms as catalysts, characterization of the enzymes responsible for the conversions and determination of reaction conditions for practical preparation.

At present, commercial production of pantothenate depends exclusively on chemical synthesis. The conventional chemical process involves reactions yielding racemic pantoyl lactone from isobutyraldehyde, formaldehyde and cyanide,

Bioproducts and Bioprocesses 2
Editors: Yoshida, Tanner
© Springer-Verlag Berlin Heidelberg 1993

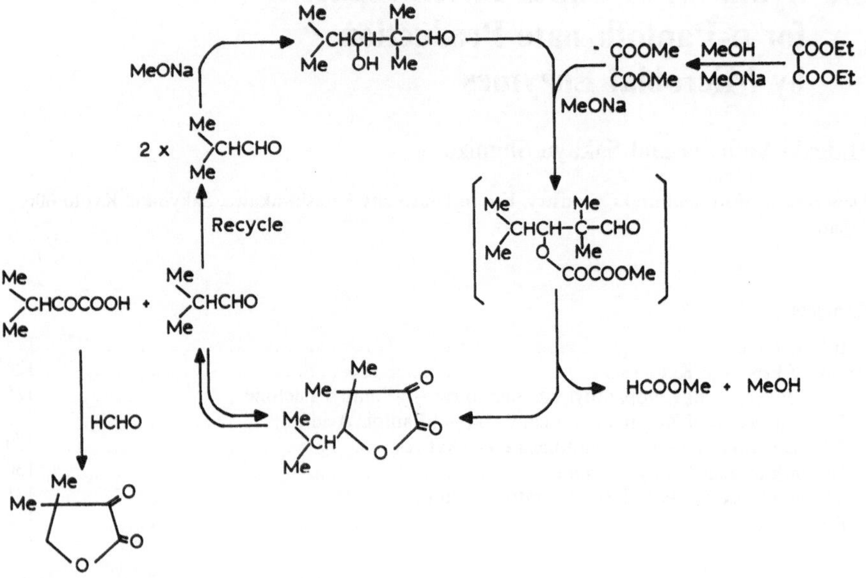

Fig. 1. The proposed reaction pathway for the one-pot synthesis of ketopantoyl lactone from isobutyraldehyde, diethyl oxalate and formalin

optical resolution of the racemic pantoyl lactone to D-(−)-pantoyl lactone with quinine, quinidine, cinchonidine, brucine and so on and condensation of D-(−)-pantoyl lactone with β-alanine. This is followed by isolation of the calcium salt and drying to obtain the final product. A problem with this chemical process, apart from the use of poisonous cyanide, is the troublesome resolution of the racemic pantoyl lactone and the racemization of the remaining L-(+)-isomer. Therefore, most of the recent studies in this area have been concentrated on the development of efficient methods to obtain D-(−)-pantoyl lactone.

To eliminate this resolution-racemization step, several microbial or enzymatic methods have been proposed. They fall roughly into two types based on the starting substrate used [1, 7].

2 Use of Prochiral Ketones

Recently, we have developed an efficient one-pot synthesis method for ketopantoyl lactone, in which it is synthesized from isobutyraldehyde, sodium methoxide, diethyl oxalate and formalin (Fig. 1). The reaction is performed in one step at room temperature with a yield of 81.0% [8]. Ketopantoyl lactone is a very promising starting material for the synthesis of D-(−)-pantoyl lactone, because it might permit several microbiological approaches leading to D-(−)-pantoyl lactone or D-(+)-pantothenate, as shown in Fig. 2. We therefore assayed a variety of microorganisms as to their reducing ability using several

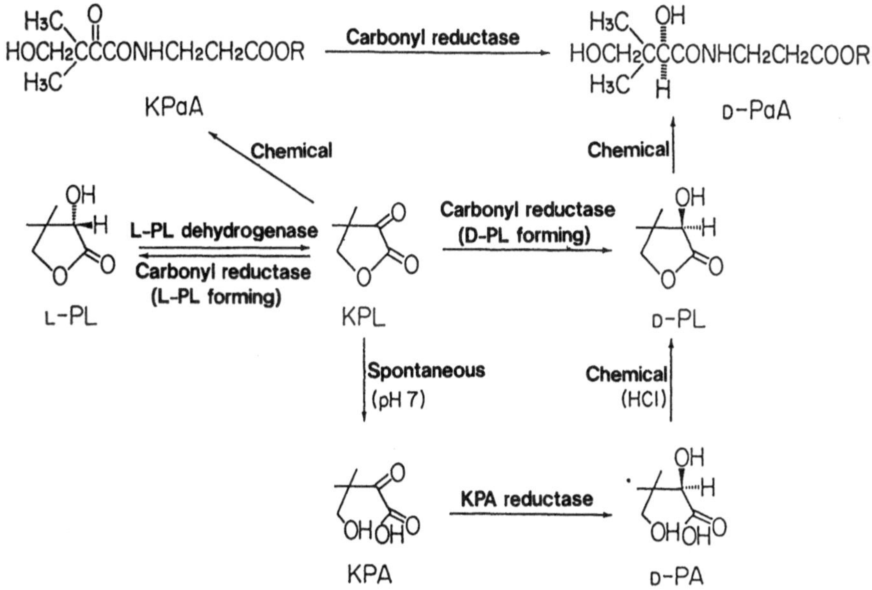

Fig. 2. Reactions involved in the enzymatic transformation to D-(−)-pantoyl lactone or D-(+)-pantothenate. *L-PL*, L-(+)-pantoyl lactone; *D-PL*, D-(−)-pantoyl lactone; *KPL*, ketopantoyl lactone; *KPA*, ketopantoic acid; *D-PA*, D-(−)-pantoic acid; *KPaA*, 2′-ketopantothenate; *D-PaA*, D-(+)-pantothenate

prochiral carbonyl compounds, such as ketopantoyl lactone, ketopantoic acid, ethyl 2′-ketopantothenate and 2′-ketopantothenonitrile.

2.1 Conversion of Ketopantoyl Lactone to D-(−)-Pantoyl Lactone

This conversion was assayed at pH 4–6 by incubating ketopantoyl lactone (10 mg/mL) in the culture broth, which had been grown with each test micro-organism, for 2 days at 28°C. Many microorganisms were found to convert the added ketopantoyl lactone to pantoyl lactone (Fig. 3a). However, the ratios of D- and L-isomers of formed pantoyl lactone were randomly distributed among the strains tested and the stereospecificity shown by the tested strains showed almost no relation to the genera or sources. For example, *Mucor racemosus* produced almost specifically the L-isomer with more than 85% molar yield. On the other hand, *Mucor javanicus* yielded a racemic mixture with 59% yield. Among 9 strains of *Rhodotorula glutinis*, which produced pantoyl lactone with greater than 90% molar yields, 5 strains gave racemic mixtures, 2 strains gave the L-isomer predominantly and the remaining 2 strains gave the D-isomer with more than 70% ee [9].

Practical stereospecific reduction of ketopantoyl lactone to D-(−)-pantoyl lactone was carried out with washed cells of *Rhodotorula minuta* or *Candida parapsilosis* as a catalyst and glucose as energy for the reduction. About 50 or 90

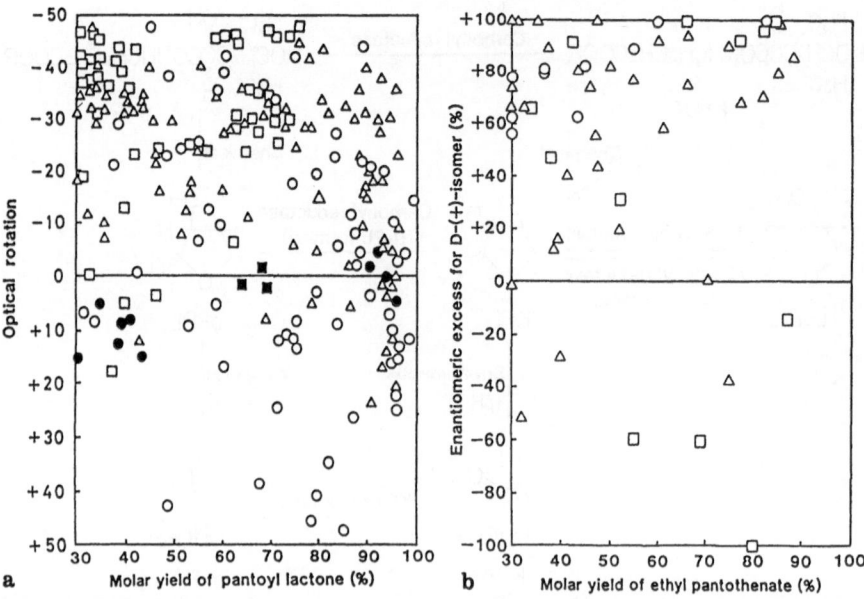

Fig. 3. Diversity of microbial reduction of ketopantoyl lactone (**a**) and ethyl 2'-ketopantothenate (**b**). △, yeasts; ○, molds; □, bacteria; ■, actinomycetes; ●, basidiomycetes

g/L of D-(−)-pantoyl lactone (94% or 98% ee, respectively) was produced with a molar yield of nearly 100% by *Rhodotorula minuta* and *Candida parapsilosis*, respectively (Fig. 4) [8, 10].

The enzyme catalyzing the asymmetric reduction of ketopantoyl lactone was isolated in a crystalline form from the cells of *Candida parapsilosis* and characterized in some detail (11, also see Table 1 and 2). It is a novel NADPH-dependent carbonyl reductase with a molecular mass of about 40 000 daltons. In addition to the reduction of ketopantoyl lactone, the enzyme catalyzes those of a variety of cyclic diketones, including derivatives of ketopantoyl lactone, isatin, camphorquinone and so on and gave the corresponding *R*-alcohols [11, 12]. We named the enzyme "conjugated polyketone reductase", since the enzyme catalyzes only the reduction of conjugated polyketones as follows,

$$
\underset{}{\overset{O\quad O}{\underset{}{-C-C-}}} \longrightarrow \underset{}{\overset{H\ OH\ O}{\underset{}{-C-C-}}}
$$

$$
\underset{}{\overset{O\ X\ Y\ O}{\underset{}{-C-C=C-C-}}} \longrightarrow \underset{}{\overset{OH\ X\ Y\ OH}{\underset{}{-C=C-C=C-}}}
$$

X, Y = H or alkyl

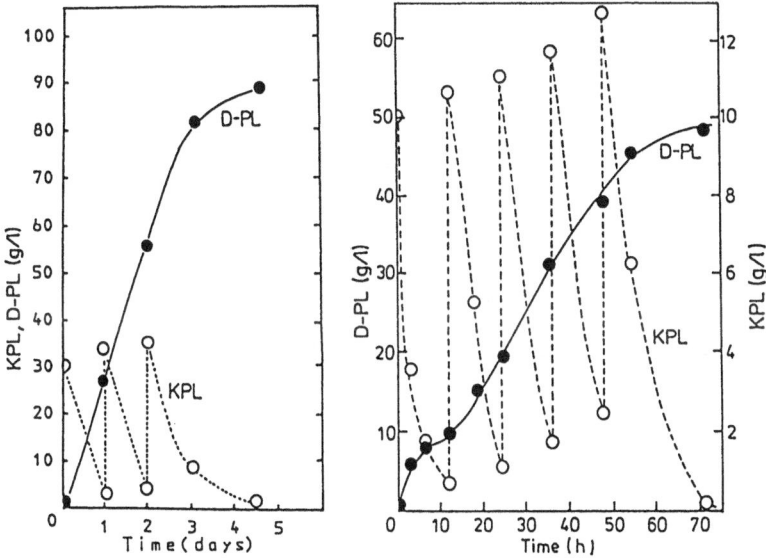

Fig. 4. Stereoselective conversion of ketopantoyl lactone (*KPL*) to D-(−)-pantoyl lactone (*D-PL*) by *Candida parapsilosis* (*left*) and *Rhodotorula minuta* (*right*)

The enzyme yielding the antipode, L-(+)-pantoyl lactone, was also isolated from *Mucor ambiguus* cells [13]. It is also a kind of "conjugated polyketone reductase" and consists of two polypeptide chains with an identical molecular mass of about 27 500 daltons (see Tables 1 and 2). The occurrence of two kinds of enzymes that show similar substrate specificity but differ from each other in their stereospecificity may be one of the possible reasons why the reduction of unnatural ketopantoyl lactone resulted in the formation of the D- and L-isomers in varying ratios as shown in Fig. 3a.

Since the above-mentioned enzymes require NADPH for the reduction of ketopantoyl lactone, there must be regenerating reaction(s) for NADPH coupled with the reduction reaction in cells. Experiments using cell-free extracts of *Candida parapsilosis* demonstrated that hexokinase, glucose 6-phosphate dehydrogenase and 6-phosphogluconate dehydrogenase were involved in the regeneration of NADPH. Cellular levels of these enzymes in *Candida parapsilosis* were almost the same as those in other common yeasts such as bakers' yeast and brewers' yeast, while considerable higher activity of "conjugated polyketone reductase" was detected in this yeast. Figure 5 outlines the mechanism for this reduction.

2.2 Conversion of Ketopantoic Acid to D-(−)-Pantoic Acid

The stereospecific reduction of ketopantoic acid to D-(−)-pantoic acid by ketopantoic acid reductase (EC 1.1.1.169), which is involved in the pantothenate

Table 1. Properties of the carbonyl reductases purified from various microorganisms

	Polyketone reductase (C. parapsilosis)	Polyketone reductase (M. ambiguus)	KPA reductase (P. maltophilia)	K-PaOEt reductase (C. macedoniensis)	L-PL dehydrogenase (N. asteroides)
Native M_r	37000	56000	116000	45000	600000
Subunit M_r	41600	27000	30500	42000	42000
$s_{20,w}$ (S)	—	—	—	—	—
pI	4.8	6.4	7.75	5.5	4.3
Absorption maximum (nm)	278	276	276	278	276, 370, 456
E (1%)	8.3	—	20.0	—	—
COOH amino acid	—	—	Phe	Thr	—
K_m (mM)	0.33 (KPL)	0.71 (KPL)	0.40 (KPA)	2.50 (K-PaOEt)	26.8 (L-PL)
V_{max} (µmol/min/mg)	481 (KPL)	541 (KPL)	1310 (KPA)	120 (K-PaOEt)	—
Cofactor	NADPH	NADPH	NADPH	NADPH	FMN
Optimum pH	7.0	6.0	6.0	6.5	9.0–9.5
Optimum temperature (°C)	40	40	37	40	40
pH stability	6.0–7.5	5.5–7.0	6.0–10	4.5–10.5	9.0–10
Thermal stability	42% (40°C, 10 min)	75% (45°C, 10 min)	90% (60°C, 10 min)	100% (55°C, 10 min)	60% (45°C, 10 min)
Inhibitor	Quercetin	Quercetin	—	—	—
Reaction mechanism	—	—	Ordered Bi-Bi	—	—
Enzyme formation	Constitutive	Constitutive	Constitutive	Constitutive	Inducible
Reference	(11)	(13)	(14)	(15)	(16)

KPA, ketopantoic acid; KPL, ketopantoyl lactone; K-PaOEt, ethyl 2'-ketopantothenate; L-PL, L-(+)-pantoyl lactone.

Table 2. Comparison of substance specificities of the carbonyl reductases purified from various microorganisms[a]

	$\begin{array}{cc} O & O \\ \parallel & \parallel \\ -C-C- \end{array}$ R-CHO	$\overset{X}{\underset{o\text{-}\quad m\text{-}\quad p\text{-}}{\bigotimes\text{-CHO}}}$	$\begin{array}{cc} O & O \\ \parallel & \parallel \\ Ha \end{array}$ OR	$\begin{array}{cc} & O \\ HO & \parallel \\ & OH \\ & O \end{array}$
Polyketone reductase				
(*C. parapsilosis*)	O			
(*M. ambiguus*)	O			
KPA reductase				
(*P. maltophilia*)				O
K-PaOEt reductase				
(*C. macedoniensis*)	O	O	O	

[a] Open circles indicate that enzymes can reduce the substrates indicated in the table. Ha = Cl, Br, F or N_3.

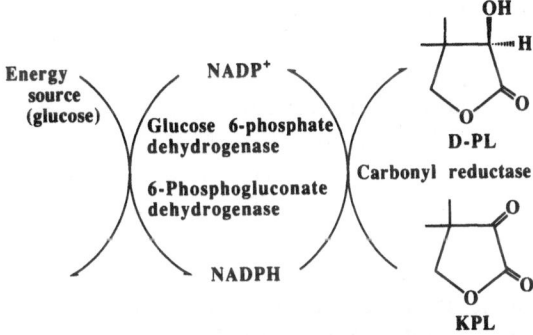

Fig. 5. The proposed mechanism for the reduction of ketopantoyl lactone (*KPL*) to D-(−)-pantoyl lactone (*D-PL*) coupled with the regeneration of NADPH

biosynthesis pathway, is also a promising reaction for the same purpose, because the ring-opening of ketopantoyl lactone to yield ketopantoic acid is easy and ketopantoic acid reductase shows absolute stereospecificity for D-(−)-pantoic acid [14]. When the same screening as described above was performed at pH 7 to 8, the added ketopantoyl lactone underwent rapid and spontaneous hydrolysis to ketopantoic acid and we observed a quite different distribution profile of the reducing activity. As expected, most of the microorganisms which showed high reducing activity were found to produce only the D-isomer. Through this screening, we found that many bacteria belonging to the genus *Agrobacterium* almost specifically produce the D-(−)-isomer (> 96% ee) in high yields. On incubation with a soil isolate, *Agrobacterium* sp. S-246, the yield of D-(−)-pantoic acid reached 119 g/L (molar yield, 90%; optical purity, 98% ee [17].

Ketopantoic acid reductase was isolated in a crystalline form from one of the potent D-(−)-pantoic acid producers, *Pseudomonas maltophilia* and characterized in some detail (14, see also Tables 1 and 2). It is an NADPH-dependent enzyme and is strictly specific to ketopantoic acid. The observation that mutants lacking this enzyme require either D-(−)-pantoic acid or pantothenic acid for growth and the revertants regain this activity indicates that it is involved in the pantothenate biosynthesis.

2.3 Reduction of 2'-Ketopantothenate Derivatives

We recently found that the rate of condensation of ketopantoyl lactone or D-(−)-pantoyl lactone with ethyl β-alanine, that yields ethyl 2'-ketopanto- thenate or ethyl D-(+)-pantothenate, respectively, is quite fast in comparison with that of the condensation of ketopantoyl lactone or D-(−)-pantoyl lactone with β-alanine, and that it proceeds more stoichiometrically [18]. Since the enzymatic hydrolysis of ethyl D-(+)-pantothenate has already been established [19], if the stereoselective reduction of ethyl 2'-ketopantothenate to ethyl D-(+)-pantothenate is possible, both the troublesome resolution and the incomplete condensation might be avoided at the same time. Thus, we assayed the reducing ability toward ethyl 2'-ketopantothenate of a variety of micro- organisms (see Fig. 3b). Several yeast strains, such as *Pichia aganobii, Hansenula miso* and *Candida macedoniensis*, exhibited high ability to reduce the 2'- ketopantothenate ester. Under the optimal conditions, *Candida macedoniensis* converted ethyl 2'-ketopantothenate (80 g/L) almost specifically to ethyl D-(+)- pantothenate (> 98% ee), with a molar yield of 97.2% [20]. In a similar manner, 2'-ketopantothenonitrile (50 g/L) was converted to D-(+)- pantothenonitrile (93.6% ee), with a molar yield of 95.6%, on incubation with *Sporidiobolus salmonicolor* cells as a catalyst [21].

The enzyme catalyzing these conversions was isolated from *Candida mace- doniensis* and characterized in some detail (see Tables 1 and 2). The enzyme shows broad substrate specificity; not only conjugated polyketones, but also aromatic aldehydes and 4-haloacetoacetic acid esters are reduced. The enzyme gives only the D-isomer on reduction of ethyl 2'-ketopantothenate, whereas it gives a mixture of the D- and L-isomers in a ratio of 4:1 on reduction of ketopantoyl lactone. The reduction product from ethyl 4-chloroacetoacetate is ethyl (S)-4-chloro-3-hydroxybutanoate (95% ee).

These enzymatic methods are simple and do not require a racemization step, which is necessary for the conventional chemical resolution.

3 Use of Racemic Pantoyl Lactone

Racemic pantoyl lactone can also be used as a starting substrate [22, 23]. We found that *Nocardia asteroides* cells specifically oxidizes the L-(+)-isomer in a racemic mixture of pantoyl lactone to ketopantoyl lactone, which is then converted to D-(−)-pantoyl lactone by the reduction with *Candida parapsilosis* cells as described above. In these two enzymatic steps, the coexisting D-(−)- isomer remains without any modification (see Fig. 2). Under suitable conditions, 72 g/L of D-(−)-pantoyl lactone was obtained from 80 g/L of DL-pantoyl lactone [22]. *Agrobacterium* sp. 246 also can be used in place of *Candida parapsilosis* [23]. Similar specific oxidation and reduction reactions can also be carried out with a single microorganism as catalyst (24). On incubation with washed cells of *Rhodococcus erythropolis*, D-(−)-pantoyl lactone in the reaction mixture reached 18.2 g/L with a molar yield of 90.5% (optical purity, 94.4% ee). This unique conversion proceeds through the successive reactions as follows:

1. the enzymatic oxidation of L-(+)-pantoyl lactone to ketopantoyl lactone, the same enzyme as that in *N. asteroides* has been suggested as being the responsible enzyme for this oxidation
2. the rapid and spontaneous hydrolysis of ketopantoyl lactone to ketopantoic acid
3. the enzymatic reduction of the ketopantoic acid to D-(−)-pantoic acid.

The enzyme catalyzing this reduction seemed to be a ketopantoic acid reductase, because *Rhodococcus erythropolis* cells could not utilize ketopantoyl lactone as substrate, different from *Candida parapsilosis*.

The enzyme catalyzing specific oxidation of L-(+)-pantoyl lactone to ketopantoyl lactone was purified from *Nocardia asteroides* cells. It is a flavoprotein with a molecular mass of about 60 000 daltons. Non-covalently bound FMN was identified as the responsible cofactor for the oxidation (see Table 1).

3.1 Stereospecific Hydrolysis of Pantoyl Lactone

Optical resolution of unmodified DL-pantoyl lactone can be carried out by a specific fungal hydrolase, as shown in Fig. 6. We found that many mold strains belonging to the genera *Fusarium*, *Giberella* and *Cylindrocarpon* specifically hydrolyze D-(−)-pantoyl lactone. When *Fusarium oxysporum* cells were incubated in 70% (w/v) aqueous solution of DL-pantoyl lactone for 24 h at 30 C with

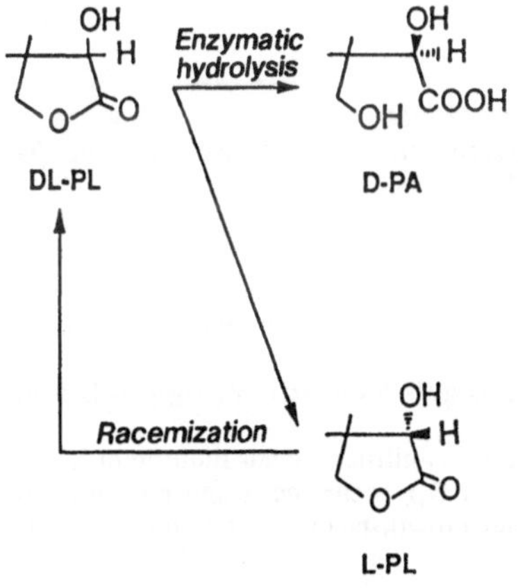

Fig. 6. Enzymatic resolution of racemic pantoyl lactone. *DL-PL*, DL-pantoyl lactone; *L-PL*, L-(+)-pantoyl lactone; *D-PA*, D-(−)-pantoic acid

Chemical resolution

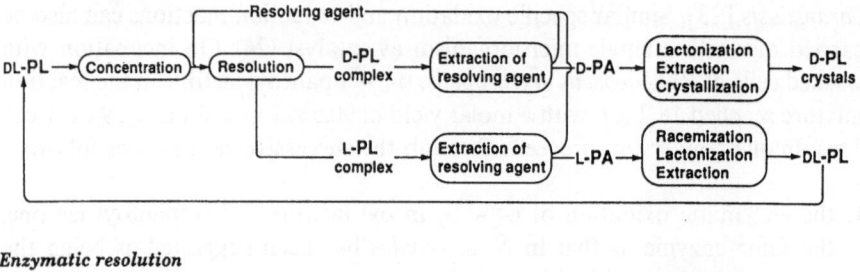

Enzymatic resolution

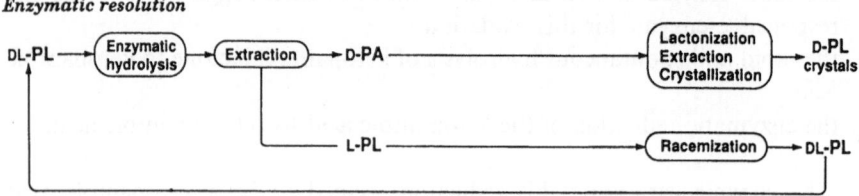

Fig. 7. Comparison of enzymatic and conventional resolution processes for DL-pantoyl lactone

automatic pH control (pH 6.8–7.2), about 90% of the D-isomer was hydrolyzed. The resultant D-(−)-pantoic acid in the reaction mixture showed a high optical purity (96% ee) and the coexisting L-isomer remained without any modification.

The enzyme responsible for this hydrolysis was isolated from *Fusarium oxysporum* cells. It is a kind of aldonolactonase with a molecular mass of 125 000 daltons. It specifically hydrolyzes several sugar lactones, such as D-galactonolactone, L-mannonolactone, and D-gulonolactone. All the substrate lactones have a downward hydroxyl group at 2-position, when the lactone rings are drawn according to the Haworth system.

A comparison of the enzymatic process proposed by the present study and the conventional chemical process for the resolution of DL-pantoyl lactone is shown in Fig. 7.

4 References

1. Yamada H, Shimizu S (1988) Angew Chem Int Ed Eng 27: 622.
2. Yamada H, Shimizu S (1985) Biotechnology—microbial conversion. In: Ullman's Encyclopedia of Industrial Chemistry, vol A4, VCH, Weinheim, p 150.
3. Shimizu S, Yamada H (1988) Microbial and enzymatic processes for the production of pharmacologically important nucleosides, Trends Biotechniol 2: 137 (1984).
4. Yamada H, Shimizu S (1985) Microbial enzymes as catalysts for synthesis of biologically useful compounds. In: Tramper J, van der Pals HC, Linko P Elsevier, (eds) Biocatalysts in Organic Synthesis, Amsterdam, p 19.
5. Shimizu S, Yamada H (1986) Coenzymes, In: "Biotechnology," Vol. 4, VCH Verlagsgesellschaft, Weinheim, pp. 159-184.
6. Nagasawa T, Yamada H (1989) Microbial transformations of nitriles, Trends Biotechnol 7: 153.
7. Shimizu S, Yamada H (1989) Pantothenic acid (vitamin B₅), coenzyme A and related compounds, In: "Biotechnology of vitamins, pigments and growth factors," EJ Vandamme, ed., Elsevier, London, pp. 199–219.

8. Shimizu S, Yamada H, Hata H, Morishita T, Akutsu S, Kawamura M (1987) Novel chemoenzy-matic production of D-(−)-pantoyl lactone, Agric Biol Chem 51: 289.

9. Shimizu S, Hata H, Yamada H (1984) Reduction of ketopantoyl lactone to D-(−)-pantoyl lactone by microorganisms, Agri Biol Chem 48: 2285.

10. Hata H, Shimizu S, Yamada H (1987) Enzymatic production D-(−)-pantoyl lactone from ketopantoyl lactone, Agric Biol Chem 51: 3011.

11. Hata H, Shimizu S, Hattori S, Yamada H (1989) Ketopantoyl-lactone reductase from *Candida parapsilosis*: purification and characterization as a conjugated polyketone reductase, Biochim Biophys Acta 990: 175.

12. Hata H, Shimizu S, Hattori S, Yamada H (1990) Stereoselective reduction of diketones by a novel carbonyl reductase from *Candida parapsilosis*, J Org Chem 55: 4377.

13. Shimizu S, Hattori S, Hata H, Yamada H (1988) A novel fungal enzyme, NADPH-dependent carbonyl reductase, showing high specificity to conjugated polyketones, purification and characterization, Eur J Biochem 174: 37.

14. Shimizu S, Kataoka M, Chung MCM, Yamada H (1988) Ketopantoic acid reductase of *Pseudomonas maltophilia* 845, purification, characterization and role in pantothenate biosyn-thesis, J Biol Chem 263: 12077.

15. Kataoka M, Doi Y, Sim T-S, Shimizu S, Yamada H (1992) A novel NADPH-dependent carbonyl reductase of *Candida macedoniensis*: purification and characterization, Arch Biochem Biophys 294: 469.

16. Kataoka M, Shimizu S, Yamada H (1992) Purification and characterization of a novel FMN-dependent enzyme, membrane-bound L-(+)-pantoyl lactone dehydrogenase from *Nocardia asteroides*, Eur J Biochem 204: 799.

17. Kataoka M, Shimizu S, Yamada H (1990) Novel enzymatic production of D-(−)-pantoyl lactone through the stereoselective reduction of ketopantoic acid, Agric Biol Chem 54: 177.

18. Sakamoto K, Kita S, Morikawa T, Shimizu S, Yamada H (1989) Preparation of ethyl 2'-ketopantothenate, Japanese patent application H1-45407.

19. Shimizu S, Sakamoto K, Yamada H (1987) Studies on the enzymatic hydrolysis of pantothenate esters (in Japanese) Nippon Nogeikagaku Kaishi 62: 283.

20. Kataoka M, Shimizu S, Doi Y, Yamada H (1990) Stereoselective reduction of ethyl 2'-ketopantothenate to ethyl D-(+)-pantothenate with microbial cells as a catalyst, Appl Environ Microbial 56: 3595.

21. Kataoka M, Shimizu S, Doi Y, Sakamoto K, Yamada H (1990) Microbial production of chiral pantothenonitrile through stereospecific reduction of 2'-ketopantothenonitrile, Biotechnol Lett 12: 357.

22. Shimizu S, Hattori S, Hata H, Yamada H (1987) Stereoselective enzymatic oxidation and reduction system for the production of D-(−)-pantoyl lactone from a racemic mixture of pantoyl lactone, Enzyme Microb Technol 9: 411.

23. Kataoka M, Shimizu S, Yamada H (1991) Stereospecific conversion of a racemic pantoyl lactone to D-(−)-pantoyl lactone through microbial oxidation and reduction reactions, Recl Trav Chim Pays-Bas 110: 155.

24. Shimizu S, Hattori S, Hata H, Yamada H (1987) One-step microbial conversion of a racemic mixture of pantoyl lactone to optically active D-(−)-pantoyl lactone, Appl Environ Microbiol 53: 519.

3.3 Mechanistic Aspects of Enzymatic Catalysis in Anhydrous Organic Solvents

Alexander M. Klibanov

Department of Chemistry, Massachusetts Institute of Technology, Cambridge, MA 02139, USA

Enzymatic catalysis in non-aqueous media has progressed from using enzymes in aqueous solutions containing relatively low fractions of water-miscible organic cosolvents to that in biphasic aqueous-organic mixtures, to that in microemulsions and reversed micelles, to that in monophasic organic media containing small amounts of water, to that in anhydrous organic solvents. It is easy to understand why enzymes retain catalytic activity in the first three types of reaction media, for in all of them the enzyme molecules are located in aqueous environments (and therefore the inherent enzymatic properties in such systems are usually not significantly different from those in water). Conversely, the phenomenon of enzymes vigorously functioning in organic solvents with no water goes against the conventional wisdom and universally accepted dogmas. Nevertheless, it has now been firmly established (see Ref. [1] for a review) that this phenomenon exists, that it is quite general and highly beneficial for bioprocessing [2], and that enzymes in organic solvents exhibit remarkable novel properties, e.g., greatly enhanced thermal stability and dramatically altered substrate specificity and stereoselectivity. In this presentation, fundamental questions concerning enzymatic catalysis in organic solvents will be addressed: How does enzymatic activity depend on the nature of the solvent and why? What physicochemical rules govern substrate, inhibitor, enantiomeric, and positional specificities of enzymes in organic solvents? What is the enzyme's structure and mechanisms of action in anhydrous media? and What additional new properties do enzymes acquire when placed in non-aqueous solvents?

References

1. AM Klibanov (1989) Trends Biochem Sci 14: 141.
2. AM Klibanov (1990) Accounts Chem Res 23: 114.

Bioproducts and Bioprocesses 2
Editors: Yoshida, Tanner
© Springer-Verlag Berlin Heidelberg 1993

1.3 Mechanistic Aspects of Enzymatic Catalysis in Anhydrous Organic Solvents

Alexander M. Klibanov

Department of Chemistry, Massachusetts Institute of Technology, Cambridge, MA 94123, USA

References

3.4 Construction of a Novel Biocatalyst Based on an Organelle Model—Expression of the Latent Function of the Enzyme by Organized Assemblies

Mitsuyoshi Ueda and Atsuo Tanaka

Laboratory of Industrial Biochemistry, Department of Industrial Chemistry,
Faculty of Engineering, Kyoto University, Yoshida, Sakyo-ku, Kyoto 606, Japan

Contents

The function of catalase was studied and it was demonstrated that catalase exclusively degrades hydrogen peroxide in a reaction mixture containing methanol and hydrogen peroxide, but when the enzyme was coupled with glucose oxidase, successful conversion of methanol to formaldehyde occurred at an optimized ratio of glucose oxidase to catalase. The ratio in the coupled system was very similar to the ratio of alcohol oxidase to catalase in peroxisomes, one of the subcellular organelles from a methanol-assimilating yeast, *Kloeckera* sp. 2201, in which these enzymes were coupled to metabolize methanol efficiently. Construction of the immobilized system of the coupled enzymes at the optimum ratio demonstrated that the oxidation of methanol through the peroxidatic function of catalase could be continuously and stably operated, the results indicating the usefulness of the system as a model of yeast peroxisomes. The coupled reaction with glucose oxidase brought out the latent function of catalase, which could not be expected in the system containing only catalase.

1 Introduction

In eukaryotic cells, some supramolecular systems composed of associated biomolecules are assembled into cellular organelles surrounded by membranes, for example mitochondria, chloroplasts, peroxisomes, lysosomes, vacuoles, endoplasmic reticula and nuclei. These organelles may become attractive models of novel biocatalysts for carrying out multistep and multifunctional reactions efficiently, although the organelles themselves seem to be inferior to enzymes and intact cells as biocatalysts because of their labile membraneous systems.

Peroxisomes, sometimes called microbodies, are one of such organelles appearing in eukaryotic cells. They are characterized biochemically by the presence of one or more hydrogen peroxide-producing oxidases and hydrogen peroxide-degrading catalase, and morphologically by a single-unit membrane with a fine granular matrix and often a crystalline core. Mammalian peroxisomes have the roles in the degradation of long-chain fatty acids, synthesis of ether lipids and metabolism of bile acid and so on. The organelles (called as glyoxysomes) of germinating seeds participate not only in the degradation of fatty acids but also in the synthesis of gluconeogenic intermediates through the glyoxylate cycle. The enzymes involved in photorespiration are localized in peroxisomes of plant leaves.

During the studies on the physiology and metabolism of alkane-assimilating and methanol-assimilating yeasts, we discovered the abundant occurrence of peroxisomes whose appearance and development were related to alkane or methanol assimilation and accompanied by a significant increase in the cellular activity of catalase (Fig. 1) [1, 2]. Yeast peroxisomes, especially those of

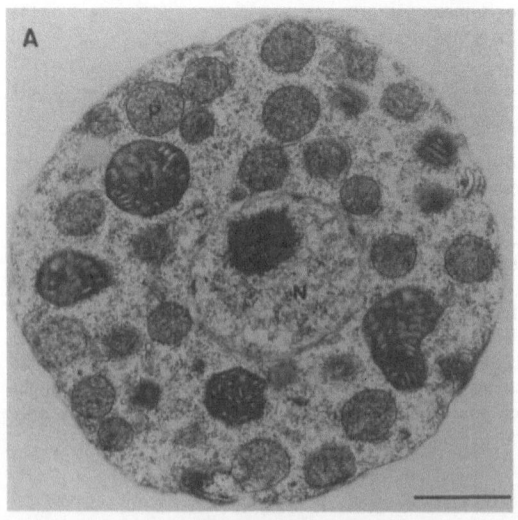

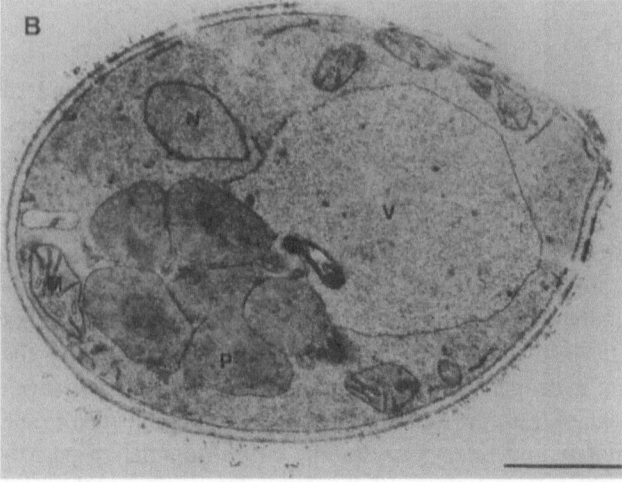

Fig. 1. Ultrastructure of *C. tropicalis* pK 233 cell (protoplast) grown on alkanes (**A**) and *Kloeckera* sp. 2201 cell grown on methanol (**B**) [1, 2]. *M*, mitochondrion; *N*, nucleus; *P*, peroxisome; *V*, vacuole. Bar: 1 μm

methanol-grown cells, seem to be the most adequate system for studying the cooperation of enzymes, because the simple synergistic action of alcohol oxidase and catalase functions in the organelle matrix. It was suggested that this coupling of alcohol oxidase and catalase in the peroxisomes repressed the catalatic activity and enhanced the peroxidatic activity of catalase. Such a synergistic reaction may bring out an alternative or latent function of enzymes coupled in the system. We describe the efficient expression of the peroxidatic activity of catalase in vitro by the construction of a coupled system with glucose oxidase as a model of peroxisomes and, subsequently, the construction of a novel oxidation reaction system.

2 Features of Yeast Peroxisomes as a Model of Organelles

Kloeckera sp. (*Candida boidinii*) 2201 grown on methanol has peroxisomes identified by cytochemical staining for catalase, the marker enzyme of peroxisomes, and, after isolation by means of differential centrifugation followed by sucrose density gradient centrifugation, it was proved that the organelles contain alcohol oxidase and catalase among the enzymes participating in the methanol oxidation, while formaldehyde dehydrogenase and formate dehydrogenase were found to be cytosolic (Fig. 2) [3]. Such organelles were also observed in the methanol-grown cells of yeasts belonging to the genera of *Hansenula*, *Pichia* and so on. Induced synthesis of catalase was observed along

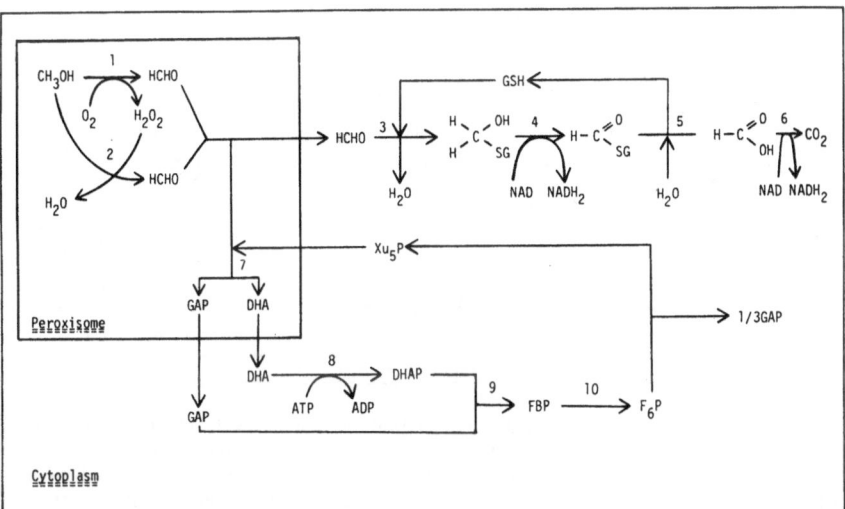

Fig. 2. Methanol metabolism in the methanol-assimilating yeast and the function of peroxisome. Enzymes: *1*, alcohol oxidase; *2*, catalase; *3*, non-enzymatic; *4*, formaldehyde dehydrogenase; *5*, S-formylglutathione hydrolase; *6*, formate dehydrogenase; *7*, dihydroxyacetone synthase; *8*, dihydroxyacetone kinase; *9*, fructose 1,6-bisphosphate aldolase; *10*, fructose 1,6-bisphosphate phosphatase. Abbreviations: *DHA*, dihydroxyacetone; *DHAP*, dihydroxyacetone phosphate; *FBP*, fructose 1,6-bisphosphate; F_6P, fructose 6-phosphate; *GAP*, glyceraldehyde-3-phosphate; *GSH*, reduced glutathione; Xu_5P, xylulose 5-phosphate

with the development of peroxisomes in methanol-assimilating cells of
Kloeckera sp. 2201. In addition to alcohol oxidase and catalase, dihydroxyacetone synthase, which participates in the first step of formaldehyde fixation, is
peroxisomal together with several oxidases. As described later, the initial
oxidation step of methanol is mediated by a synergistic action of alcohol oxidase
and catalase localized in peroxisomes. Thus, peroxisomes of methanol-utilizing
yeast cells serve not only to separate the hydrogen peroxide-forming system
from the cytosol but also to oxidize methanol efficiently.

On the other hand, the peroxisomes isolated from the alkane-grown cells of
Candida tropicalis pK 233 were shown to contain various enzymes related to
fatty acid metabolism [4]. The results of enzyme localization indicate that fatty
acids derived from alkanes are degraded to acetyl-CoA via β-oxidation exclusively in peroxisomes and the peroxisomes and mitochondria cooperate in
the synthesis of gluconeogenic intermediates, such as malate and succinate,
through the glyoxylate cycle since isocitrate lyase and malate synthase are
peroxisomal but other enzymes common to the TCA cycle are mitochondrial
(Fig. 3). Acetyl-CoA, which is necessary for the operation of the TCA cycle, is

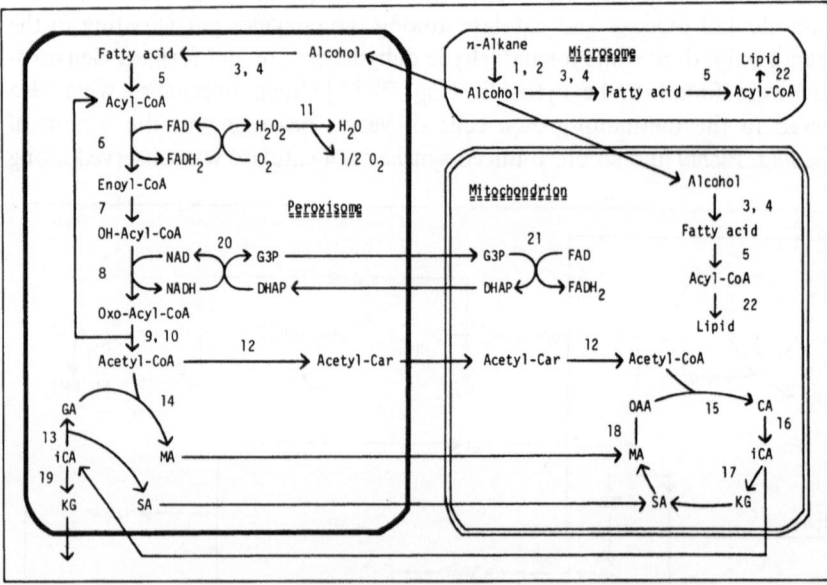

Fig. 3. Alkane metabolism in the alkane-assimilating yeast and the function of peroxisome.
Enzymes: *1*, cytochrome P-450; *2*, NADPH-cytochrome P-450 (cytochrome c) reductase; *3*, long-
chain alcohol dehydrogenase; *4*, long-chain aldehyde dehydrogenase; *5*, acyl-CoA synthetase; *6*,
acyl-CoA oxidase; *7*, bifunctional enzyme (enoyl-CoA hydratase); *8*, bifunctional enzyme (3-
hydroxyacyl-CoA dehydrogenase); *9*, 3-ketoacyl-CoA thiolase; *10*, acetoacetyl-CoA thiolase; *11*,
catalase; *12*, carnitine acetyltransferase; *13*, isocitrate lyase; *14*, malate synthase; *15*, citrate synthase;
16, aconitase; *17*, NAD-linked isocitrate dehydrogenase; *18*, malate dehydrogenase; *19*, NADP-
linked isocitrate dehydrogenase; *20*, NAD-linked glycerol-3-phosphate dehydrogenase; *21*, FAD-
linked glycerol-3-phosphate dehydrogenase; *22*, glycerol-3-phosphate acyltransferase. Abbreviations: *Acetyl-Car*, acetylcarnitine; *CA*, citrate; *DHAP*, dihydroxyacetone phosphate; *GA*, glyoxylate; *G3P*, glycerol-3-phosphate; *iCA*, isocitrate; *KG*, 2-ketoglutarate; *MA*, malate; *OAA*, oxalacetate; *OH-Acyl-CoA*, 3-hydroxyacyl-CoA; *Oxo-Acyl-CoA*, 3-ketoacyl-CoA; *SA*, succinate

assumed to be transported from peroxisomes to mitochondria via acetylcarnitine shuttle. The first step of β-oxidation is catalyzed by a flavoprotein, acyl-CoA oxidase (hydrogen peroxide-producing), linking to catalase (hydrogen peroxide-degrading). The second and third steps are mediated by a bifunctional enzyme having the activities of enoyl-CoA hydratase and 3-hydroxyacyl-CoA dehydrogenase. Two types of 3-ketoacyl-CoA thiolases, one being specific to acetoacetyl-CoA and the other to longer-chain 3-ketoacyl-CoAs, are responsible for the last step [5]. These enzymes are localized in peroxisomes but not in mitochondria. The results support the conclusion that fatty acids are completely degraded to acetyl-CoA (and propionyl-CoA from odd-chain fatty acids) in yeast peroxisomes. Accordingly, peroxisomes from the alkane-grown yeast cells could also be an excellent model for constructing a system with multistep reactions.

3 Application of Immobilized Yeast Peroxisomes

Yeast peroxisomes isolated from the methanol-grown cells of *Kloeckera* sp. 2201 were immobilized intact in matrices of photo-crosslinked resin gel, polyurethane resin gel, proteinic polymer and polyacrylamide gel. These entrapped preparations of peroxisomes showed relatively high activities of alcohol oxidase, catalase and D-amino acid oxidase (Table 1) [6]. The immobilized peroxisomes oxidized two moles of methanol to yield two moles of formaldehyde with consumption of one mole of molecular oxygen under both isotonic and hypotonic conditions. Addition of 3-amino-1,2,4-triazole, an inhibitor of catalase, decreased the formation of formaldehyde from methanol by a half without affecting the oxygen consumption. These results clearly reveal the synergistic reaction of alcohol oxidase and catalase in the peroxisomes. That is, hydrogen peroxide formed at the methanol oxidation by alcohol oxidase serves for the oxidation of another mole of methanol through peroxidatic action of catalase. This synergistic action of these enzymes in peroxisomes was observed even when

Table 1. Immobilization of *Kloeckera* peroxisomes [6]

Peroxisomes	Relative enzyme activity (%)		
	Catalase	Alcohol oxidase	D-Amino acid oxidase
Free	100	100	100
ENT-4000-entrapped[a]	46–55	79–80	49
PEGM-2000-entrapped[a]	36	76	31
PU-9-entrapped[b]	46	58	—
Albumin-entrapped	48	73	—
Polyacrylamide-entrapped	74	49	—

[a] Photo-crosslinked resin gel.
[b] Polyurethane resin gel.

Table 2. Substrate specificity of alcohol oxidase in free and immobilized yeast peroxisomes [6]

Substrate	Relative enzyme activity (%)		
	Free peroxisomes	ENT-4000-entrapped peroxisomes	Albumin-entrapped peroxisomes
Methanol	100	100	100
Ethanol	80	96	120
n-Propanol	56	60	50
n-Butanol	43	49	47
n-Amyl alcohol	39	35	14
Benzyl alcohol	28	11	8

the entrapped peroxisomes were burst inside gel matrices, indicating the utility of immobilized organelles for the study of their function in vivo. Alcohol oxidase of the entrapped peroxisomes showed substrate specificity similar to the free counterpart (Table 2) [6]. Not only methanol but also primary alcohols with two to five carbons were oxidized at significant rates. Stability of alcohol oxidase and catalase was improved to some extent by entrapment, but not satisfactorily so far as we have tested. However, immobilized peroxisomes are applicable to continuous or repeated assays or treatments of hydrogen peroxide, alcohols and D-amino acids as a multifunctional biocatalyst. Furthermore, this system is easily applicable to constructing a model of synergistic enzyme reactions.

4 Construction and Application of a Peroxisome Model

In the synergistic reaction of alcohol oxidase and catalase in peroxisomes of *Kloeckera* sp. 2201, hydrogen peroxide produced by alcohol oxidase is utilized for the oxidation of another mole of methanol through the peroxidatic action of catalase. When the peroxisomes were disintegrated and the resulting homogenate was diluted, no synergistic action of alcohol oxidase and catalase was observed. Furthermore, catalase itself did not exhibit the peroxidatic activity, especially in the presence of a high concentration of hydrogen peroxide. Exogenous supply of high concentrations of hydrogen peroxide may disturb the peroxidatic function of catalase due to the preferential reaction of a catalase-hydrogen peroxide complex with hydrogen peroxide. Probably, an endogenous supply of hydrogen peroxide is necessary to bring out the peroxidatic function of catalase as in the peroxisomes in vivo. Catalase (from bovine liver) and glucose oxidase (from *Aspergillus niger*), which supplies hydrogen peroxide through the oxidation of glucose, were selected to construct a model of the peroxisomal system (Fig. 4) [7]. We calculated the ratio of the total content of catalase and alcohol oxidase proteins in peroxisomes based on the results of the subcellular fractionation and their yields through purification [8, 9], the results indicating that alcohol oxidase and catalase were functioning at the ratios of alcohol

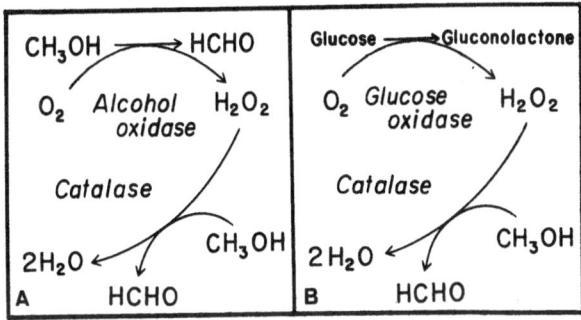

Fig. 4. Function of peroxisomes in a methanol-grown yeast, *Kloeckera* sp. 2201 **(A)** and their model **(B)** [7]

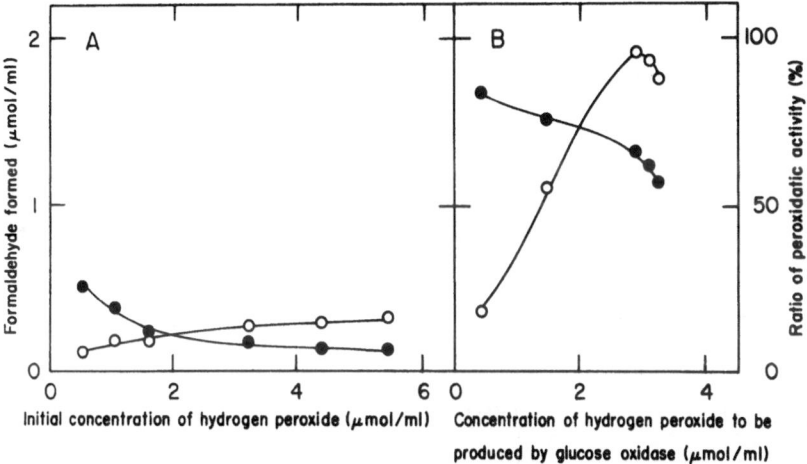

Fig. 5. Oxidation of methanol to formaldehyde with the system containing only catalase **(A)** and coupled catalase and glucose oxidase **(B)** [7]. (○), Formaldehyde formation; (●), ratio of the peroxidatic activity to total activity of catalase

oxidase to catalase: activity, 3.3×10^{-4}; number of molecules, 2–3; protein content, 5–8. We attempted to obtain optimum conditions to induce the peroxidatic activity of catalase in the catalase–glucose oxidase coupled system. By measuring the consumption of glucose, the endogenous production of hydrogen peroxide and the formation of formaldehyde, the peroxidatic activity of catalase was evaluated stoichiometrically. As shown in Fig. 5, the coupled system **(B)** revealed the peroxidatic activity of catalase, unlike the non-coupled system **(A)**. Furthermore, for the effective expression of such activity, the optimum conditions were found as follows on the ratios of glucose oxidase to catalase: activity, 1.0×10^{-3}; number of molecules, 1.3; protein content, 1 (Fig. 6). These facts were very similar to the results obtained in vivo with yeast peroxisomes and were consistent with those from the following theoretical analysis.

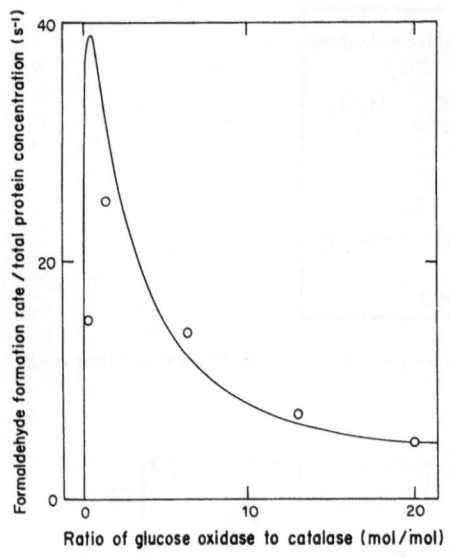

Fig. 6. Optimization of formaldehyde formation in the catalase–glucose oxidase coupled system [7]. (○), Experimental values obtained; (———), theoretical curve

The reactions in the catalase–glucose oxidase coupled system can be expressed as follows.

$$\text{Catalase} + H_2O_2 \underset{k_{-1}}{\overset{k_1}{\rightleftharpoons}} [\text{Catalase-}H_2O_2]$$

(E)   (H)   (EH)

$$[\text{Catalase-}H_2O_2] + CH_3OH \xrightarrow{k_2} \text{Catalase} + HCHO + 2H_2O$$

(EH)   (M)   (E)   (F)·   (W)

$$[\text{Catalase-}H_2O_2] + H_2O_2 \xrightarrow{k_3} \text{Catalase} + 2H_2O + O_2$$

(EH)   (H)   (E)   (W)   (O)

$$\text{Glucose oxidase-FAD} + \text{Glucose} \underset{k_{-1}}{\overset{k_4}{\rightleftharpoons}} [\text{Glucose oxidase-FAD-Glucose}]$$

(E_{ox})   (G)   (E_{ox}G)

$$[\text{Glucose oxidase-FAD-Glucose}] \xrightarrow{k_5} \text{Glucose oxidase-FADH}_2$$

(E_{ox}G)   (E_{red})

$$+ \delta\text{-Gluconolactone}$$
$$(\delta\text{-G})$$

$$\text{Glucose oxidase-FADH}_2 + O_2 \xrightarrow{\ k_6\ } \text{Glucose oxidase-FAD} + H_2O_2$$

$$(E_{red}) \qquad\qquad (O) \qquad\qquad\qquad (E_{ox}) \qquad\qquad (H)$$

As the result of the analysis with the steady-state approximation, the relationship between [H] and $[E_G]/[E_K]$ was calculated as follows,

$$[H] = \frac{1}{4k_1 k_3}\left[\left\{k_5(k_1 + k_3)\left(\frac{[E_G]}{[E_K]}\right) - k_1 k_2[M]\right\}\right.$$

$$+ \left\{k_5^2\,(k_1 + k_3)^2\left(\frac{[E_G]}{[E_K]}\right)^2 + 2k_1 k_2 k_5(3k_3 - k_1)[M]\left(\frac{[E_G]}{[E_K]}\right)\right.$$

$$\left.\left. + (k_1 k_2[M])^2\right\}^{1/2}\right],$$

where $[E_K]$ and $[E_G]$ are the total concentration of catalase and glucose oxidase, respectively. The values of k_1 and k_2 were reported on ethanol as $k_1 = 5.6 \times 10^6$ $(M^{-1}\ s^{-1})$ (for bovine liver catalase) and $k_2 = 1.8 \times 10^2$ $(M^{-1}\ s^{-1})$ (for bovine liver catalase). Although the values for methanol were not reported with bovine liver catalase, the values for ethanol and methanol were demonstrated to be very similar with horse liver catalase. The value of k_3 is 1.2×10^7 $(M^{-1}\ s^{-1})$ (for bovine liver catalase) and that of k_5, 4.0×10^2 (s^{-1}) (for *Aspergillus niger* glucose oxidase). In this reaction system, the concentration of methanol [M] was 1.6 M. Under the conditions of excess methanol, [M] can be assumed to be constant. Finally, the rate of formaldehyde formation per total protein can be expressed as follows,

$$\frac{d[F]}{dt}\bigg/[E_K] + [E_G] = \left(\frac{1}{A + \dfrac{B}{[H]}}\right)\left(\frac{1}{1 + \dfrac{[E_G]}{[E_K]}}\right)$$

$$A = \frac{k_1 + k_3}{k_1 k_2[M]},$$

$$B = \frac{k_{-1} + k_2[M]}{k_1 k_2[M]} = \frac{1}{k_1} \qquad (k_{-1},\ \text{negligible})$$

(A, B; constants).

Then, the theoretical curve was illustrated in Fig. 6. This kinetic analysis of the reaction in the catalase–glucose oxidase coupled system has demonstrated that the optimum relationship between the rate of formaldehyde formation and the ratio of glucose oxidase to catalase is present. The experimental results mentioned before were clearly confirmed.

By the immobilization of catalase and glucose oxidase with a photo-crosslinkable resin prepolymer ENT-4000, the coupled system is expected to be operated continuously. Figure 7 shows that the immobilized coupled enzyme system was more stable than the free coupled enzyme system in the formation of formaldehyde under the optimized conditions of the ratio of glucose oxidase to

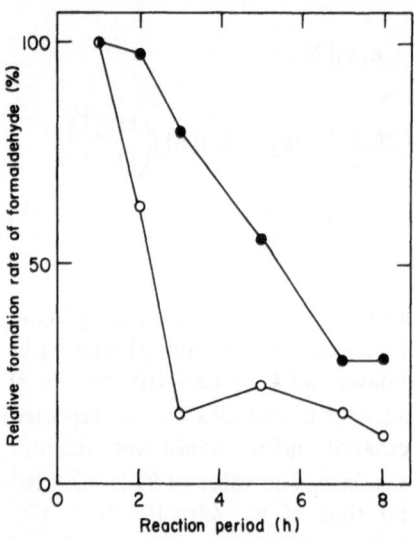

Fig. 7. Relative formation rate of formaldehyde with the free (○) and the immobilized (●) coupled enzyme system at the optimum ratio of the enzymes [7]

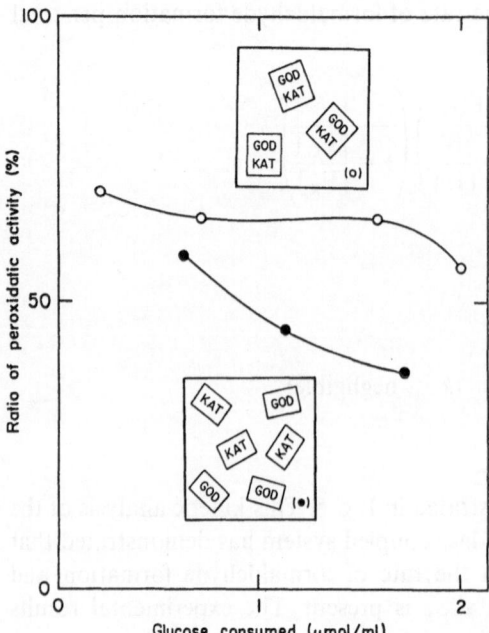

Fig. 8. Comparison of the ratio of per-oxidatic activity with the immobilized systems at the optimum ratio of enzymes [7]. (○), Co-immobilized catalase and glucose oxidase; (●), the system mixed with individually immobilized catalase and glucose oxidase

catalase. Even after 5 h, the immobilized system retained about 55% of the initial activity. In immobilizing both enzymes, the co-immobilized system at the optimum ratio exhibited a higher level of formation of formaldehyde than the mixed system of the individually immobilized enzymes at the optimum ratio as illustrated in Fig. 8. This result confirmed that endogenous production and supply of hydrogen peroxide were important to the synergistic action. Furthermore, the repeated production of formaldehyde was performed in a batch system at 30°C for 30 min. The immobilized coupled system exhibited the formation of formaldehyde even after 15 repeated uses, maintaining the activity of about 46% of that of the initial value.

5 Conclusion

As described above, a novel system with the immobilized coupled enzymes, in which the peroxidatic activity of catalase was expressed efficiently, could be operated continuously. This system will be applicable to the production of various aldehydes from alcohol substrates.

6 References

1. Osumi M, Miwa N, Teranishi Y, Tanaka A, Fukui S (1974) Arch Microbiol 99: 181.
2. Fukui S, Tanaka A, Kawamoto S, Yasuhara S, Teranishi Y, Osumi M (1975) J. Bacteriol 123: 317
3. Fukui S, Kawamoto S, Yasuhara S, Tanaka A, Osumi M, Imaizumi F (1975) Eur J Biochem 59: 561.
4. Tanaka A, Fukui S, (1989) The Yeasts, 2nd edn 3: 261.
5. Kurihara T, Ueda M, Tanaka A (1989) J Biochem 106: 474.
6. Fukui S, Tanaka A (1979) J Appl Biochem 1: 171.
7. Kawaguchi T, Ueda M, Tanaka A, Teramoto K (1989) Biocatalysis 2: 273.
8. Kato N, Omori Y, Tani Y, Ogata K (1976) Eur J Biochem 64: 341.
9. Mozaffar S, Ueda M, Kitatsuji K, Shimizu S, Osumi M, Tanaka A (1986) Eur J Biochem 155: 527.

3.5 Enzyme Catalysis in Unusual Environments

Harvey W. Blanch

Department of Chemical Engineering, University of California at Berkeley, Berkeley, CA 94720, USA

Although enzymes possess many unique catalytic properties, combining substrate and reaction specificity and high activity at ambient temperatures and pressures, they have not been widely applied in industrial chemical processes. Enzymes generally require an aqueous environment to function and thus cannot be used effectively to synthesize organic chemicals which have limited solubility in water. They may also be inhibited by reaction substrates or products, and in some cases production of the desired product requires the enzyme to operate in the "reverse" direction. Recently enzymes have been shown to function in selected organic solvents, to a large extent overcoming these obstacles. Several examples of enzyme reactions under conditions of low water content will be discussed.

One approach to overcoming substrate or product inhibition is to employ enzymes in reverse micellar liquid-membrane systems. The synthesis of tryptophan from indole is an example of the reversal of the direction of *tryptophanase*. A second model to be described is chymotrypsin activity in reverse micelles. The kinetics of the reaction and the influence of the aqueous "mini-phase" on catalysis will be discussed. The conformation of the enzyme and surrounding water has been probed by EPR spectroscopy.

The second approach we will examine is alcohol dehydrogenase in microcapsules. Microcapsules dispersed in organic solvents provide a suitable environment for conducting enzyme reactions involving cofactors and hydrophobic substrates. Encapsulated NADH is active and stable in cyclohexane provided the pH is adjusted appropriately. Mass transfer does not influence batch reaction rates. Conversion in a fluidized-bed reactor containing encapsulated NADH/NAD$^+$ depends strongly on residence time and inlet cinnamyl alcohol concentration employing cyclohexane as the continuous phase. However, analysis of these data is complicated by enzyme inactivation and interference from residual encapsulating agents.

We will also examine enzyme-catalyzed reactions in supercritical CO_2; the transesterification of triacylglycerides catalyzed by lipase. Kinetics of this reaction will be compared with those in aqueous–organic mixtures and the merits of this unusual solvent will be discussed.

Bioproducts and Bioprocesses 2
Editors: Yoshida, Tanner
© Springer-Verlag Berlin Heidelberg 1993

3.6 Preparation and Characterization of Semisynthetic Oxidases

Tetsuya Yomo, Itaru Urabe* and Hirosuke Okada

Department of Biotechnology, Faculty of Engineering, Osaka University,
2-1 Yamada-oka, Suita-shi, Osaka 565, Japan

Contents

5-Ethylphenazine-poly(ethylene glycol)-glutamate dehydrogenase conjugate (EP-PEG-GltDH) was prepared by linking poly(ethylene glycol)-bound 5-ethylphenazine to glutamate dehydrogenase. The average number of the ethylphenazine moieties bound per enzyme subunit was 0.7. This conjugate is a semisynthetic enzyme having NADH oxidase activity; the ethylphenazine moiety acts as a catalytic group, and the coenzyme-binding site of glutamate dehydrogenase acts as a substrate-binding site. The results of the kinetic analysis of the activity of EP-PEG-GltDH show that the apparent turnover number of the active site is 0.38 s^{-1}, which corresponds to the apparent intramolecular rate constant of the oxidation of NADH bound in the active site. 5-Ethylphenazine-glucose dehydrogenase-NAD conjugate (EP-GlcDH-NAD) was prepared by linking both poly(ethylene glycol)-bound 5-ethyl-phenazine and poly(ethylene glycol)-bound NAD to glucose dehydrogenase. The average number of the ethylphenazine moieties bound per enzyme subunit was 0.8, and that of the NAD moiety was 1.2. This conjugate is a semisynthetic enzyme having glucose oxidase activity. The catalytic cycle of the semisynthetic oxidase has two catalytic steps: reduction of the NAD moiety by the active site of the glucose dehydrogenase moiety and oxidation of the NADH moiety by another catalytic site of the ethylphenazine moiety. Strategies for designing enzyme-like catalysts are presented.

1 Introduction

There are three main characters in an enzyme reaction: a substrate-binding site, a catalytic site, and a substrate. If we prepare a new set of these players, the set will show a new catalytic activity and become an artificial enzyme. The investigation of the kinetic properties of these artificial enzymes will enable us to understand the rate-acceleration mechanisms used in natural enzyme reactions.

---

* To whom all correspondence should be addressed.

Bioproducts and Bioprocesses 2
Editors: Yoshida, Tanner
© Springer-Verlag Berlin Heidelberg 1993

Recently, we have prepared 1-(3-carboxypropyloxy)-5-ethylphenazine (Fig. 1), a stable derivative of 5-alkylphenazine with a carboxyl group at the end of the side chain at postion 1 [1]. This ethylphenazine (EP) derivative is an electron mediator that accepts two electrons form NAD(P)H and is then reoxidized with oxygen or other electron acceptors, such as 3-(4',5'-dimethylthiazole-2-yl)-2,5-diphenyltetrazolium bromide (MTT). Therefore, the EP derivative works as a catalyst that oxidizes NAD(P)H with oxygen or MTT by the reaction cycle as shown in Fig. 1. In addition, the carboxyl group of the EP derivative can be used for linking it to other molecules. Thus, we decided to prepare new artificial enzymes using EP as a catalytic group.

When EP is a catalytic group, the substrate should be NAD(P)H. As for the substrate-binding site, we decided to use the coenzyme-binding site of a dehydrogenase. This set shows an NAD(P)H oxidase activity, and becomes a semisynthetic NAD(P)H oxidase. In this set, we can also use the catalytic groups in the active site of the dehydrogenase, and the dehydrogenase is converted into the corresponding oxidase.

In this work, we prepared the following conjugates: 5-ethylphenazine-poly(ethylene glycol)-glutamate dehydrogenase conjugate (EP-PEG-GltDH) and 5-ethylphenazine-glucose dehydrogenase-NAD conjugate (EP-GlcDH-NAD). These conjugates show new catalytic activities, and the EP moiety acts as an artificial catalytic group for the oxidation of NADH (or the NADH moiety) with oxygen or MTT in the new catalytic reactions. These conjugates are unique enzyme-like catalysts of semisynthetic oxidases, and provide us with kinetic basis for designing artificial enzymes.

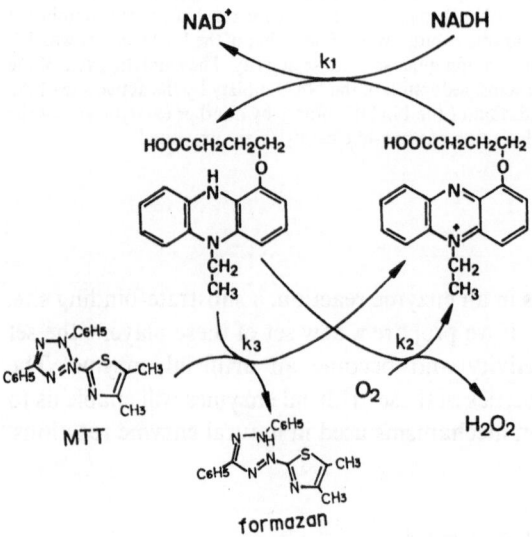

Fig. 1. NADH oxidation catalyzed by an ethylphenazine derivative

2 Semisynthetic NADH Oxidase [2]

2.1 Reactivity of NADH Bound at a Coenzyme-Binding Site

At first, we planned to link a catalytic group to a substrate-binding site; that is, covalently linking the EP group to a dehydrogenase with the long, flexible, and hydrophilic linker poly(ethylene glycol) (PEG). The conjugate of this type is expected to work as an NADH oxidase, if the bound NADH reacts well with the EP moiety. Figure 2 shows all the possible reactions catalyzed by EP-PEG-dehydrogenase.

The reactivity of the bound NADH may be different depending on the structure of the binding site, and we must select an appropriate dehydrogenase for this purpose. The reactivity of NADH bound to the site can be examined by measuring the rate of oxidation of NADH with an EP derivative in the presence of various concentration of the dehydrogenase (Fig. 3). In the presence of an excess amount of MTT, all the EP moiety is in the oxidized form, and the initial reaction rate (v) at the steady state is:

$$v = k_1[EP]_t[NADH] + k_2[EP]_t[site \cdot NADH], \tag{1}$$

where $[EP]_t$ is the total concentration of the EP moiety, $[NADH]$ is the concentration of free NADH, $[site \cdot NADH]$ is the concentration of NADH bound to the site (and that of the site having NADH), and k_1 and k_2 are the second-order rate constants of the reactions of the EP moiety with free and bound NADH, respectively.

The mass balances for this reaction system at the initial steady state are:

$$[NADH]_t = [NADH] + [site \cdot NADH], \tag{2}$$

$$[site]_t = [site] + [site \cdot NADH], \tag{3}$$

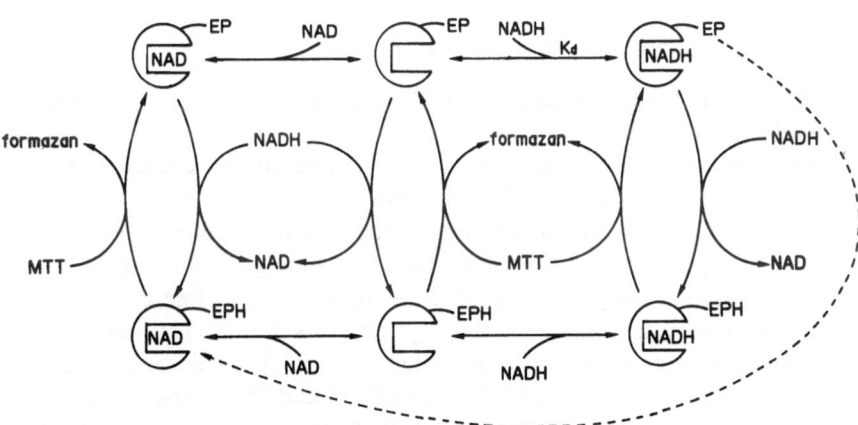

Fig. 2. Reactions catalyzed by EP-PEG-dehydrogenase

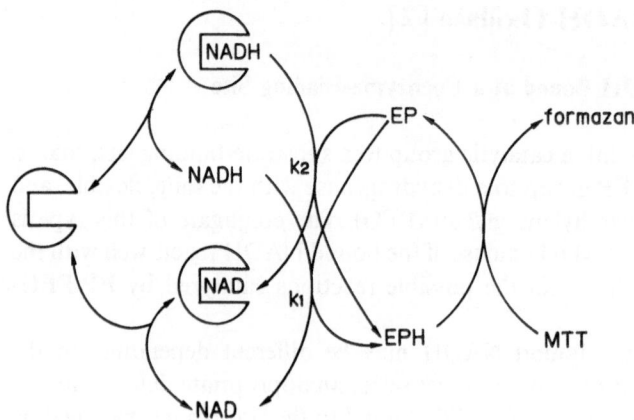

Fig. 3. Reaction system for determination of reactivity of NADH bound to dehydrogenase

where $[NADH]_t$ and $[site]_t$ are the total concentrations of NADH and the binding sites, respectively, and $[site]$ is the concentration of the free sites.

The dissociation constant (K_d) of the site $\cdot$ NADH complex and the fraction (α) of NADH bound to the site are expressed as:

$$K_d = [site][NADH]/[site \cdot NADH], \tag{4}$$

$$\alpha = [site \cdot NADH]/[NADH]_t. \tag{5}$$

From Eqs. (2)–(5), α is:

$$\alpha = ([site]_t + [NADH]_t + k_d)/(2[NADH]_t) - \{([site]_t$$
$$+ [NADH]_t + K_d)^2 - 4[NADH]_t[site]_t\}^{1/2}/(2[NADH]_t). \tag{6}$$

From Eqs. (1), (2), and (5),

$$v = [EP]_t[NADH]_t\{k_1 + (k_2 - k_1)\alpha\}. \tag{7}$$

Equation (6) shows that α changes depending on $[site]_t$, and Eq. (7) shows that v is linearly related to α.

We examined the reactivity of NADH bound to the coenzyme site of glutamate dehydrogenase (GltDH) and also of lactate dehydrogenase (LDH). The values of v were measured at different concentrations of one of the enzymes, and the α values were calculated from Eq. (6) assuming an arbitarary value of K_d. The plot of v vs α thus obtained generally shows a curved line, and the shape of the line varies depending on the K_d value. As Eq. (7) requires a linear relationship between v and α, the K_d value that satisfy the linear relationship were calculated for GltDH and LDH to be $1.0\,\mu M$ and $23\,\mu m$, respectively, so as to make the correlation coefficient maximum by the least-squares method. Figure 4 shows the straight lines obtained for the two enzymes.

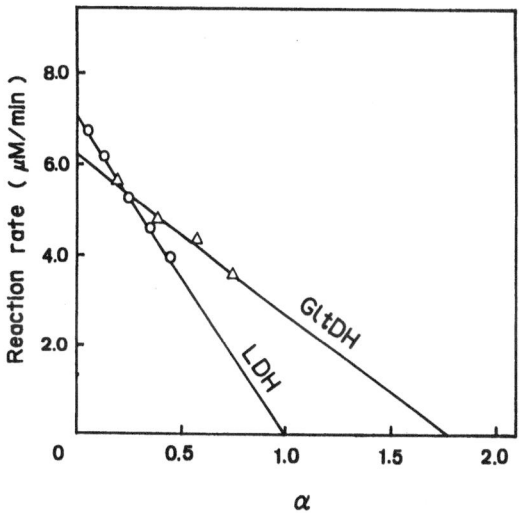

Fig. 4. Plot of Eq. (7)

Equation (7) indicates that the slope of the lines in Fig. 4 corresponds to the difference between k_2 and k_1. This means that the slope is negative when the bound NADH has lower reactivity, the slope is zero when the site has no effect on the reactivity of NADH, and the slope is positive when the bound NADH has higher reactivity than the free coenzyme. Equation (7) also shows that the value of k_2 can be obtained from the value of v at $\alpha = 1$ or from the α-intercept (α_0) of the straight line from the following equation:

$$k_2 = k_1(1 - 1/\alpha_0). \tag{8}$$

Thus the k_2 values for GltDH and LDH is calculated to be 1.20 mM^{-1} s^{-1} and 0, respectively, using a k_1 value of 2.78 mM^{-1} s^{-1} for the reaction of NADH with EP-PEG [3]. These results show that NADH bound to the site of GltDH reacts with EP-PEG at a rate 43% of that of free NADH, but NADH bound to the site of LDH cannot react. Therefore, we selected GltDH as a substrate-binding site for preparing a semisynthetic NADH oxidase.

2.2 Preparation of EP-PEG-GltDH

EP-PEG was covalently linked to GltDH by the procedure previously described for the preparation of dehydrogenase-PEG-NAD conjugates [4, 5]. The terminal amino group of poly(ethylene glycol)-bound ethylphenazine (EP-PEG) was activated with a bifunctional reagent [3,3'-(1,6-dioxo-1,6-hexanediyl)bis-2-thiazolidinethione, DHBT], and the activated EP-PEG was linked to GltDH (Fig. 5). The conjugate thus prepared had 0.7 mol of the EP moiety per mol of enzyme subunit in average. The GltDH activity of the conjugate was 39% of that of the native enzyme.

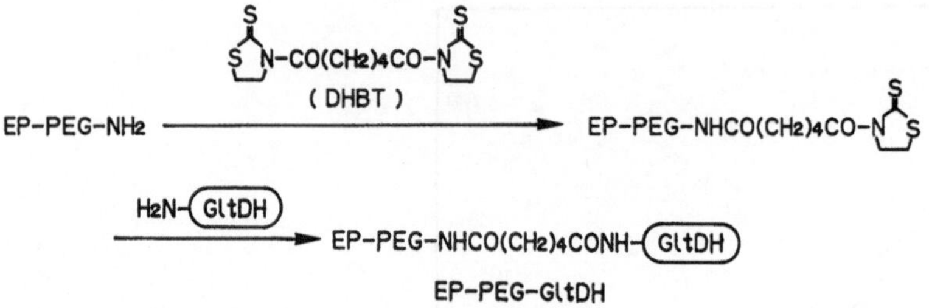

Fig. 5. Scheme for the preparation of EP-PEG-G1tDH

2.3 NADH Oxidase Activity of EP-PEG-GltDH

The active site of EP-PEG-GltDH is composed of one NADH-binding site and some catalytic groups of the EP moieties. The number of the EP moieties may be different for each active site, the distance between the binding site and the EP moieties may be different for each EP moiety, and the affinity of the binding site for NADH may be different for each active site because of the effects of the chemical modification. Therefore, these active sites are grouped by the affinity of the substrate-binding site and the turnover number: the ith active site has a dissociation constant of $K_{d,i}$ and a turnover number of $k_{cat,i}$. In this case, the initial steady-state rate of the NADH oxidase reaction catalyzed by EP-PEG-GltDH is expressed by the following equation (if the NADH concentration is much larger than the active-site concentration, and the MTT concentration is in excess):

$$d[\text{formazan}]/dt = 0.7 \cdot k_1' \, [\text{EP-PEG-GltDH}]_t [\text{NADH}]_t$$

$$+ \sum_i \{ k_{cat,i} \cdot \gamma_i [\text{EP-PEG-GltDH}]_t [\text{NADH}]_t / (K_{d,i}$$

$$+ [\text{NADH}]_t) \}, \tag{9}$$

where k_1' is the second-order rate constant of the reaction between the EP moiety and free NADH (the value is assumed to be the same for all the EP moieties), [EP-PEG-GltDH]$_t$ is the total concentration of the conjugate expressed as the subunit (i.e. active site) concentration, γ_i is the fraction of the ith active site. The first term of Eq. (9) is the intermolecular reaction rate between free NADH and the EP moiety. The second term is the intramolecular reaction rate between bound NADH and the EP moiety, i.e. the activity of the active site (see Fig. 2). The effects of the presence of the NADH-binding site are expressed by this second term. It should be noted that for a dehydrogenase with an oligomeric structure, one EP moiety may be used commonly by two or more active sites on the enzyme, but the form of Eq. (9) does not change. It should also be noted that almost all the EP moieties are in the oxidized form and are always ready to react with NADH in the presence of excess MTT.

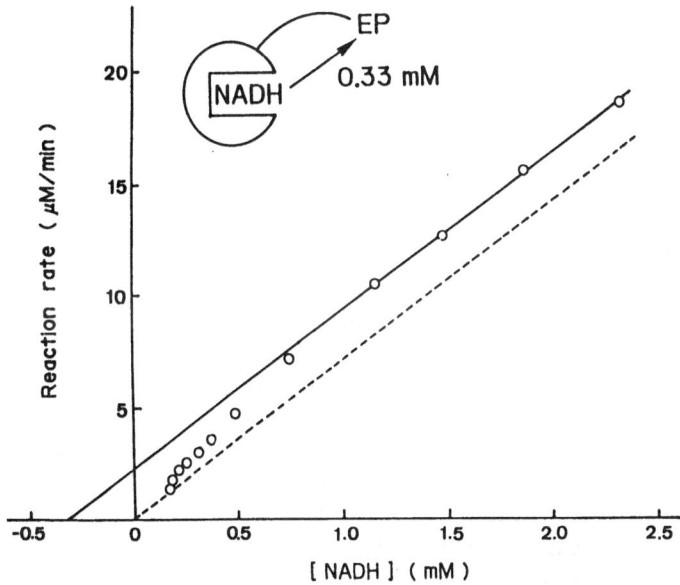

Fig. 6. Effect of NADH concentration on the activity of EP-PEG-G1tDH

The NADH oxidase activity of EP-PEG-G1tDH was measured at different initial concentrations of NADH, and the results are shown in Fig. 6. The reaction rate increases with the increase in the NADH concentration, and reaches a straight line with a slope of $1.94 \times 10^{-1}\,\mathrm{s}^{-1}$. When $[\mathrm{NADH}]_t \gg K_{d,i}$, Eq. (9) is simplified to:

$$d[\text{formazan}]/dt = 0.7 \cdot k_1'[\text{EP-PEG-GltDH}]_t[\text{NADH}]_t$$
$$+ k_{cat}'[\text{EP-PEG-GltDH}]_t, \qquad (10)$$

where k_{cat}' is the apparent k_{cat} of this preparation of the semisynthetic oxidase, and corresponds to $\sum\limits_i k_{cat,i} \cdot \gamma_i$ in Eq. (9). The values of k_1' and k_{cat}' were found to be $1.60\,\mathrm{mM}^{-1}\mathrm{s}^{-1}$ and $0.38\,\mathrm{s}^{-1}$, respectively, from the straight line shown in Fig. 6 and by using Eq. (10). This k_1 value is 58% of the second-order rate constant (k_1) of the reaction between EP-PEG and NADH; probably reflecting some effect of the enzyme molecule on the reactivity of the EP moiety.

If the NADH-binding site of EP-PEG-GltDH is not competent (i.e. for all $i, k_{cat,i} = 0$), the reaction rate is expressed by the first term of Eq. (9) and by the broken line in Fig. 6 with a slope of $0.7 \cdot k_1'[\text{EP-PEG-GltDH}]_t$ ($= 1.94 \times 10^{-3}\,\mathrm{s}^{-1}$). Thus, the rate-acceleration effect of the NADH-binding site (i.e. the second term of Eq. (9)) is shown in Fig. 6 as the difference between the data points and the broken line. As expected from Eqs. (9) and (10), the difference increases with the increase in $[\mathrm{NADH}]_t$ and reaches a constant value of $k_{cat}'[\text{EP-PEG-GltDH}]_t$ (0.65 μM s^{-1}) when $[\mathrm{NADH}]_t \geqq 1.1\,\mathrm{mM}$. In this constant region, almost all the binding sites are occupied with NADH.

2.4 Effective Concentration

EP-PEG-GltDH worked as an NADH oxidase; that is, GltDH was converted to NADH oxidase just by linking with EP-PEG. In addition, the oxidase activity was higher than the intermolecular reaction rate between the EP moiety and NADH, and the difference between them corresponds to the rate acceleration due to the presence of the substrate-binding site near the catalytic group. This rate-acceleration effect can be explained by the increase in the effective concentration of NADH around the catalytic group due to the presence of the bound NADH. The apparent effective concentration of bound NADH for the catalytic group can be calculated as the ratio of the rate constants (based on the EP moiety) of the intramolecular and intermolecular reactions, i.e. $k'_{cat}/(0.7 \cdot k')$; this value has been found to be 0.33 mM. It is also obtained from Fig. 6 as the intercept at $d[\text{formazan}]/dt = 0$ of the straight line. This means that when all the binding sites are occupied by NADH, the EP moiety acts as if it were in a solution with an NADH concentration of $([\text{NADH}]_t + 0.33\,\text{mM})$. This increase in the local NADH concentration is the effect of the presence of the substrate-binding site near the catalytic group. This is one of the rate acceleration mechanisms used by native enzymes. The effective concentration of 0.33 mM is not large compared to $[\text{NADH}]_t$, but this effective concentration is due only to the bound NADH molecules whose actual concentration is at most 1.73 μM $([\text{EP-PEG-GltDH}]_t)$. Therefore, the reactivity of NADH is enhanced about 200 times by this binding to the site. It should also be noted that this rate-acceleration mechanism is effective even when the binding sites are not saturated with NADH. In this case, the apparent effective concentration varies depending on the NADH concentration, in contrast to the constant effective concentration obtained for EP-PEG-NAD [3] this difference is due to the difference between the absence and the presence of the covalent linkage between the two reactants.

The apparent effective concentration of the NADH-binding site for the EP moiety is calculated to be 0.77 mM from the apparent effective concentration (0.33 mM) and the relative reactivity (0.43) of bound NADH to free NADH. The effective concentration of 0.77 mM is about twice the effective concentration (0.4 mM) of the NAD moiety for the EP moiety of EP-PEG-NAD [3]), and this corresponds to the situation that the EP moiety is linked with two NADH-binding sites by the spacer of PEG (M_r 3000, about 25 nm). In reality. the EP moiety is linked by the spacer at a point on the surface of GltDH, and the point is generally not at the binding site. In this case, the maximum distance between the binding site and the EP moiety is longer than the length of the PEG spacer, and the effective concentration of one NADH-binding site must be less than 0.4 mM. Therefore, the EP moiety can interact with more than three binding sites of a hexameric GltDH molecule. The apparent effective concentration of the EP moiety for the NADH-binding site, on the other hand, is 0.54 mM. The effective concentration of 0.54 mM means that one NAD-binding site can interact with more than two EP moieties. It should be noted that the effective concentrations obtained here are apparent values for this sample, because the attachment points and the number of EP-PEG on the enzyme molecule are not constant for each subunit.

3 Semisynthetic Glucose Oxidase [6]

3.1 Preparation of EP-GlcDH-NAD

Glucose dehydrogenase (GlcDH) catalyzes the oxidation of glucose to gluconolactone using NAD or NADP as a coenzyme (Fig. 7). In our previous work, we found that this enzyme shows very low activity for NAD derivatives, such as PEG-NAD, but when PEG-NAD is covalently linked to the enzyme, the

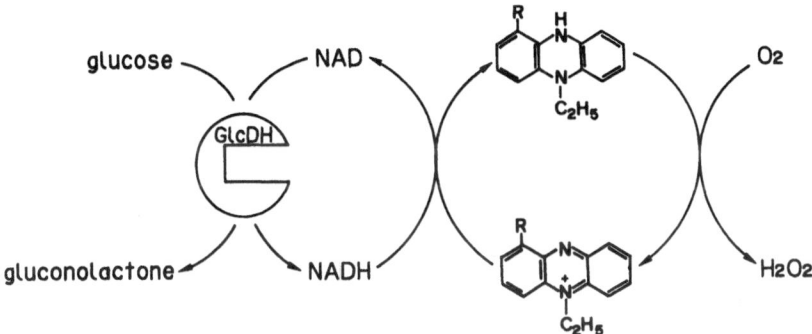

Fig. 7. Coupling of the G1cDH reaction and the EP reaction

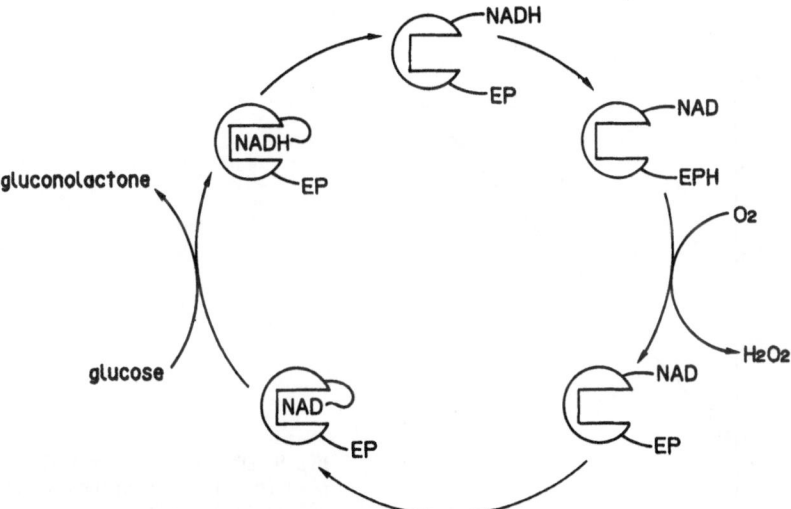

Fig. 8. Reaction scheme of semisynthetic glucose oxidase

conjugate shows good activity due to the increase in the effective concentration of the NAD moiety [5]. When GlcDH reaction is coupled with the reaction cycle of EP, as shown in Fig. 7, the overall reaction becomes the oxidation of glucose with oxygen, i.e. glucose oxidase. Therefore, we linked both EP and NAD to GlcDH using PEG as a spacer by the methods used for the preparation of EP-PEG-GltDH.

EP-GlcDH-NAD thus prepared has the following characteristics. The average number of the EP moieties bound per molecule of enzyme subunit was 0.8, and that of NAD was 1.2. The GlcDH activity of EP-GlcDH-NAD in the presence of exogenous NAD and EP was 53% of that of the native enzyme.

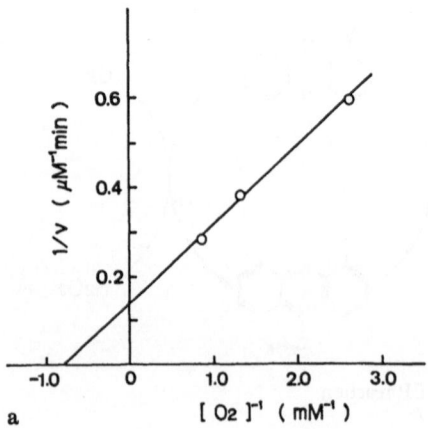

a

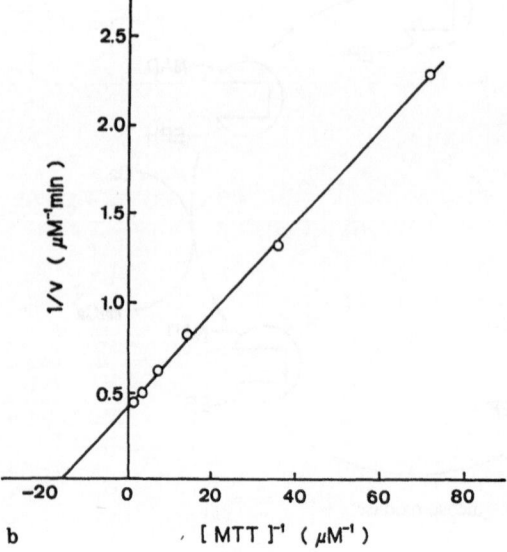

b

Fig. 9. Effect of **a** oxygen and **b.** MTT concentrations on the activity of EP-G1cGH-NAD

3.2 Glucose Oxidase Activity of EP-GlcDH-NAD

EP-GlcDH-NAD catalyzes the oxidation of glucose by oxygen or MTT, and therefore, works as a glucose oxidase by the catalytic cycle shown in Fig. 8. This means that natural GlcDH is artificially converted to glucose oxidase.

Figure 9 shows the effects of oxygen and MTT concentrations on the activity of the semisynthetic oxidase. From these results, the values of the parameters for the Michaelis–Menten equation were obtained as follows: for oxygen, V_{max} $= 8.2\,\mu M/min$ and $K_m = 1.6\,mM$; for MTT, $V_{max} = 2.5\,\mu m/min$ and K_m $= 0.07\,mM$. It is interesting that EP-GlcDH-NAD has no binding site for oxygen or MTT but has K_m values for them. This is because the EP moiety takes two forms, i.e. oxidized (EP) and reduced (EPH) forms, in the catalytic cycle of the semisynthetic oxidase (see Fig. 8). In this case, the reaction kinetics fit the Michaelis–Menten equation, even if the catalyst has no binding site for the substrates. Generally, K_m is thought to be a dissociation constant of an enzyme–substrate complex, but these findings clearly show it is not always true.

In the reaction cycle of EP-GlcDH-NAD, the following two catalytic reactions are connected by intramolecular coupling: the reduction of the NAD moiety by the active site of GlcDH and the reoxidation of the NADH moiety by the EP moiety. The importance of the intramolecular coupling of these two reactions is expressed as follows. When the EP moiety is separated from the conjugate, the V_{max} value decreases to 0.14% due to the decrease in the rate of the second catalytic reaction. When the first reaction is made into intermolecular type keeping the second reaction intramolecular type, the V_{max} value also decreases to 0.13%. Therefore, this intramolecular coupling of the two successive reactions is essential to the glucose oxidase activity of EP-GlcDH-NAD.

4 Conclusions

We have transformed dehydrogenases into oxidases. This method seems to be generally applicable to preparing new oxidases. On the basis of the above results, the following strategies for designing enzyme-like catalysts are presented [7]. For simple catalysts having one substrate-binding site and one catalytic group, one should use a binding site with higher affinity for the substrate and make the ratio of the intramolecular and intermolecular rate constants as large as possible. To increase the intramolecular rate constant, one should select a good binding site and use a flexible linker with an appropriate length. A catalyst with a subunit structure is also effective in enhancing the specific activity.

For a more complex catalyst, having one substrate-binding site and two kinds of catalytic groups, it is important to couple the two catalytic reactions intramolecularly. For the reaction system with coupling, covalent linking of the intermediate is the best way. If the intermediates are covalently linked, the catalyst does not need to have the binding sites for the intermediates.

5 Abbreviations

EP, 5-ethylphenazine
PEG, Poly(ethylene glycol)
GltDH, glutamate dehydrogenase
LDH, lactate dehydrogenase
GlcDH, glucose dehydrogenase
EP-PEG, poly(ethylene glycol)-bound 5-ethylphenazine
EP-PEG-NAD, 5-ethylphenazine-poly(ethylene glycol)-NAD conjugate
EP-PEG-GltDH, 5-ethylphenazine-poly(ethylene glycol)-glutamate dehydro-genase conjugate
EP-GlcDH-NAD, 5-ethylphenazine-glucose dehydrogenase-NAD conjugate
MTT, 3-(4′,5′-dimethylthiazole-2-yl)-2,5-diphenyltetrazolium bromide
DHBT, 3,3′-(1,6-dioxo-1,6-hexanediyl)bis-2-thiazolidinethione.

6 References

1. Yomo T, Sawai H, Urabe I, Yamada Y, Okada H (1989) Eur J Biochem 179:293.
2. Yomo T, Urabe I, Okada H (1991) Eur J Biochem 196: 343.
3. Yomo T, Sawai H, Urabe I, Okada H (1989) Eur J Biochem 179: 299.
4. Eguchi T, Iizuka T, Kagotani T, Lee J:H, Urabe I, Okada H (1986) Eur J Biochem 155: 415.
5. Nakamura A, Urabe I, Okada H (1986) J Biol Chem 261: 16792.
6. Yomo T, Urabe I, Okada H (1991) Eur J Biochem 200: 759.
7. Yomo T, Urabe I, Okada H (1992) Eur J Biochem 203: 543.

4 Bioprocess Engineering

4 Bioprocess Engineering

4.1 On-line Diagnosing System for Fed-Batch Fermentation

Isao Endo, Hajime Asama, Mikio Nakajima, Teruyuki Nagamune, Terhi Siimes* and Pekka Linko*

Chemical Eng. Lab., The Institute of Physical and Chemical Research 2-1 Hirosawa, Wako-shi, Saitama 351-01, Japan

An on-line diagnosing system for a fed-batch fermentation process has been developed. The system is based on the physiological activities of a microorganism which are represented by specific rates like cellular growth, substrate consumption and production of an aiming substance. These process variables can be measured by an on-line system named Bio Advanced Control System (BIOACS) which we developed in 1985.

Recently, we have expanded the function of this BIOACS and established an expert system for diagnosing bioreaction processes on the basis of a database in which the physiological activities of microorganisms are stored. The system was implemented in Smalltalk/V on IBM PC-AT or Macintosh II, and was developed by Aarts et al. The software shell was also developed and it can handle uncertainties both in the measurements and knowledge by fuzzy logic. By defining fuzzy sets like "normal", "low" or "high" to each specific rate, transfer of human expert knowledge and handling of imprecise data was facilitated. The knowledge is represented as a network of rules together with their truth values.

We applied this system to diagnosing a batch fermentation process of lactic acid and obtained a good result. We are now aiming to apply this expert system to a fed-batch process. The key technology is how to introduce a "defuzzification" program in order to control the process precisely according to the counter measures which are obtained from the diagnoses. We have no software applicable to "defuzzification"so far, but we think it should be like a pharmacist compounding a medicine according to a prescription from a doctor. Timing of control is important too. The control of a bioprocess is quite similar to therapeutics in so far as we take account of the physiological state of a microorganism.

* Lab. Biotechnol. Food Eng., Helsinki University of Technology SF 02150 Espoo, Finland

4.2 Sensors for Bioprocess Monitoring and Control

Daniel I.C. Wang

Department of Chemical Engineering and Biotechnology Process Engineering Center, M.I.T., Cambridge, MA 02139, USA

Abstract

The need for robust sensors for bioprocess monitoring leading to effective control strategies is well recognized by biochemical engineers. Firstly, the ability to monitor on-line cell mass concentrations has been the goal of many researchers. Secondly, the added complexity of many fermentation media containing suspended solids renders the cell concentration measurement in situ in a fermenter even more challenging. Lastly, the ability to differentiate the total cell mass from the viable fraction in situ could lead to some very exciting on-line control and optimization strategies.

This paper will address these three goals in our on-line sensor research program. The measurements on the cell concentration are achieved through an integrated light-scatter spectra analysis. The cell concentrations are performed using a fiber optic probe during fermentation in the presence of medium solids such as fish meal and soybean meal. The on-line measurements of the cell viability are achieved through the use of recombinant DNA technology.

Bioproducts and Bioprocesses 2
Editors: Yoshida, Tanner
© Springer-Verlag Berlin Heidelberg 1993

4.3 Development of Micro—Biosensors for Brain Research

Eiichi Tamiya* and Isao Karube**

*Japan Advanced Institute of Science and Technology, 15 Asahidai, Tatsunokuchi, Ishikawa 923-12, Japan
**Research Center for Advanced Science and Technology, University of Tokyo, 4-6-1 Komaba, Meguro-ku, Tokyo 153, Japan

Contents

Carbon fibers are used to construct ultramicroelectrodes with 7µm diameter. When a triangular potential (-2 to $+2$V vs Ag/AgCl) was applied before every measuring double pulse (first pulse: 750 mV, second pulse: 1100 mV), the determination limit was 0.1µM of hydrogen peroxide. A micro-acetylcholine sensor was fabricated by immobilizing acetylcholine esterase and choline oxidase on the carbon fiber by entrapment with PVA-SbQ. This sensor gave a linear calibration plot for the range from 0.1 to 1.0 mM with a linear correlation coefficient of 0.9842. A micro-glutamate sensor consists of a platinized carbon fiber disk electrode modified with the immobilized glutamate oxidase membrane. This sensor gave a linear calibration for the range 2 µM to 1.2 mM. Release of glutamate in the cerebellar cortex was detected after pottassium stimulation.

1 Introduction

Neurotransmitters play important roles in the brain since they are the key link in communication between neurons. Acetylcholine is the transmitter of motor neurons in the spinal cord as well as in all the nerve-sketal junctions in vertebrates. It is located diffusely throughout the brain. The activity of the enzyme choline acetyltransferase in the brain decreases often significantly with age. This is the enzyme responsible for synthesis of acetylcholine. Glutamate is also the transmitter which is related to expression of long term memory. Mechanisms of long-term potentiation (LTP) in the hippocampus and long-term depression (LTD) in the cerebellar cortex are unclear because of difficulty in measuring glutamate release from the presynaptic membrane and the sensitivity of the glutamate receptor in the postsynaptic membrane. *In vivo* acetylcholine and glutamate sensors are powerful tools for elucidating the sites of action in the brain where they operate. A smaller electrode does less damage to tissue during insertion into brain or nerve tissue. An extremely small environment can be examined with a microelectrode whose diameter is of the order of a few micrometers. The carbon fiber electrode is considered to be one of the most useful transducers for *in vivo* biosensors, because carbon fibers provide us with

Bioproducts and Bioprocesses 2
Editors: Yoshida, Tanner
© Springer-Verlag Berlin Heidelberg 1993

ultramicroelectrodes with high strength and electrochemically pretreated fibers have a much greater sensitivity to dopamine and catecholamine. Acetylcholine and glutamate cannot normally be directly oxidized on the electrode, however the application of carbon fibers has proven otherwise using immobilized enzymes. Several acetylcholine sensors have been developed based on enzyme electrodes with a potentiometric detector or amperometric detection of oxygen and hydrogen peroxide. However these sensors are not suitable for *in vivo* analysis. The detection of hydrogen peroxide is an established approach to the construction of biosensors based on electrodes containing immobilized oxidases. In a report by Akiyama it was shown that it is possible to obtain a high sensitivity and an excellent reproducibility to dopamine using microcomputer-controlled potentiostatic pulse polarization techniques [1].

In this study, electrochemical operations for preelectrolysis and measurement were employed for sensitive determination of hydrogen peroxide. Then carbon fiber electrodes were modified with acetylcholine esterase and choline oxidase for an acetylcholine sensor and glutamate oxidase for a glutamate sensor.

2 Fabrication and Measurement of Micro-Biosensors

Choline oxidase, acetylcholine esterase and acetylcholine were purchased from Sigma. Glutamate oxidase was donated by the Yamasa Company. Poly(vinyl alcohol)-styrylpyridinium (PVA–SbQ, degree of polymerization 1700, degree of saponification 88, SbQ content 1.3 mol%, solid content 11%, pH 7) was kindly donated by the Toyo Kagaku Co. Other reagents were of analytical grade.

The material and fabrication of carbon fiber electrodes were described in our previous paper [4]. The diameter of the electrode was 7 μm. All electrochemical experiments were done with a three-electrode system enclosed in a steel Faraday cage. A microcomputer-controlled potentiostat system was used for precise control of the applied potential. This system was newly designed and constructed for determination of H_2O_2. The basic method for the construction of this system was shown in our previous paper [1]. Our polarization technique was applied to analyze H_2O_2. An activation signal consisted of an anodic–cathodic triangular wave was introduced immediately before every measuring pulse. The parameters for measuring pulses were set as follows. The length of first pulse (750 mV) was 1000 ms. The current was measured every 625 μs for 160 ms to the end of the first pulse. The average of 256 measured points for 160 ms was employed for calculation of current intensity. The length of second pulse (1100 mV) was also 1000 ms. The average value was obtained in a similar manner as that above. The 32 measured points were obtained for 20 ms which is one cycle of the electric line (50 cycles/s). The operation of averaging improves sensitivity and reproducibility because of noise reduction. The current intensity was defined as the difference in currents between the first and second pulses. A platinized carbon fiber disk (PCD) electrode was also used for a transducer of a micro glutamate sensor.

3 Response to Hydrogen Peroxide

Oxidation of H_2O_2 was observed over 850 mV vs Ag/AgCl based on voltammogram of H_2O_2 in a phosphate buffer (Fig. 1). When 1100 mV of potential was applied to the carbon electrode, a stationary current was obtained within 1 s. Therefore 1 s is suitable for the length of pulse potential. Figure 2 showed relationships between the current intensity and H_2O_2 concentration. A single pulse of 750 mV was not available for H_2O_2 determination because the current fluctuated randomly against H_2O_2 concentration. Pulse operation shown by No. 4 in the Fig. 2 gave an excellent calibration for H_2O_2. Activation triangular potential is necessary to obtain a reproducible response. The subsequent response was gradually reduced without activation potential (Fig. 3). Aoki et al. indicated that oxidation of H_2O_2 began at 0.5 V based on a cyclic voltammograph and the peak current was related to the H_2O_2 concentration higher than 0.1 mM [3]. A more sensitive determination was not possible because the

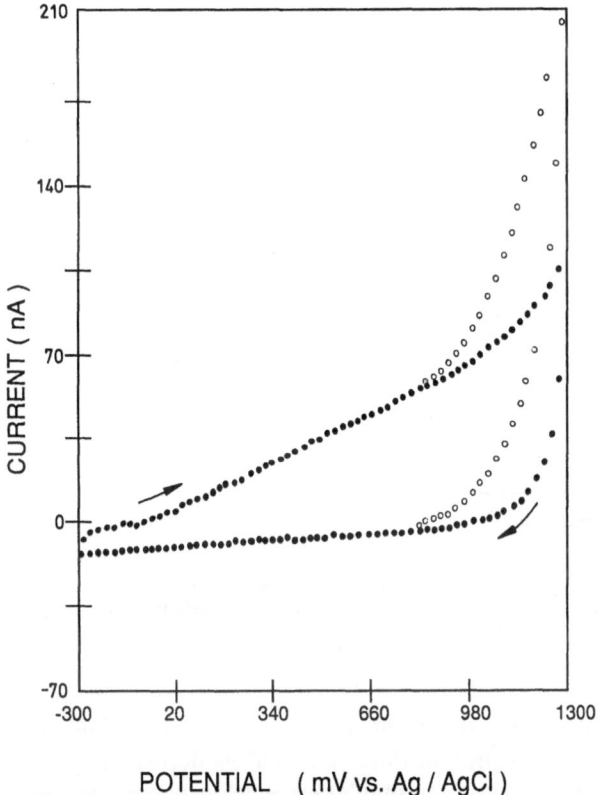

Fig. 1. Voltammograms of hydrogen peroxide
The potential was increased in 25 mV steps from -300 to 1300 mV vs Ag/AgCl electrode and followed by a decrease to -300 mV. Currents were measured after 30 ms of each potential. *Arrows* indicate directions of potential increase and decrease. ($\bigcirc$) and ($\bullet$) show voltammograms with and without hydrogen peroxide (final conc. 1 mM) respectively

172   Eiichi Tamiya and Isao Karube

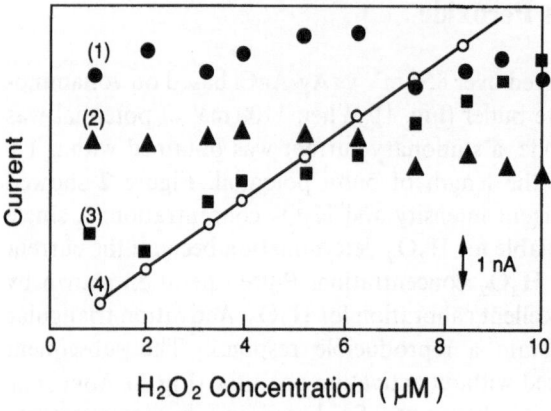

Fig. 2. Relationship between intensity and hydrogen peroxide concentration under various kinds of pulse operation
Single pulse (1): $(P_1: 750\,mV, T_1: 0.5\,s)$.
Double pulse (2): $(P_1: 750\,mV, P_2: 850\,mV, T_1: 0.5\,s, T_2: 0.5\,s)$
(3): $(P_1: 600\,mV, P_2: 950\,mV, T_1: 1\,s, T_2: 1\,s)$
(4): $(P_1: 750\,mV, P_2: 1100\,mV, T_1: 1\,s, T_2: 1\,s)$

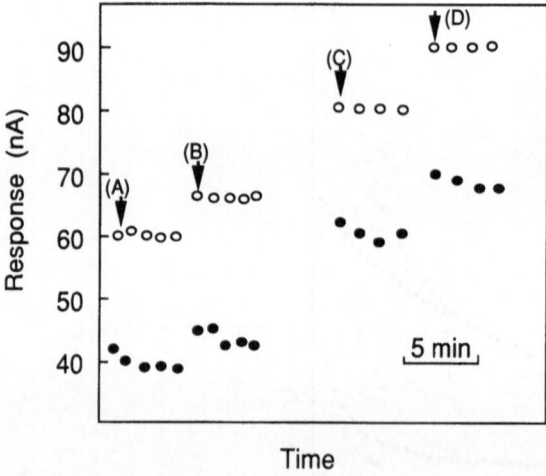

Fig. 3. Timecourse of current intensity after addition of hydrogen peroxide. Double pulse (4) in Fig. 2 was used with (○) and without (●) activation wave. A,B,C,D indicate addition of hydrogen peroxide of 40, 50, 80 and 100 μM (final conc.) respectively

oxidation was masked by background current. The polarization technique has been established for sensitive detection of dopamine and its derivative [1, 2]. This technique is based on the activation triangular wave before each measuring pulse. The resting time (T_r) between the activation wave and the measuring pulse was related to current intensity. 2 s of T_r was employed for this study since longer T_r caused a further decrease of the current intensity. Activation wave included negative potential where oxygen was electrochemically reduced possibly to produce hydrogen peroxide. However, current intensity was not affected by dissolved oxygen concentration, because background current based on oxygen was compensated by our double pulse method. The detection limit is less than 0.1 μM. The sensitivity of our carbon fiber is better than that of micro-gold electrodes reported previously [4].

The use of the activation potential was effective for the improvement of sensitivity. Linear relationship was obtained between current and hydrogen peroxide concentration in the range from 0.1 μM to 1 mM; slope, y-intercept and linear correlation coefficient were 0.638 nA/μM, 10.3 nA and 0.9997, respectively. Reproducibility was tested by repeated 25-μl injections of 50 mM hydrogen peroxide solutions; the relative standard deviation for a set of 10 injections was 0.68%. The current value became the saturated level above 4 mM which is the detectable upper limit of this method. This sensor is stable for at least 1 month. Slope and background currents were different between electrodes probably because of different surface areas and microenvironmental diffusion conditions.

4 Effects of Albumin and Ascorbic Acid

Albumin protein which is abundant in body fluids and adsorbed nonspecifically onto the surface of the electrode causing a reduction in the active surface area. In our electrode system, reproducible determination of H_2O_2 is possible even in samples containing albumin. Our triangular wave produces oxygen and hydrogen gases and etches the surface of the carbon fiber. These phenomena may strip adsorbed protein from the electrode surface. The slope value in albumin solution was reduced to 73% of that in albumin-free solution. The diffusion speed of hydrogen peroxide became low due to the high viscosity of a 0.5 g/mL albumin solution. Oxidation of ascorbic acid occurred over 10 mV of potential according to the voltammogram of ascorbic acid. The separation of H_2O_2 from ascorbic acid was possible because of the large difference of oxidation potential between ascorbic acid and hydrogen peroxide. The double pulse method compared with the single pulse method at the various concentrations of ascorbic acid in Fig. 4. Current intensity of single pulse indicated total current in ascorbic acid but without hydrogen peroxide. The double pulse method was not affected by ascorbic acid concentration because it subtracted the oxidation current from ascorbic acid. A large standard deviation of current values was obtained when the double pulse operation (P_1 : 750 mV, P_2 : 1100 mV) was employed for determination of ascorbic acid. Lower potentials of the double pulse (P_1 : 100 mV, P_2 : 150 mV) were useful for a stable response to ascorbic acid. Higher overpotential increases the rate of ascorbic acid oxidation, and a mixing condition has a more serious influence on the sensor response because of a thinner diffusion layer. Another possibility of slightly poor linearity for calibration is due to the formation of slimy electrode deposits on oxidation.

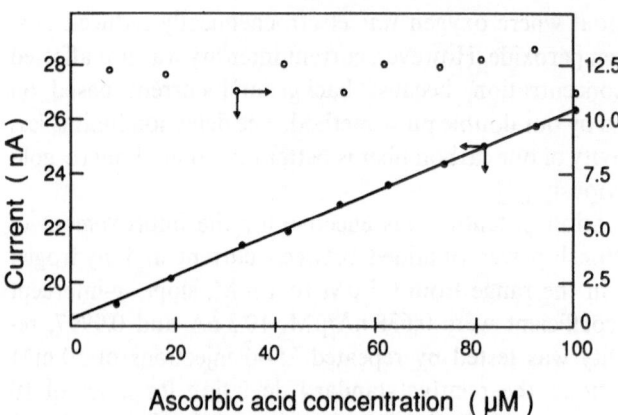

Fig. 4. Comparison of the double pulse method with the signal pulse method at various concentration levels of ascorbic acid. (○) and (●) show double pulse (4) and single pulse (1) respectively as used in Fig. 2. Hydrogen peroxide concentration is 1 μM

5 Characteristics of Micro-Acetylcholine Sensors

Choline oxidase (1 unit) and acetylcholine esterase (0.5 unit) dissolved in 0.5 mL of water were added to 0.2 mg of 11% aqueous solution of PVA-SbQ (enzyme-PVA mixture) [5]. A carbon fiber electrode was dipped in the enzyme–PVA mixture, and air dried for 5 h at room temperature. The air-dried carbon fiber was exposed to normal fluorescent room light for 1 h.

The micro-acetylcholine sensor gave a linear calibration plot for the range 0.1–1.0 mM with a linear correlation coefficient of 0.9842. The sensitivity with a PVA–SbQ membrane is better than that obtained by using immobilized enzymes by crosslinking with albumin and glutaraldehyde. The determination time for acetylcholine was 5s. Previous acetylcholine sensors needed at least 1 min of response time because diffusion of acetylcholine and hydrogen peroxide diffused slowly in a thick enzyme-immobilized membrane [6–8]. The determination limit of our sensor is 0.05 mM (S/N = 2) [9]. This value is higher than those obtained by the other groups using the same type of detection [6, 8]. Sensitivity is strongly related to amount of enzymes immobilized in the membrane which covers the electrode surface. Scanning electron micrographs indicated that the immobilized enzyme membrane is very thin. So, loading of the enzyme is not sufficient to obtain optimal sensitivity of immobilized acetylcholine esterase or choline oxidase is more than 1 month according to previous papers [6–8]. The half life of our micro-acetylcholine sensor was approximately 1 week which was dependent on the use and storage conditions.

6 Characteristics of Micro-Glutamate Sensors

A microglutamate sensor consists of a platinized carbon fiber disk (PCD) electrode and an immobilized glutamate oxidase membrane. A PCD electrode

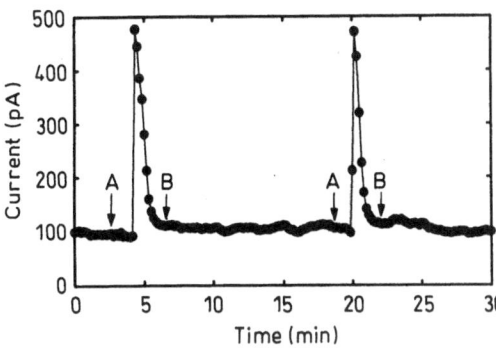

Fig. 5. Response of the sensor on applying high pottassium stimulation to a cerebellar slice

At A, the perfusate was changed from normal Ringer solution (N.R.) to high potassium containing Ringer solution (high K^+), and at B was changed from high K^+ to N.R.

whose diameter is 15 μm gave a more stable response to hydrogen peroxide because the background current of the PCD electrode is smaller than that of a carbon fiber electrode. Glutamate oxidase was immobilized using a photo-crosslinking method with PVA–SbQ. The micro-glutamate sensor gave a linear calibration for the range from 2 μM to 1.2 mM. This sensor was applied to monitor the release of glutamate from the synapse in the cerebellar cortex (Fig. 5). The amount of glutamate depended on the calcium concentration in the perfusate. Release of glutamate after electrical stimulation was also monitored by our sensor.

7 References

1. Akiyama A, Kato T, Yasuda E (1985) Anal Chem 57: 1518.
2. Nakazato T, Akiyama A (1988) J Neurochemistry 51: 1007.
3. Aoki K, Ishida M, Tokuda K, Hasebe K (1988) J Electroanal Chem 251: 63.
4. Tamiya E, Karube I (1988) Sensors & Actuators 15: 199.
5. Ichimura K (1987) Makromol Chem 188: 763.
6. Mascini M, Moscone D (1986) Anal Chim Acta (1986) 179: 439.
7. Sode K, Marty J, Karube I (1989) Anal Chim Acta 228: 49.
8. Morelis RM, Coulet PR (1990) Anal Chim Acta 231: 27.
9. Tamiya E, Sugiura Y, Navera EN, Mizoshita S, Nakajima K, Akiyama A, Karube I (1991) Anal Chim Acta 251: 129.

4.4 Application of 2-D Gel Electrophoresis to Study Intracellular Events in Industrial Fermentations

Henry Y. Wang and Kevin H. Dykstra

The University of Michigan, Department of Chemical Engineering, Ann Arbor, MI 48109, USA

Contents

High resolution two-dimensional polyacrylamide gel electrophoresis (2–D PAGE) is an extremely powerful technique for separating complex protein mixtures. It allows visualization of thousands of cellular proteins and has been used to study a variety of bacterial and eucaryotic systems. However, it has not yet been used systematically to study or to monitor industrially significant bioprocesses. Until recently, a major limitation to the application of 2-D PAGE for bioprocess monitoring is the time required to process and evaluate the 2-D gels. Recent advances in micro-2-D gel electrophoresis, which have shortened processing time down to 2–4 hs, and sophisticated image analysis have the potential to break down this barrier even further. Our preliminary experimental results indicate that this approach for the detection of intracellular events is viable and deserves further investigation. *Streptomyces* fermentations are used to serve as model bioprocesses for this evaluation. More specifically, we employed cycloheximide producing mutants of *Streptomyces griseus* to establish and evaluate the utility of 2-D gel electrophoresis as a bioprocess monitoring tool. We have also initiated the compilation of a computerized protein data bank on this industrially important organism.

1 Introduction

Most industrial bioprocesses are generally conducted in batch or fed batch modes with limited process control capabilities. The cell is essentially treated as a black box which takes up nutrients and produces the desired product. Improvements in product yield and productivity have been achieved primarily by random media manipulation and shotgun mutation programs. Although traditional biochemical and genetic methods have resulted in dramatic improvements in production since the late 1940s, they are not based on fundamental understanding of cellular metabolism. Attempts to optimize product yield and nutrient availability have met with only limited success because of a lack of

178   Henry Y. Wang and Kevin H. Dykstra

knowledge concerning regulatory metabolism in these organisms. Environmental factors which trigger the onset of bioproducts production are not
completely understood, and means to sustain biosynthesis of these bioproducts
are usually unknown. Catabolite repression, feedback regulation of product
synthesis, degeneration of synthesis, and degradation of the extracellular product are some problems frequently associated with these industrial fermentations
[3]. In addition, the diversity of industrial significant metabolites and the
organisms which produce them makes it even more difficult to draw general
conclusions about cellular regulation. A general and convenient means is needed
to study and to monitor intracellular events of cellular metabolism associated
with product biosynthesis of industrial significant organisms at a fundamental
level. Such a method would allow meaningful comparisons among different
industrial significant bioproduct producing cells.

High resolution two-dimensional polyacrylamide gel electrophoresis (2-D
PAGE) has been developed as a reliable analytical technique for resolving
complex protein mixtures such as cellular extracts into individual components
(Fig. 1). Proteins can be visualized by a variety of means such as silver staining
(actual amount of protein) or radioactive labeling (rate of synthesis). 2-D PAGE
is capable of clearly and reproducibly resolving thousands of individual proteins
on the basis of isoelectric point and molecular weight. It has been used by
biochemists and microbiologists to analyze the intracellular protein profiles of
Escherichia coli and of mammalian cells. Coupled with computerized image
analysis, it is a powerful tool for analyzing and comparing cellular proteins from
various hosts. In addition, two-dimensional electrophoresis presents an opportunity to look inside the cells to assess their metabolic state [28]. However, it has
not yet been systematically applied to industrially important organisms or for
possible process monitoring. We are interested in adapting this technology for

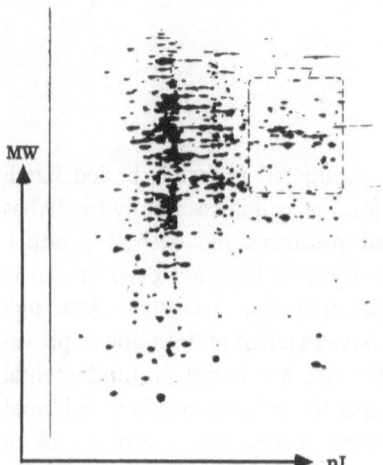

Fig. 1. Two-dimensional electrophoretogram
of *Streptomyces griseus*, UC-2132. The portion
of the complete protein profile to be analyzed
with a micro 2-D gel (outlined by *dotted line*).
By using narrow range ampholytes and gels of
appropriate acrylamide concentration, it
should be possible to focus on a limited portion of the complete profile without loss of
resolution

direct monitoring the metabolic status of various industrial fermentation processes. However, in order for this approach to be useful, the time required to run 2-D gels must be reduced substantially so that reproducible results can be achieved in relatively short time. Recent developments in miniaturization of 2-D gels have reduced the time down below 4h.

Even though little is known about the biosynthesis of cycloheximide by *Streptomyces griseus*, its production has been demonstrated to be feedback regulated, and thus, continuous removal of the antibiotic from the fermentation broth leads to tremendous increases in production [6, 29, 43]. In studying these antibiotic fermentations using both complex and synthetic media, we would like to learn what cellular functions are required for or associated with production of cycloheximide and other antibiotics. We are also interested in studying the coordination of the appearance of these antibiotics with other cellular functions in order to learn more about the scope of the changes involved in secondary metabolism. We are particularly interested in finding common characteristics in the regulation of different antibiotic syntheses. Secondly, a catalog of the intracellular proteins of *Streptomyces griseus* has been initiated. Previous experience in developing a protein database for *Escherichia coli* in Professor F.C. Neidhardt's laboratories at the University of Michigan has helped us to initiate this task. Our ultimate goal is to develop a centralized information bank on *Streptomyces* that will be available to the scientific community and to industry. Thirdly, because the ability to determine the metabolic state of industrial organisms at a fundamental level is important for monitoring and control of biological processes, a process monitoring system based on both mini- and micro 2-D electrophoresis and computerized image analysis system should be developed. Standardized procedures will be defined for fast sample preparation and gel staining on a micro scale. Data analysis can be based upon the newly developed *Streptomyces* protein database. Finally, information gained can potentially be used to construct a structured model of secondary metabolism which will be useful for process control and optimization.

2 Two-Dimensional Polyacrylamide Gel Electrophoresis

2.1 Conventional 2-D PAGE

Although several two-dimensional techniques had been proposed for the separation of cell proteins, the first truly two-dimensional, high resolution technique was that of O'Farrell [28]. By combining isoelectric focusing (IEF) in the first dimension with SDS electrophoresis in the second dimension, O'Farrell was able to resolve proteins on the basis of two independent properties: isoelectric point and molecular weight (Fig. 2). This technique has the ability to resolve thousands of proteins from whole cell extracts clearly and reproducibly [28].

Protein samples are prepared for two-dimensional polyacrylamide gel electrophoresis (2-D PAGE) by first disrupting the cell sample. New approaches

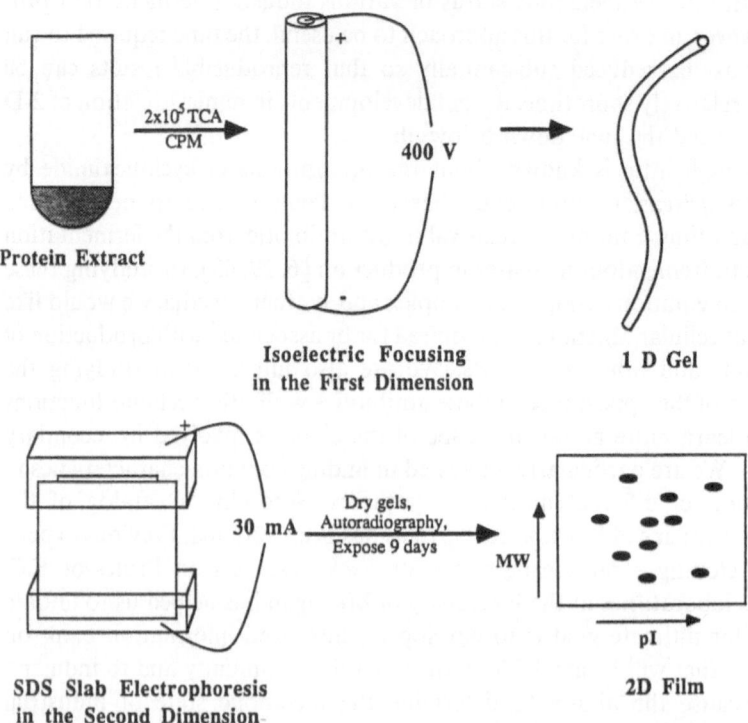

Fig. 2. Schematic diagram of experimental procedure for a conventional 2-D PAGE

developed in our laboratories to selectively permeabilize various cells for protein release may simplify this step significantly. Cell proteins are then applied to the first dimension tube gel for isoelectric focusing. In isoelectric focusing, the polyacrylamide support matrix is polymerized in the presence of a mixture of highly mobile amphoteric compounds with different isoelectric points, usually polyamino–polycarboxylic acids. When a voltage gradient is applied across the gel, these ampholytes quickly migrate to the point where their net charge is zero, forming a pH gradient within the matrix. Next, the various proteins in the sample begin migrating to their respective isoelectric points, and at equilibrium, proteins with similar isoelectric points are focused into sharp bands (Fig. 2). This step in conventional 2-D PAGE techniques is extremely slow, requiring 15–20 h. However, miniaturization can lead to substantial savings in time by reducing the distance proteins must travel to their isoelectric points and by allowing the strength of the electric field to be increased without distorting the protein bands due to inadequate heat dissipation in the gels.

In the second dimension, separation is achieved on the basis of molecular weight. The 1-D gel (a spaghetti-like strand of polyacrylamide) containing the IEF-separated proteins is dialysed in a solution containing SDS. This compound, combined with a reducing agent (β-mercaptoethanol), disrupts intraprotein interactions, wrapping around the polypeptide backbone and imparting each amino acid residue with a negative charge. The proteins are thus transformed into negatively charged rods having a roughly constant charge to mass ratio. In an electric field, the proteins' ability to migrate therefore depends on the size of the charged rod and the "molecular sieving" effect of the gel matrix. Following dialysis, the 1-D gel is laid across a Laemmli slab gel and the proteins electrophoresed for 3–6 h, migrating in logarithmic proportion to their molecular weights. At this point, the proteins are present as a two-dimensional array of spots in the gel (Fig. 1). As with the first dimension, substantial time savings may be achieved by miniaturization, with the separation being complete in about 30 min for a commercial micro electrophoresis system [30].

Following 2-D PAGE, proteins may be visualized in a number of ways. With Coomassie Blue staining, optical density varies linearly with the amount of protein, but this method has relatively low sensitivity. Recently, reduction of silver complexed with protein has been exploited to give a more sensitive staining method [22, 23]. The fact that many proteins stained by these methods have a characteristic color makes it a simple matter to distinguish and track specific proteins on several gels. However, it also means that the changes in the absolute concentration of a protein can only be measured relative to some reference state, and determination of a protein's concentration as a fraction of total cellular protein must be accomplished by other means. A variety of immunological and enzymatic methods have also been used to visualize proteins in gels, and give very specific biochemical methods for determining the identity of proteins in the gels [4, 34]. Finally, radioactive labeling of proteins followed by autoradiographic detection offers a powerful array of sensitive techniques to visualize protein spots [1, 25].

One of the chief problems with 2-D PAGE is assimilating the tremendous amount of information on each gel into a cohesive form for interpretation of results. A variety of computer methods have been used to analyze and compare different gels [14, 18, 35, 36, 40, 42]. These methods generally first include image acquisition, via either video camera [35], rotating drum densitometer [14], or laser densitometer [40]. Protein spots are defined using various segmentation algorithms and assigned coordinate values. Different gels are then compared and individual spots evaluated for matching and relative protein concentration. Finally, the spot measurements become part of a database of similar measurements used for future analysis and comparison [19, 38].

2.2 Micro 2-D PAGE

SDS polyacrylamide gel electrophoresis on a miniature scale was first performed in 1964. With the advent of conventional 2-D PAGE techniques, there have also been various attempts at using micro 2-D PAGE [12, 20, 21, 27, 31–34]. The chief advantages of miniaturization are the small size of the protein sample required and the shorter time necessary for separation. Samples of less than 1 µg total protein have been shown to give acceptable results with silver staining [2], and micro gels generally take roughly one tenth the time required for conventional electrophoresis procedures [26]. However, most of these systems could only be used semi-quantitatively to assimilate all the information on the gel.

There have been several descriptions of 2-D PAGE using micro systems giving gels which are roughly postage stamp size [12, 20, 27, 32–33]. In these systems, the first dimension isoelectric focusing is carried out in glass capillary tubes (5–25 µL), requiring 15–80 min [27, 32]. The second dimension gels are generally formed using microscope slides. The proteins are usually visualized using silver staining [21, 27], or autoradiography [31], although enzyme assays have been used to measure some proteins in gels prepared under non-denaturing conditions [34]. Recently, a commercial micro electrophoresis system has become available which uses prepared media and is capable of performing 2-D electrophoresis in less than 4 h [13, 30]. With this system both isoelectric focusing and SDS electrophoresis are performed in a horizontal slab configuration. However, following IEF, strips of the IEF gel are cut and laid directly upon the second dimension SDS gel. With this system, the 2-D separation can be accomplished in roughly 3.5 h, including staining.

From the above discussion, it is clear that 2-D PAGE is an extremely general technique for the analysis of complex protein mixtures. Its use to study the effects of heat shock in *E. coli* and various mammalian cell lines has recently been reviewed by Neidhardt and Van Bogelen [24], and Goochee [10], respectively. It has also been used to measure the rate of synthesis of various proteins in *S. cerevisiae* [8, 39], to examine mutation rates in the human genome [22], and to study protein changes due to disease [9], among many other applications. However, 2-dimensional electrophoresis has not been used for analysis of industrially important bioprocesses, and it has not been used to study secondary metabolism. With a few exceptions [11, 15, 37], the protein patterns of *Streptomyces* sp.have not been examined in detail. In addition, the application of micro 2-D methods for routine analyses of protein samples and application to bioprocess monitoring and control have not been exploited.

3 Model Antibiotic Fermentation: Cycloheximide Production by *Streptomyces griseus*

The discovery of an antifungal agent produced by *Streptomyces griseus* was reported by Whiffen and coworkers in 1946 [17, 44]. At first called actidione, then later renamed cycloheximide, it acts on 80 S eucaryotic ribosomes to block peptidyl transferase activity, thereby stopping translation. Cycloheximide is primarily used in agriculture to control mildew and other fungal infestations. It is also a valuable research tool for studying protein synthesis in eucaryotes.

The biochemical pathway for cycloheximide synthesis is largely unknown. Although studies suggest that the cycloheximide molecule is assembled via the condensation of acetate or malonate units [41], the actual order in which the units are assembled is not known. Furthermore, no metabolic intermediates in the process of cycloheximide biogenesis have ever been identified.

Batch fermentation of *Streptomyces griseus* for the production of cycloheximide is similar to many other secondary metabolite fermentations. The process is characterized by rapid mycelial growth during the first 24 h. The cessation of mycelial growth and the first significant accumulation of cycloheximide appear to coincide, although the exact mechanism by which cycloheximide production is initiated is not well understood. The production of cycloheximide, as shown in Fig. 3, is closely linked to the consumption of glucose, and cycloheximide accumulation ceases almost immediately after depletion of glucose in the medium [16]. Also, high levels of cycloheximide in the fermentation medium have also been shown to stop further accumulation of the antibiotic [16]. Computer modelling of the batch fermentation with and without continuous product removal has suggested that cycloheximide synthesis is regulated by feedback repression, that is, cycloheximide decreases the *amount* of its biosynthetic enzymes [5–6].

Feedback regulation can be alleviated by maintaining the broth concentration of cycloheximide at a low level. As seen in Fig. 3, this results in a dramatic increase in both the rate of production and overall yield of cycloheximide. In addition, an on-line extraction system based on the use of neutral polymeric resin was developed in our research laboratories [29, 43]. Rohm and Haas XAD-4 resin particles were immobilized in hydrogel beads to facilitate recovery of the product. The adsorbent beads were placed in an extraction vessel, and

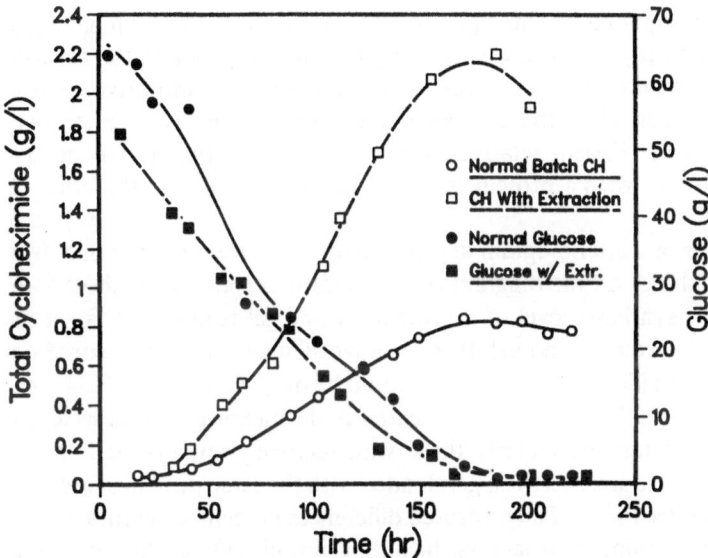

Fig. 3. Typical fermentation profiles for cycloheximide production both with and without resin addition

whole fermentation broth was circulated through the extractor. The cyclo-heximide concentration in the fermentation broth was maintained at around 300 μg/L, and total antibiotic production was enhanced to 2000–2500 μg/l (2–3 times the normal total product yield).

4 Application of 2-D Gel Electrophoresis to Study Intracellular Events in Cycloheximide Fermentations

Our initial work has focused on the cycloheximide fermentation, and on the synthesis rates of various proteins over the course of the normal fermentation and under various operating conditions. We have therefore prepared auto-radiograms of gels loaded with pulse-labeled protein extracts. We first wanted to assess the repeatability of the technique, and then look at the changes in the protein profile over the normal course of the fermentation. The effects of feedback regulation and continuous product removal on the intracellular proteins were then examined.

4.1 Reproducibility in Quantitative and Qualitative Comparisons

A typical 2-D electrophoretogram of *Streptomyces griseus* is shown in Fig. 1. There are obviously hundreds of proteins clearly resolved in these images. Since we are interested in changes in the proteins, the first task was to assess what constitutes a significant change in the protein pattern. In order to accomplish this, 4 samples were taken simultaneously from the same fermentation, and prepared for electrophoresis. Three gels of each sample were then run, giving a total of 12 theoretically identical gels. These gels were compared on the basis of quantitative and qualitative differences in their images. A quantitative change was defined to occur when the corresponding protein spot was present two different gels, but at different relative levels of synthesis. A qualitative change was defined to occur when a protein spot was missing from one or the other of the two gels.

Figure 4 shows a scatter diagram depicting a quantitative comparison of two of these gels. Each dot on the diagram corresponds to a spot matched on both gels. The relative synthesis level of a protein on one gel is plotted against its relative synthesis on the second gel. If the relative synthesis levels measured on each of the gels were identical, all the dots would lie on the line $x = y$. However, as can be clearly seen, there is significant scatter in the measured synthesis levels of the various proteins, particularly those with relatively low synthesis rates. Similar diagrams were seen between gels loaded with the same protein sample or those with different samples. The measured differences in relative synthesis rates could be due to a number of factors, including insufficient resolution of the scanner, inaccurate modeling of spot shape by the spot detection algorithm, crowding of spots on the gels, variation in the level of different proteins entering

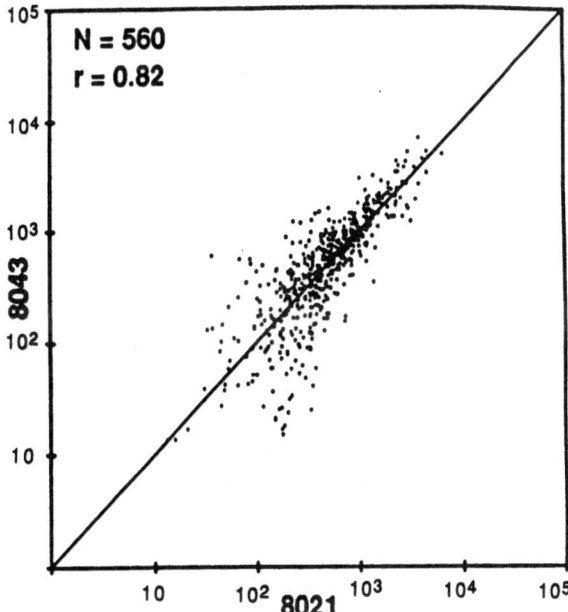

Fig. 4. Scatter diagram showing the quantitative reproducibility between two identically prepared gels. Quantitation values for matched spots on gel #8043 are plotted against the values for the corresponding spots on gel #8021

the gels, etc. The important point here is that the significance of any quantitative differences must be assessed on a case by case basis using appropriate statistical comparisons and replicate gels.

In Table 1 the average number of qualitative differences occuring between pairs of replicate gels from the same protein sample and those derived from different simultaneously prepared protein samples is shown. It is important to note that in each case, roughly 600 proteins were identified on each gel, so that the number of computer-detected qualitative differences comprised no more than 5% of the total number of proteins on each gel. However, after visual inspection of the autoradiograms at the positions of the computer-detected qualitative differences, no actual differences could be found in the protein

Table 1. Qualitative differences detected in identically prepared samples and gels

	Average number of computer detected differences	Visually confirmed differences
Different gels from same extract	23	0
Gels from different extracts	24	0

patterns. Most of the spurious differences were found in crowded regions of the protein pattern, where the spot detection algorithm had interpreted too crowded or overlapping protein spots as a single protein on one gel, and as two proteins on the second. From these results, it can be seen that the existence of the various proteins in the pattern can be very repeatably detected.

4.2 Intracellular Protein Changes During the Normal Fermentation

In the next experiment, it was desired to look at the way the protein pattern changes over the course of a normal fermentation. In this experiment, cell samples were taken from shake flask fermentations at various times during the growth phase, during the production phase, and after the exhaustion of glucose in the medium. At each time point, 2-D gels were prepared from the cells, and the cycloheximide and glucose levels assayed. Figure 5 is a graph of the number of qualitative protein changes occuring between each of the sample points. It should be noted that roughly 500 spots were analyzed on each gel. These data very clearly show that there are substantial changes in the proteins synthesized by the cells after the change from the innoculum medium to the production medium, at the transition from growth to stationary phase, and with the exhaustion of glucose in the medium. There are thus a relatively large number of proteins whose presence or absence is specific to one or another of the phases in the growth cycle of *Streptomyces griseus*. These data mark the first time that global changes in the intracellular makeup of the bacterial cell have been examined during different growth phases.

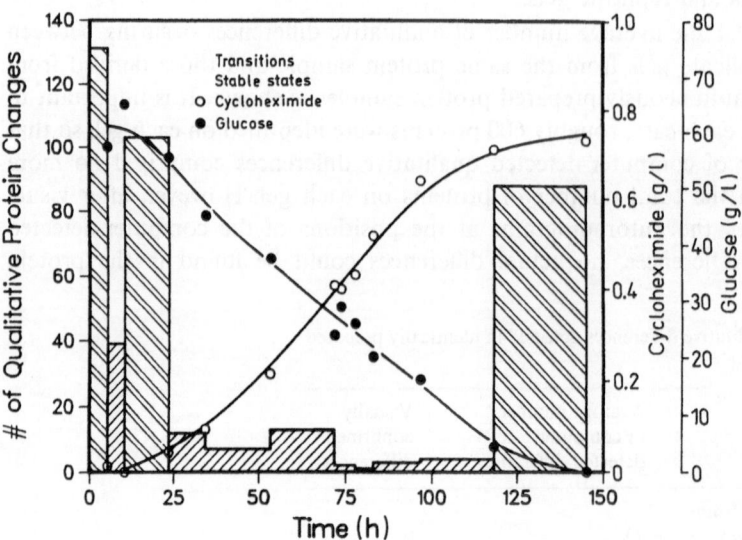

Fig. 5. Graph of the number of qualitative protein changes occurring between sample points during a normal cycloheximide fermentation

4.3 The Effect of Feedback Regulation

We next wished to examine the effect of cycloheximide addition and removal on the protein pattern. In the first experiment, 1 g/L of cycloheximide was added to an idiophase fermentation, and the cells sampled periodically thereafter. After antibiotic addition, further accumulation of the product ceases. Cycloheximide addition also immediately and specifically depresses the synthesis of at least 4 proteins (CR1–CR4) [7]. It is interesting to note that synthesis of all of these proteins is specific to the stationary phase, and ceases after the exhaustion of glucose (data not shown).

Addition of XAD-4 resin to the fermentation removes the antibiotic from the fermentation broth, and causes a sharp increase in the production rate of cycloheximide (Fig. 3). Addition of resin also causes a dramatic increase in the synthesis rate of two of the same proteins whose synthesis was depressed by cycloheximide addition [7]. We have thus identified a group of proteins whose synthesis coincides with antibiotic production in *S. griseus* [7]. Two of these proteins exhibit behavior consistent with what would be expected if they were responsible for rate limiting steps in the biochemical pathway of cycloheximide production (Fig. 6).

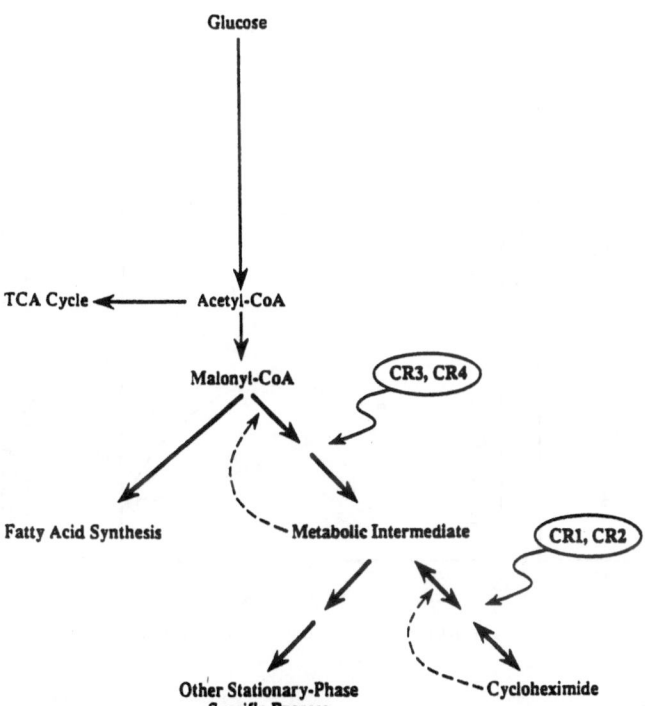

Fig. 6. Proposed pathway for cycloheximide biosynthesis showing the hypothetical positions of proteins CR1–CR4. The *dashed lines* show the effects of feedback repression of the synthesis of these proteins

188   Henry Y. Wang and Kevin H. Dykstra

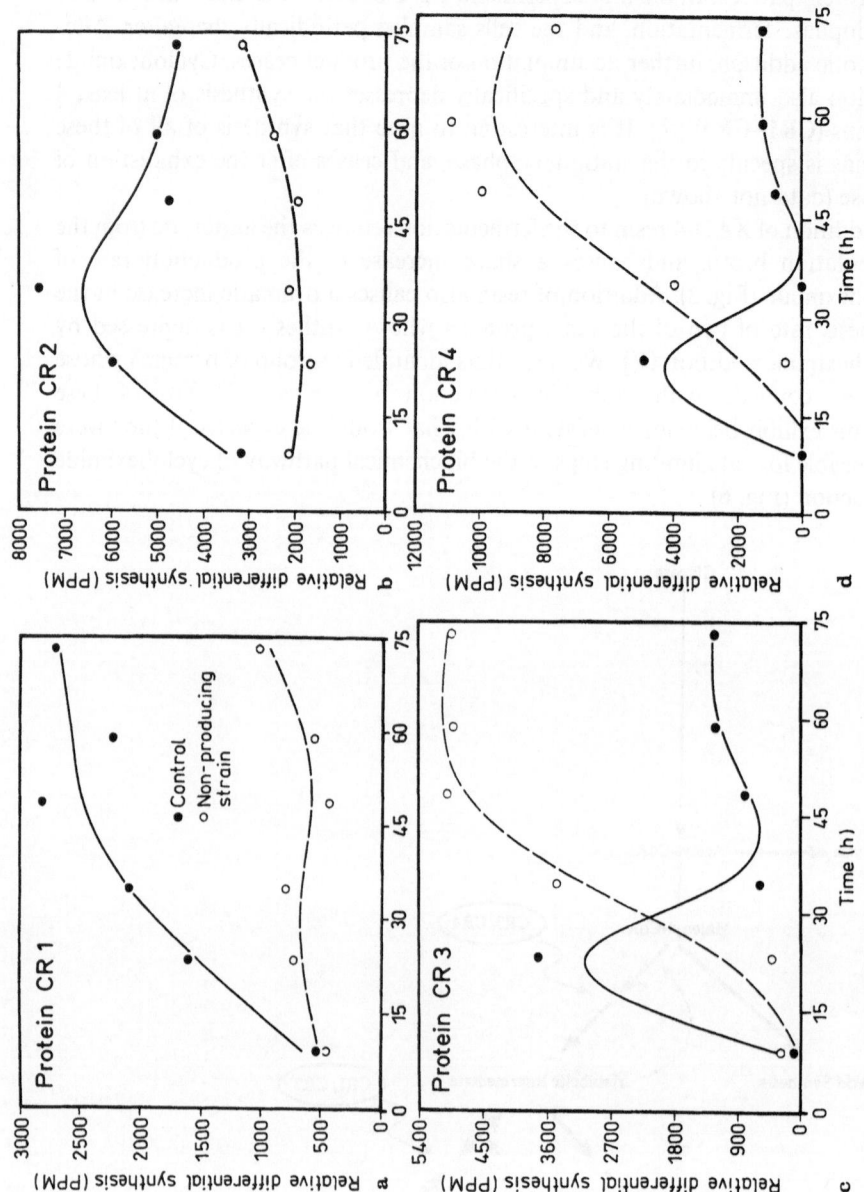

Fig. 7. Time profiles of the synthesis of proteins CR1–CR4 in the non-producing strain (BEL-1) in comparison with VC-2132 producing strain **a, b.** Synthesis of CR1–CR2 are markedly lower than in the producing strain **c, d.** Production of proteins CR3 and CR4 are substantially delayed compared to the producing strain

Work on this project continues in this laboratory. When cycloheximide is added from the beginning of the fermentation, the level of antibiotic falls over time, and then slowly rises again. However, preliminary results indicate that when resin added to such a fermentation, the increase in production rate is even more dramatic than when resin is added to the normal fermentation. It will be interesting to see the effect of these operating modes on the intracellular protein pattern.

4.4 Comparison of Producing and Non-Producing Cultures

We have also examined the differences in the protein patterns between the production strain, UC-2132, and a non–producing strain (BEL-1). Besides producing no cycloheximide, this non-producing strain (BEL-1) sporulates profusely during the stationary phase, while very few spores can be detected in producing cultures of UC-2132 at any time during the fermentation. More than 99% of the intracellular proteins produced by this non-producing strain could be matched to corresponding proteins in the producing strain, UC-2132 during the growth phase. With the onset of the stationary phase of the fermentation (which occurred simultaneously in multiple comparisons, as evidenced by reduced levels of protein synthesis), the amount of homology between the two strains fell to around 90%. However, throughout the fermentation there was substantial correspondence in the protein pattern of the two strains. In the first experiment, synthesis of some of the confirmed proteins was examined over the course of the normal fermentation. As shown in Figs. 7a and 7b, synthesis of proteins CR1 and CR2 is substantially lower in the non-producing strain than in UC-2132 throughout the fermentation. Thus, in every case examined where antibiotic production was unambiguously reduced, synthesis of these two proteins was also at a substantially reduced level.

In Figs. 7c and 7d, the time course of synthesis for proteins CR3 and CR4 in BEL-1 is shown respectively. It can be seen that compared to UC-2132, significant synthesis of these proteins is delayed from 14–26 h. In the case of UC-2132, synthesis of these two proteins goes through a maximum value following induction. However, in BEL-1, following delayed induction, the synthesis rate is maintained at a relatively high, constant value. Again, in both strains, synthesis of these proteins appears to be regulated in a coordinated manner, and their time courses are qualitatively similar. Added cycloheximide reduced production of proteins CR1 and CR2 in BEL-1, but had no appreciable effect on CR3 and CR4. Resin addition to the non-producing strain had no noticeable effect on the intracellular protein profile of BEL-1.

5 Conclusions

These data show first of all that 2-D PAGE is an extremely powerful means with which to study secondary metabolism. It can be clearly seen that a large number of changes in the cellular pattern of gene expression accompany the process of differentiation. Even though very little is known about the biochemistry of cycloheximide production, it has been possible to identify several proteins which may be linked to antibiotic production, and to study the regulation of their synthesis. Finally, we have seen that by manipulating the fermentation conditions with an eye toward maximizing the production of these proteins, it may be possible to further improve the production of various antibiotics. The fundamental knowledge gained about regulation of secondary metabolism in *Streptomyces griseus* will also provide a basis for comparison with other organisms that produce different types of industrially important secondary metabolites. A more complete understanding of the regulatory mechanisms associated with secondary metabolism will permit rational design of better production media and improved process control and feeding strategies. The strategy developed here to couple an automated 2-D gel electrophoresis with a complete protein database of an industrially important organism could have a significant impact on the development and optimization of other bioprocesses using these and other cells, including mammalian cells.

Acknowledgements. Financial support of this work by the National Science Foundation is gratefully acknowledged.

6 References

1. Bonner WM, Laskey RA (1974) Eur J Biochem 46: 83.
2. Condeelis JS (1977) Anal Biochem 77: 195.
3. Demain AL (1984) Adv Biochem Engin 1: 113–202.
4. Dunbar BS (1987) BioTechn 5(3): 218–226.
5. Dykstra KH, Wang HY (1987) Ann NYAS 506: 511–522.
6. Dykstra KH, Li X-M, Wang HY (1988) Biotech Bioeng 32: 356–362.
7. Dykstra KH, Wang HY (1990) Appl Microbiol Biotechnol 34: 191–197.
8. Elliot SG, McLaughlin CS (1978) J Bact 137(3): 1185–1190.
9. Goldman D, Merril CR, Polinsky RJ, Ebert MH (1982) Clin Chem 28(4): 1021–1025.
10. Goochee CG, Passini CA (1988) Biotech Prog 4(4): 189–201.
11. Gräfe U, Sarfert E (1985) FEMS Micro 28: 249–253.
12. Heukeshoven J, Dernick R (1988) Electrophoresis 9: 28–32.
13. Jägersten C, Edström A (1987) Pharmacia Miniposter, Pharmacia AB Biotechnology, Uppsala, Sweden.
14. Jansson PA, Grimm LB, Elias JG, Bagley EA, Longberg-Holm KK (1983) Electrophoresis 4: 82–91.
15. Kendrick KE, Ensign JC (1983) J Bact 155(1): 357–366.
16. Kominek LA (1975) Antimicrob Agents Chemother 7(6): 856.
17. Kornfeld EC, Jones RG, Parke TV (1949) J Amer Chem Soc 71: 150.
18. Lemkin PF, Lipkin LE, Lester EP (1982) Clin Chem 28(4): 840–849.
19. Lemkin PF, Lipkin LE (1983) Electrophoresis 4: 71–81.
20. Manabe T, Hayama E, Okuyama T (1982) Clin Chem 28(4): 824–827.
21. Manabe T, Visvikis S, Steinmetz J, Galteau MM, Okuyama T, Siest G (1987) Electrophoresis 8: 325–330.
22. Merril CR, Goldman D, Ebert MH (1981) Proc Nat Acad Sci USA, 78: 6471–6475.
23. Merril CR, Goldman D, Van Keuren ML (1982) Electrophoresis 3: 17–22.
24. Neidhardt FC, Van Bogelen RA (1981) Biochem Biophys Res Comm 100: 894–900.

25. Neidhardt FC, Phillips TA (1983) The protein catalogue of *Escherichia coli*, In: Celis JE, Bravo R (eds) Two-dimensional gel electrophoresis of proteins: methods and applications. Academic, New York, p 417.
26. Neuhoff V (1980) In: Radola BJ (ed) Electrophoresis '79. Walter de Gruyter, New York, p 203.
27. Neukirchen RO, Schlosshauer B, Baars S, Jäckle H, Schwarz U (1987) J Biol Chem 257: 15229–15234.
28. O'Farrell PH (1975) J Biol Chem 250: 4007.
29. Payne GF, Wang HY (1989) Arch Microbiol 151: 331–335.
30. Pharmacia Phast System, Product Sales Literature, Pharmacia, Uppsala, Sweden (1987).
31. Pipkin JL, Anson JF, Hinson WG, Burns ER, Casciano DA (1986) Electrophoresis 7: 463–471.
32. Poehling HM, Neuhoff V (1980) Electrophoresis 1: 90–102.
33. Rüchel R (1977) J Chromatog 132: 451–468.
34. Sanderink G-JCM, Artur Y, Galteau M-M, Wellman-Bednawska M, Siest G (1986) Electrophoresis 7: 471–475.
35. Skolnick MM, Sternberg SR, Neel JV (1982) Clin Chem 28: 969–978.
36. Skolnick MM (1982) Clin Chem 28: 979–986.
37. Sugiyama M, Katoh T, Mochizuki H, Nimi O, Nomi R, (1983) J Ferment Technol 61(4): 347–351.
38. Taylor JN, Anderson L, Scandra AE, Willard KE, Anderson NG (1982) Clin Chem 28(4): 861–866.
39. Trew BJ, Frieson JK, Moens PB (1979) J Bact 138(1): 60–69.
40. Tyson JJ, Haralick RH (1986) Electrophoresis 7: 107–113.
41. Vanek Z, et al. (1964) Biochem Biophys Res Comm 17(5): 532.
42. Vo K-P, Miller MJ (1981) Anal Biochem 112: 258.
43. Wang HY (1983) Ann NY Aca Sci 431: 313.
44. Whiffen AJ, Bohonos N, Emerson RL (1946) J Bact 52: 610.

4.5 A Knowledge-Based Approach for Control of Phenylalanine Production by a Recombinant *Escherichia coli*

Konstantin B. Konstantinov and Toshiomi Yoshida

International Center of Cooperative Research in Biotechnology,
Faculty of Engineering, Osaka University, 2-1, Yamada-Oka, Suita, Osaka 565, Japan

Contents

A newly developed knowledge-based approach for diagnosis and control of variable structure processes was applied to the fed-batch cultivation of a recombinant *Escherichia coli* for phenylalanine production. To accumulate a consistent amount of expert knowledge required for the practical development of the control system, the general characteristics of the process were investigated including the structural transformations of the microbial system during the cultivation, as well as several other phenomena which are directly related to the control of the process, e.g. the excretion of acetic acid, the gradual decrease of the oxidative capacity of the cell, the dependence of CO_2 evolution on the glucose feeding, etc. Practical development and functioning of the knowledge-based part and the algorithmic modules are described in relation to the physiological pecularities of the microbial system. In the last part, the results from a cultivation conducted under control of the developed system are presented and commented on.

1 Introduction

Already for two decades interest in the automatic control of fermentation processes has been growing rapidly because of its theoretical and practical importance. However, despite some achievements, there are still many unsolved problems ensuing mainly from the great complexity and the unusual characteristics of the living systems [1–5]. The serious difficulties encountered call for revision of the methods and techniques used in modeling and control of fermentation processes [6–8]. To compensate for the shortcomings of the traditional control concepts, some new trends have recently arisen; among which most significant is the application of modern AI methods to the process control [9–14]. Our work has been dedicated to the development and application of a system for control of fermentation processes utilizing such techniques, particularly the knowledge-based control approach.

We developed a knowledge-based approach for real-time diagnosis of the microbial population and control of fermentation processes focussing on the knowledge management mechanism, the capability of handling uncertain and fuzzy information and learning functions [15–24]. The proposed control system was applied on the process for phenylalanine production by the fed-batch cultivation for recombinant *Escherichia coli*. From these works, this paper summarizes investigation of the general characteristics of the process to accumulate expert knowledge, the practical development and functioning of the knowledge-based part and the algorithmic modules and the results from a cultivation conducted under control of the developed system.

2 Materials and Methods

2.1 Strain

A tyrosine and thiamine deficient mutant—*Escherichia coli* AT2471 was used as a host strain. It had been modified [25] by inclusion of plasmid pSY130–14 carrying the genes *aroF*^{FR} and *pheA*^{FR} for enzymes involved in the key steps in the phenylalanine synthesis pathway. The plasmid also included the gene for kanamycin resistance, and this antibiotic was used to suppress the growth of plasmid free cells.

2.2 Cultivation

Two hundred ml LB medium together with 40 mg L^{-1} kanamycin, 20 g L^{-1} glucose and 1 drop of antifoam (Adecanol 109) was used for preculture. The pH was initially adjusted to 7.5 with NaoH. One ml of the stock culture stored at − 80°C in LB medium containing 20% glycerol and 25 mg L^{-1} kanamycin was inoculated and cultivated for 12 h in a 2 L Sakaguchi flask at 37°C on a reciprocating shaker. For phenylalanine production, a synthetic medium with the following composition was used (in g L^{-1}): 7.0 Na$_2$HPO$_4$·12H$_2$O, 4.0 KH$_2$PO$_4$, 1.0 NaCl, 5.0 NH$_4$Cl, 2.0 K$_2$SO$_4$, 3.0 MgSO$_4$·7H$_2$O, 10.0 Na-citrate, 0.55 Na-glutamate, 0.06 CaCl$_2$·2H$_2$O, 0.6 tyrosine, 0.05 thiamine and 0.04 kanamycin. The initial glucose concentration was 20 g L^{-1}. After depletion 60% (w/v) glucose solution was fed. All cultivations were conducted in a Chemap FZ-2000 fermenter under the following conditions: temperature 38.5°C, pH 7.0 (controlled as indicated in the text), initial cultivation volume 3.0 L and air flow rate 3.0 L min^{-1}. A newly developed computer–fermenter complex was used for on-line data measurement, calculation and control. An IBM AT (286, 10 Mhz) compatible machine equipped with a floating point co-processor was linked with the fermenter through an ADC-200 interface card (CONTEC, Japan) with 16 analog inputs, 2 analog outputs and 12 bit resolution. This complex worked under the control of the software system described elsewhere.

2.3 On-line Measurements and Calculations

The following variables were measured on-line: temperature, pH, dissolved oxygen (DO), agitation speed, air flow rate, O_2 and CO_2 concentrations in the outlet gas, glucose and ammonia feed rates, cultivation volume, and optical density of the culture. To fill out the picture of the process, several additional variables were calculated on-line by the computer system.

2.4 Off-line Analyses and Data Processing

Samples were taken regularly at 2 h intervals for off-line analyses. Cell concentration was measured by a spectrophotometer at 660 nm after proper dilution. Glucose was assayed enzymatically by a Glucostat. Acetic acid concentration was measured by a gas chromatograph with a 2 m glass column packed with PORAPAK-Q. Phenylalanine was assayed using the characteristic reaction with 2,4,6–trinitrobenzene-sulfonic acid (TNBS) [26]. It was performed after careful evaporation of the ammonia existing in the culture supernatant by mixing 1 ml of a 50 times diluted samples with 1 ml 2 N NaOH and boiling this for 90 min. The ammonia free residual was diluted additionally in 100 L water; 0.5 ml from this solution was mixed with 5 ml 0.2 M $Na_2B_4O_7$ and 1 ml freshly prepared TNBS reagent (0.5 g L^{-1} TNBS and 0.625 g L^{-1} Na_2SO_3), and incubated in a water bath for 30 min at 40°C. The phenylalanine concentration was measured photometrically at 420 nm. This procedure provided perfect linearity for phenylalanine concentration up to 10 g L^{-1}.

To improve the quality, some noisy data obtained by off-line analysis were passed through a spline-smoothing procedure using the program complex SMOOTH [27]. The program was slightly modified and recorded in the C language to best suit our application.

3 Results and Discussion

3.1 Study of Physiological Phenomena and their Effect on the Phenylalanine Production Process

During the phenylalanine production process, glucose is transformed into four main products—biomass, phenylalanine, acetic acid and carbon dioxide. Initially due to the inadequate distribution of the carbon between these products, the efficiency of the cultivation was not sufficiently high. The physiological mechanisms governing the balance between their synthesis, and the possibility to control this balance by manipulation of some process factors were experimentally studied.

3.1.1 Acetic Acid Excretion and its Effect on Cultivation Pattern

The large amount of acetic acid and the negligible amount of ethanol were detected during the fed-batch cultivation of the recombinant of *E. coli*. Acetic acid is well-known as a growth inhibitory substance [28–36]. Inhibition by acetic acid depends on the oxygen supply and on the pH, being higher under conditions of oxygen deficiency [35] and low pH values [34].

A series of batch experiments were carried out in order to investigate the effects of acetic acid inhibition on our strain. As shown in Fig. 1, the acetic acid concentration which completely inhibited the cell growth was about 15 g L^{-1}. No inhibition was observed at 2 g L^{-1} of acetic acid. To investigate the effect of glucose limitation, several fed-batch cultivations were made under sufficient

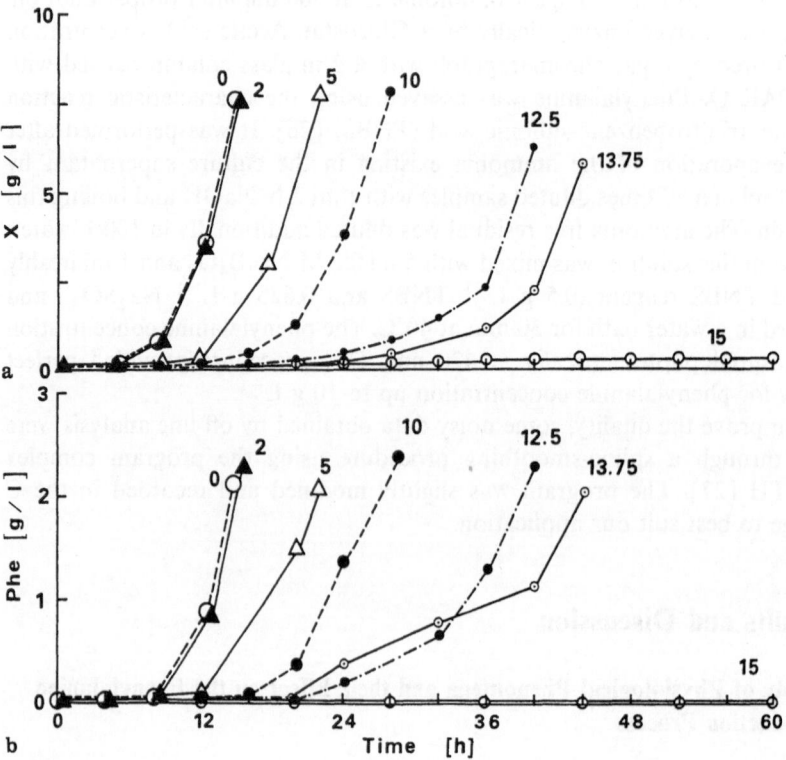

Fig. 1. Batch experiments at different initial concentrations of acetic acid. (a) Cell growth; (b) phenylalanine production

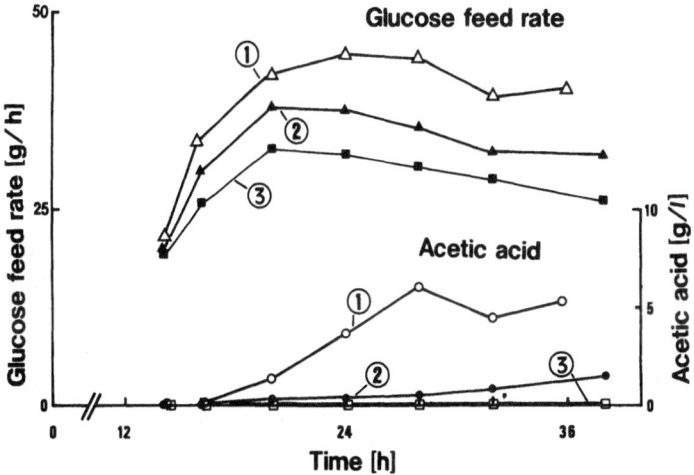

Fig. 2. Acetic acid excretion in three experiments, conducted with continuous glucose feeding under different glucose limitation. The limitation was lowest in exp. 1 and highest in exp. 3

oxygen supply and continuous glucose feeding with several feed rates. In all cases the glucose concentration in the broth was zero. The results (Fig. 2) showed that the progressive decrease of the glucose feed rate results in suppression of the acetic acid formation. When the glucose feed rate was sufficiently low, acetic acid excretion was completely prevented.

These results proved that not only oxygen deficiency but also glucose overfeeding also contribute to fast acetic acid excretion. Consequently, the well-known and widely used DO-stat strategy with intermittent glucose feeding [31] is not suitable for phenylalanine production by our strain.

3.1.2 Decreasing Oxidative Capacity of Cell Population

Another phenomenon closely related to the development of an efficient glucose feeding policy is the gradual decrease of the oxidative capacity of the cell population, which was observed during the phenylalanine production process. It was found that during the prolonged fed-batch cultivation for phenylalanine production the metabolic activity of the population undergoes dramatic transformations.

A large base of experimental data was created by collecting information from many processes conducted under identical conditions, except for differences in glucose feed rates. These data showed that acetic acid excretion may begin at different cultivation stages. The moment of initiation of excretion appeared to be random and chaotic, unless a reasonable physiological explanation can be found.

Oxygen limitation is an inconsistent reason because all experiments were conducted under strict aerobic condition, and a glucose effect can be eliminated because the cultivations were run under glucose limitation, and the glucose concentration was practically zero. The potential possibility for hidden limitation by other medium components was also excluded, since sufficient amounts of all nutrients were fed.

An acceptable explanation of these data can be found in the hypothesis of limited cell respiratory capacity [37–39]. This postulates a relation between the glucose feed rate and fermentative by-product excretion under conditions of glucose limitation and sufficient oxygen supply. To test this hypothesis, the data for the specific glucose uptake rate GUR^* obtained in different experiments against time. The results obtained are shown in Fig. 3, where three distinct types of experimental points are plotted: points of intensive acetate excretion; points where acetate excretion had just been initiated and its concentration was still very low (less than 0.1 g L^{-1}); and points where acetate was not excreted at all.

The distribution of the randomly obtained points in the figure roughly implies that the suspected relation between GUR^* and the acetic acid excretion really exists. There are different ways to separate analytically the points into distinct classes according to the excretion of acetate. First, one may follow the discrimanant analysis technique and build a borderline separating the points of intensive acetate excretion from those where excretion was not detected. However, the border has apparently a nonlinear shape, and due to the unsolved difficulties in identification of nonlinear discriminant functions, this approach is hardly applicable. A more practical method is to use only the set of points corresponding to very low acetate concentration, which have a medial position

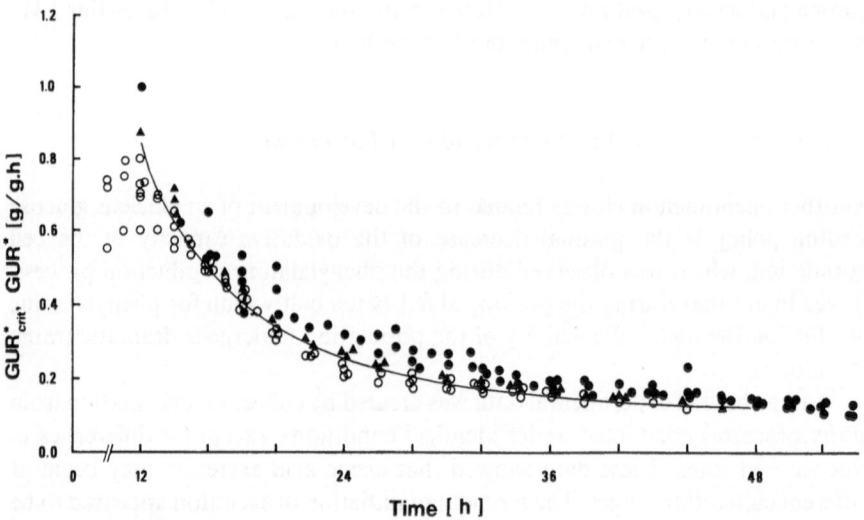

Fig. 3. GUR* versus time. The points represent experiment data collected in different cultivations, conducted under identical condition except for the glucose feed rates. Symbols indicate; (●) excretion of acetic acid; (○) no acetic acid excretion; (▲) very low acetic acid concentration. The continuous line represents GUR^*_{crit}.

between the other two sets. The corresponding time-course may be approximated by a nonlinear function, which can be considered also as a discriminant function.

Among different approximating functions tested, the best fitting was obtained by the declining exponential function. It was interpreted as the time-profile of the so called critical specific glucose uptake rate GUR^*_{crit}, and used in the separation of the GUR^*-time field into distinct regions (Fig. 4). In the first region the oxidative capability of the cells was not exceeded by the glucose feed rate, glucose was completely oxidized and acetate was not formed. Within this "region of subcritical feed rates" the degree of saturation of respiration, defined as the ratio between GUR^* and GUR^*_{crit}, is restricted between zero and unity. In the second region, referred to as the "region of supracritical feed rates", the glucose feed rates exceed the oxidative potential of the population. The glucose fed cannot be completely oxidized and the excess will be directed through the glycolytic pathway, resulting in by-product excretion. Here, the degree of saturation of respiration is higher than one.

Because it is impractical to consider that a clear single borderline GUR^*_{crit} exists between these regions, it is useful to introduce a third region—a narrow band around GUR^*_{crit} discriminating the first two area. This is referred to as the "region of critical feed rates" and is described by the approximate equality. Here in this region glucose feed rates fit with the oxidative capacity of the cells, and the degree of saturation of the respiration is near to unity. These results showed that in fed-batch cultivation for phenylalanine production, GUR^*_{crit} is not a constant, but its value decreases noticeably, following a near-exponential pattern.

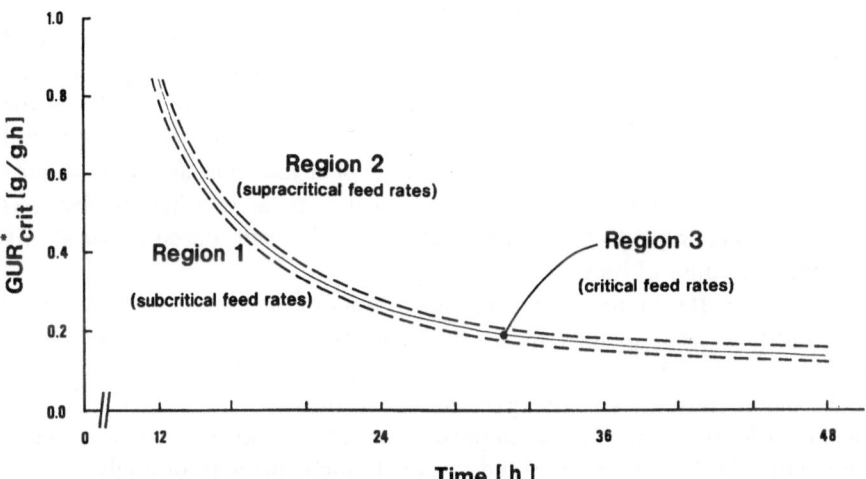

Fig. 4. Discrimination of the time-field of fermentation based on acetic acid excretion. The *continuous line* represent GUR^*_{crit}. *Dotted lines* represent the borders of the region of critical feed rates

3.1.3 CO$_2$ Evolution

More than 50% of the carbon in this cultivation was released in the form of CO$_2$. It was observed that the CO$_2$ evolution is also related to the glucose feed rate. Under conditions of strong glucose limitation, the relative CO$_2$ production was intensified. In contrast, when the cells were satisfied with glucose, the CO$_2$ production was considerably lower. This was clearly indicated by the instantaneous CO$_2$ yield from glucose, which was calculated on-line by the computer. Its value quickly increased after reduction of the glucose feed rate. The dependence of the CO$_2$ yield on the glucose feeding was probably related to the cell maintenance. Since the required energy for this remains independent of the glucose feeding, the relative share of CO$_2$ in the overall carbon flow will increase under conditions of insufficient glucose supply. Therefore, unnecessarily strong glucose limitation will worsen the process performance by increasing the CO$_2$ yield, resulting in reduced phenylalanine yield. Consequently, the process should run with the highest glucose feed rate, which still provides a limitation sufficient for prevention of the acetic acid excretion.

3.1.4 Physiological Alteration of Cell Population

During the cultivation, the cell population chronologically passes through four physiological stages. In the first several hours after inoculation, the cells grew under conditions of excessive supply of all nutrients. During this period cells were not able to produce acetate regardless of the glucose concentration. After this period, however, the population acquired the capability of excreting acetate and the process entered its second stage. Accordingly, glucose feeding strategy satisfying the required limitation should be activated in this stage. During this period cells still grow actively and form the by product. After the depletion of tyrosine, growth rapidly ceased and the process entered its third stage. During this period cells produced phenylalanine without active growth. With the advance of the process, however, the production rate decreased, and the cell population gradually lost activity. Finally, phenylalanine synthesis almost completely stopped and the process entered the last stage when glucose was transformed mainly into acetic acid and CO$_2$, and the cell concentration slowly decreased because of lysis.

Most important point here is that the transfer of the process from one stage to another is accompanied by drastic changes in the characteristics of the biological plant. These alterations expressed typical variable structure behavior and required flexible alteration of the control strategies, particularly the strategy of glucose feeding. Therefore, an important task of the control system is to detect automatically these alterations and to modify the control accordingly.

In addition to the chronological changes, there are some other phenomena which affect the cell characteristics, but are not strictly time associated. Some of these events appear randomly during the second and the third process stages.

They might indicate drift from the normal process course which can result from errors in the measurements and in the control, faults in the technical equipment, or unknown physiological phenomena. Examples are premature initiation of acetic acid excretion, increase of the background glucose concentration, too strong glucose limitation, secondary growth during the third process stage, etc. Such events also require reliable diagnosis and adequate reaction of the control system.

3.2 Control of the Fed-batch Process for Phenylalanine Production

The accumulated knowledge of the phenylalanine production process allowed application of the proposed knowledge-based control methodology. The synthesis of the control system is logically divided into two parts: first, the design of the higher system level (diagnosis subsystem) and, second, the design of the lower system level (control subsystem). They are relatively independent and can be discussed separately.

3.2.1 Design of the Higher System Level: the Diagnosis Subsystem

As shown in Fig. 5, the higher level of the control system is composed of a procedure for calculating the structural variables of the fermentation plant which represent adequately the physiological state of the cell population, and procedure for real-time diagnosis of the cell population.

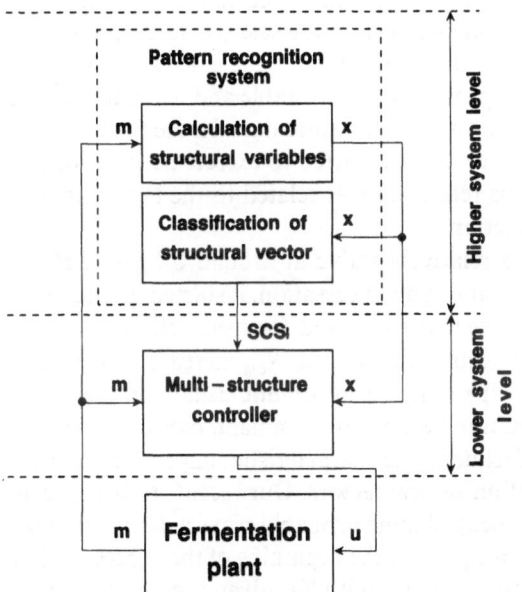

Fig. 5. Structure of a two-level hierarchical system for control of fermentation processes as variable structure plants. Notations: m, x and u are measurement, structural and control vectors, respectively, and SCS_i is the ith subspace of constant structure

On-line monitoring of representative structural variables. At the present, there is no clear analytical theory for the selection of the process structural variables. Here we propose a solution for the phenylalanine production process. Although the results presented are strictly related to this cultivation, we hope that the techniques, along with some more general discussion, may be useful in the case of other processes.

Before the selection of structural variables, some general considerations about the dimension of the structural vector should be pointed out. First of all, this dimension must be more than one. A single variable is never enough to describe complex biological phenomena; it is always ambiguous and does not supply clear information, unless combined with other variables. Furthermore, in highly changeable and noisy environments, it is reasonable to use a redundant set of variables. Redundancy is an important tool to enhance reliability: even if a certain variable is wrongly calculated, the recognition procedure will be carried out correctly, supported by the remaining variables.

By means of the computer system, the following variables were calculated on-line: O_2 uptake rate (OUR), CO_2 evolution rate (CER), respiratory quotient (RQ), glucose uptake rate (GUR) which was equal to the glucose feed rate (GFR) under conditions of glucose limitation, ammonia feed rate (AFR), ratio of the OUR to the GUR ($R_{o/g}$), ratio of the AFR to the GUR ($R_{a/g}$), CO_2 yield ($Y_{c/g}$), total O_2 consumption, total CO_2 consumption, marker of glucose accumulation (MGA, defined below), specific growth rate (SGR), increment of DO (ΔDO), specific glucose uptake rate (GUR*). However, not all of these are clearly related to the plant structural transfers, and only the followings qualify as structural variables: RQ, $R_{o/g}$ (or $Y_{c/g}$), MGA, SGR, ΔDO, $R_{a/g}$ and GUR*. All these variables belong to the class of structure-reflecting variables, have a qualitative nature and important physiological meaning, and are calculable on-line.

The characteristics and typical time-courses of the selected variables, shown in Fig. 6, will be discussed below. In the initial stage of the cultivation RQ was rather low. In the main part of the process, RQ was stable and fluctuated in the narrow range of 1.0–1.07, in accord with the literature data. In the final stage of the process, when cells lost their activity and started to excrete acetic acid, RQ values decreased to 0.95–0.98. This reduction was related to the last structural transfer and can facilitate its detection.

During growth, $R_{o/g}$ remained relatively stable at around 0.4 g g^{-1}. However, after the depletion of tyrosine and growth cessation, it started gradually to increase, indicating that more oxygen was required for metabolization of the same amount of glucose. During normal cultivation, $R_{o/g}$ increased more than two times. Around the 50th hour it reached a maximum, usually in the range 0.8–0.9 g g^{-1}. Thereafter, fermentation became predominant, oxygen consumption decreased, and $R_{o/g}$ rapidly declined. The characteristic shape of $R_{o/g}$ during the 3rd stage might be dependent on several factors. Our recent study suggests that the increasing inhibition by phenylalanine probably plays a dominant role. The $R_{a/g}$ value was high (about 0.06 g g^{-1}), at the beginning of the process, being relating to the active synthesis of the biomass. With the advance of the process, it

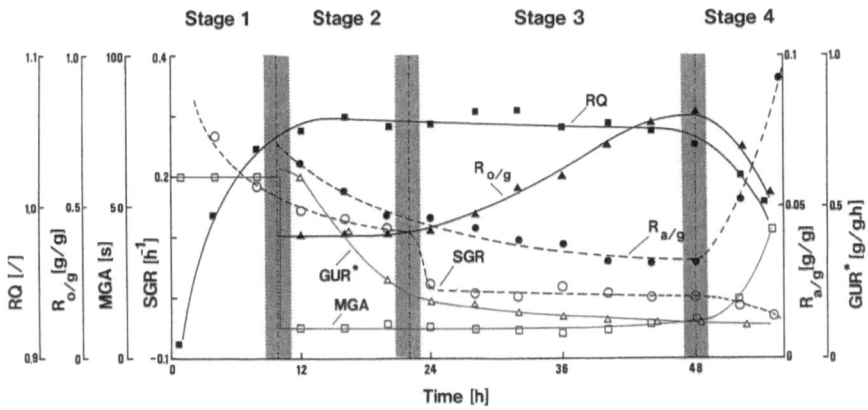

Fig. 6. Typical time-courses of selected structural variables: (■) RQ, (▲) $R_{o/g}$, (□) MGA, (○) SGR, (●) $R_{a/g}$, (△) GUR*. The *shadowed bands* indicate the periods when the structural transfers usually occur

slowly decreased. After cessation of growth, ammonia was used mainly for phenylalanine synthesis which also progressively slowed down. Upon entering the last stage, phenylalanine formation stopped and $R_{a/g}$ exhibited a minimum, usually in the range 0.02–0.03 $g g^{-1}$. Finally, it rapidly increased due to the pumping of ammonia for neutralization of the excreted acetic acid.

MGA is a special variable, and was not calculated directly from the available measurements. For its monitoring a dedicated algorithm which manipulates the glucose feed rate was developed. The procedure is based on short-time interruption of the glucose feed rate, cyclic monitoring of the DO, and detection of the moment of its jump, which indicates glucose depletion. MGA represents the delay (in seconds) from the moment of the feed rate interruption to the moment of the DO increase. The algorithm was encoded as a part of the software system; it worked completely automatically, and MGA was updated by the computer system as a normal variable. Typical shapes of the variables related to MGA estimation are shown in Fig. 7.

The MGA time-course during normal fermentation was characterized by low values of 8–12 s in the main part of the cultivation. With the advance of the process MGA slowly increased, which becomes more profound at the end of the cultivation when the cells, due to the loss of activity, could not withstand even low glucose feed rates. Although the procedure of MGA estimation seems complex, it proved to be extremely reliable and accurate. Our experience showed that it is not harmful to the cell population and does not have any negative side effects. Using this variable, it was possible to detect the early stages of glucose accumulation (MGA values around 15–20 s), which was a potential threat for acetic acid excretion.

The value of SGR dropped from the value in the normal range between 0.2 and 0.1 to about 0.01 h^{-1} when OD stopped increasing. This allowed for

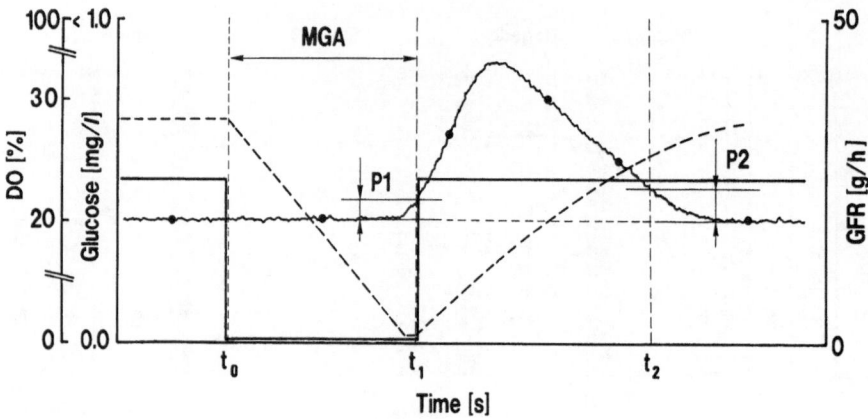

Fig. 7. Time courses of variables related to MGA estimation: (●) DO, (– – –) hypothetical glucose concentration and (——) glucose feed rate. P_1 is the minimal DO increment to consider glucose depletion (default 0.5%); and P_2 is a zone above the initial DO value in which when entered, control will be resumed (default 1.5%)

reliable detection of the moment of tyrosine depletion and good timing for its addition. The SGR was the variable which most reliably contributed to the detection of structural transfer at the end of the stage of healthy growth.

The GUR* is a fundamental variable in the phenylalanine production process. As discussed previously, it should follow a nearly-exponential declining shape to prevent fermentative glucose metabolism. This condition is related to the decreasing oxidative capacity of the cell population, caused probably by product inhibition. Unfortunately, the optimal GUR* shape is not exactly clear; it depends on many factors and practically should be considered as a "fuzzy" band, rather than as a single line. If the GUR* value is becoming high, its time-course will cross this band, leaving the so-called "region of subcritical feed rate" and entering the "region of supracritical feed rate". This will result in the GUR* exceeding the oxidative potential of the population and fermentative glucose metabolism, with a negative effect on the process efficiency. Therefore, the GUR* value can be used to roughly determine the current position of the process with respect to these regions.

Design of the diagnosis procedure. As global rules for detection of the chronological (stage-associated) plant alterations, the following rules were formulated to describe the plant structures, corresponding to the natural stages of the phenylalanine production process

IF (RQ is LOW) and (MAG is VERY HIGH) and (SGR is HIGH), THEN (the process structural state is SCS_1: "STAGE 1: BALANCED GROWTH"); $dc_1 = 1.0$.

IF (RQ is NORMAL) and ($R_{o/g}$ is LOW) and (SGR is HIGH) and (GUR* is HIGH),
THEN (the process structural state is SCS_2: "STAGE 2: UNBALANCED GROWTH"); $dc_2 = 1.0$.

IF (RQ is NORMAL) and ($R_{o/g}$) is MIDDLE-HIGH) and (SGR is LOW) and (GUR* is LOW),
THEN (the process structural state is SCS_3: "STAGE 3: UNBALANCED PRODUCTION"); $dc_3 = 1.0$.

IF (RQ is LOW) and ($R_{o/g}$ is MIDDLE) and (MGA is FINAL-HIGH) and (SGR is NEGATIVE) and (GUR* is VERY LOW),
THEN (the process structural state is SCS_4: "STAGE 4: FINAL LOSS OF ACTIVITY"); $dc_4 = 1.0$,

where SCS_i and dc_i are the abbreviations of subspace of constant structure and degree of certainty, respectively.

Their translation into a "transparent" algorithm yields a set of numerical equations in the following form,

$$
\begin{aligned}
w_{11} \cdot \mu_{11}(RQ) + w_{12} \cdot \mu_{12}(MGA) & \\
+ w_{14} \cdot \mu_{14}(SGR) &= dc_1 \\[2mm]
w_{21} \cdot \mu_{21}(RQ) \qquad\qquad + w_{23} \cdot \mu_{23}(R_{o/g}) & \\
+ w_{24} \cdot \mu_{24}(SGR) \quad + w_{25} \cdot \mu_{25}(GUR^*) &= dc_2 \\[2mm]
w_{31} \cdot \mu_{31}(RQ) \qquad\qquad + w_{33} \cdot \mu_{33}(R_{o/g}) & \\
+ w_{34} \cdot \mu_{34}(SGR) \quad + w_{35} \cdot \mu_{35}(GUR^*) &= dc_3 \\[2mm]
w_{41} \cdot \mu_{41}(RQ) + w_{42} \cdot \mu_{42}(MGA) + w_{43} \cdot \mu_{43}(R_{o/g}) & \\
+ w_{44} \cdot \mu_{44}(SGR) \quad + w_{45} \cdot \mu_{45}(GUR^*) &= dc_4
\end{aligned}
\tag{1}
$$

where the weight coefficients were obtained by a supervised learning procedure. The scale of the variables were properly segmentalized by fuzzy membership functions with a piecewise-linear plateau form, corresponding to the fuzzy expressions used in the rules' linguistic terms.

In addition to the main group of rules for diagnosis of the process stages, a second group was introduced to detect some events with a local character which may, or may not happen during the process. The rules from this group will be "fired" selectively to the process stage, i.e. according to the result obtained from the first group of rules. These additional rules have simple conditions, which makes their processing easier. Some of them are used only to advise the operator, and not for supervising of the lower system level.

206 Konstantin B. Konstantinov and Toshiomi Yoshida

*Rules for stage 1:
Rule 1.1 (for detection of the glucose depletion in the broth; ΔDO is the DO increment):
IF (ΔDO is HIGH),
THEN (activate continuous glucose feeding);
THEN (deactivate this rule).

Rule 1.2 (for detection of the entering of the DO into the set interval):
IF ($DO \leq DO_{sp}$),
THEN (activate DO control);
THEN (deactivate this rule).

*Rules for stage 2:
Rules 2.1 (for detection of abrupt interruptions in the glucose consumption):
IF (ΔDO is HIGH),
THEN (glucose consumption was interrupted).

Rule 2.2 (for detection of the entering of the N into the set interval, correspond-
ing to the maximum allowable OTR = OUR):
IF ($N \geq N_{sp}$),
THEN (do not increase GFR anymore);
THEN (deactivate this rule).

Rules for stage 3:
Rule 3.1 (for detection of premature acetic acid excretion):
IF ($R_{a/g}$ is HIGH),
THEN (premature acetic acid excretion).

Rule 3.2 (for detection of abrupt interruptions in the glucose consumption):
IF (ΔDO is HIGH),
THEN (glucose consumption interrupted).

Rule 3.3 (for detection of cell growth after the tyrosine depletion):
IF (SGR is MIDDLE),
THEN (secondary growth under tyrosine limitation).

*Rules for stage 4:
Rule 4.1 (for the process termination):
IF ($R_{a/g}$ is HIGH),
THEN (runaway acetic acid excretion);
THEN (terminate the process).

3.2.2 Design of the Lower System Level: the Control Subsystem

The design of the lower system level was performed on the basis of the
experimental study described above. Particular attention was focused on the

problem for control of the glucose feed rate which appeared to be crucial for the phenylalanine production process.

DO control. DO was maintained within the range [20–22]%. The dissolved oxygen control loop was activated after dissolved oxygen had decreased to 20% (rule 1.2). This happened during the first process stage, usually between the 8th and the 12th hour after inoculation. Thereafter, DO was maintained through all process stages within the range [20–22]% by manipulation of the agitation speed in a structure-invariant control loop. This was realized by the incremental control algorithm shown in Fig. 8. It was selected because of its simplicity, reliability and accurate performance. During the third process stage, however, the DO control loop was incorporated into a more complicated control scheme, which is explained below.

Control of the glucose feed rate. The glucose feed rate is a key factor in the phenylalanine production process. It should be controlled throughout the cultivation to prevent acetic acid excretion and to reduce excessive CO_2 evolution. However, the characteristics of the culture in respect to the glucose feeding are not constant during the process. This is consequence of the serious physiological alterations which the cell population undergo upon transfer through the consecutive cultivation stages, and which require alteration of the glucose feeding policy.

*Glucose feeding in stage 1: During the first several hours after inoculations, excretion of acetic acid was never observed, despite the glucose concentration in the broth. High glucose concentration were not dangerous during this period and there were no reasons to keep the cells under glucose limitation. Consequently, the total amount of glucose provided for this stage was supplied to the fermenter before inoculation resulting in the initial concentration of $20\,g\,L^{-1}$.

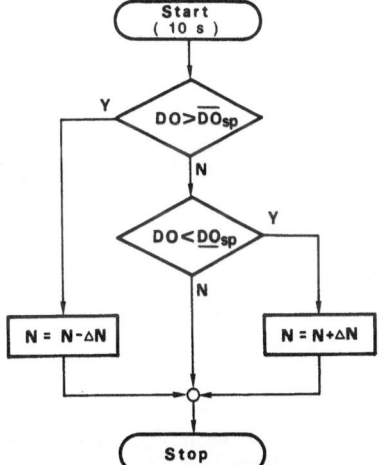

Fig. 8. Incremental algorithm for DO control. It was activated every 10 s

*Glucose feeding in stage 2: After 10 to 14 h of cultivation, due to unclear reasons, the cells acquired the capability of producing acetic acid, and excessive glucose feeding resulted in its excretion. Thus, from this period, continuous feeding of glucose should be initiated to provide adequate glucose limitation. At the same time cells still grow actively, and the total amount of biomass increases together with the overall metabolic potential of the population. The glucose feeding control should account for the growth of cells by corresponding increase of the feed rate.

To satisfy these requirements, the control scheme shown in Fig. 9, was developed. It manipulated the glucose feed rate to maintain at a desired value keeping the specific variable MGA which reflected the degree of glucose limitation. The selection of its set point (MGA_{sp}) is very important because it

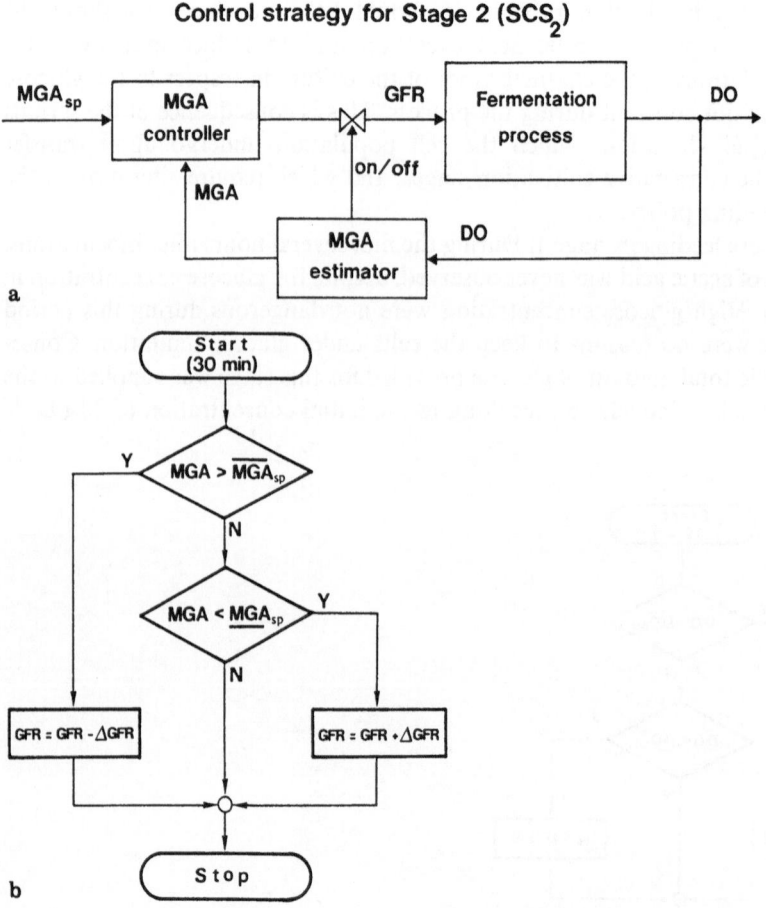

Control strategy for Stage 2 (SCS_2)

Fig. 9. The MGA controller used to manipulation of the glucose feed rate in stage 2. (**a**) control scheme; (**b**) control algorithm. The procedure was activated every 30 min

should provide the feed rate required for the compromisation, achieving simultaneously the prevention of acetic acid excretion and reduction of CO_2 evolution. A value of MGA_{sp} that is too small leads to poor feed rates, glucose overlimitation and intensive CO_2 evolution. A large MGA_{sp} value results in high glucose feed rates, insufficient limitation, and acetic acid excretion. It was found that MGA_{sp} values in the range [10–16] provided reasonable glucose limitation, and this range was used in our experiments. The corresponding logic of the control algorithm is shown in Fig. 9.

*Glucose feeding in stage 3: The active growth terminated sharply after the tyrosine depletion. The growth cessation due to tyrosine starvation leads to drastic changes in the cell physiology, initiating several phenomena, e.g. morphological alteration, darkening of the culture medium, increase in the relative oxygen consumption. Most important is, however, that the total metabolic capacity of the cell population progressively decreased so that at the final stages of the process, cells were able to oxidize much lower amounts of glucose than at the time of the growth cessation. Hence, during the non-growing period the population should be kept under adequate glucose limitation, accounting for the decrease of its metabolic capacity. To meet this requirement, a control scheme called "balanced DO-stat", was applied. It manipulated both the agitation speed and the glucose feed rate, providing a nearly-exponential declining pattern of the specific glucose uptake rate.

The proposed control scheme consists of two interconnected loops as shown in Fig. 10. The first loop, where DO was controlled at a fixed set point DO_{sp} by altering agitation speed, was the same as in other process stage. In the second loop agitation speed was also kept at a fixed set point N_{sp} by manipulating the glucose feed rate. Practically, both set points DO_{sp} and N_{sp} were substituted by narrow ranges. After detection of the transfer to the third stage, only the second loop was activated because the first one was already in operation. The link between the two loops provided an accurate dynamic balance between oxygen transfer and feeding, so that sufficient oxygen supply was guaranteed while the cells were protected from excess glucose supply. This new feature was the reason for naming the proposed strategy "balanced DO-stat."

Important feature of this control strategy, particularly with regard to the phenylalanine production process, is that it provides the gradually decreasing glucose feed rate with a proper time-profile. The mechanism of the automatic reduction of the glucose feed rate relied on a useful empirical relationship. Experiments showed that the decrease of GUR^*_{crit} was accompanied by another phenomenon: the oxygen demands of the cells to metabolize glucose, expressed by the ratio of the oxygen uptake rate to the glucose uptake rate

$$R_{o/g} = \frac{OUR}{GUR} \tag{2}$$

increased progressively with time, and more oxygen was needed for metabolization of equal amounts of glucose. Under conditions of glucose limitation, the

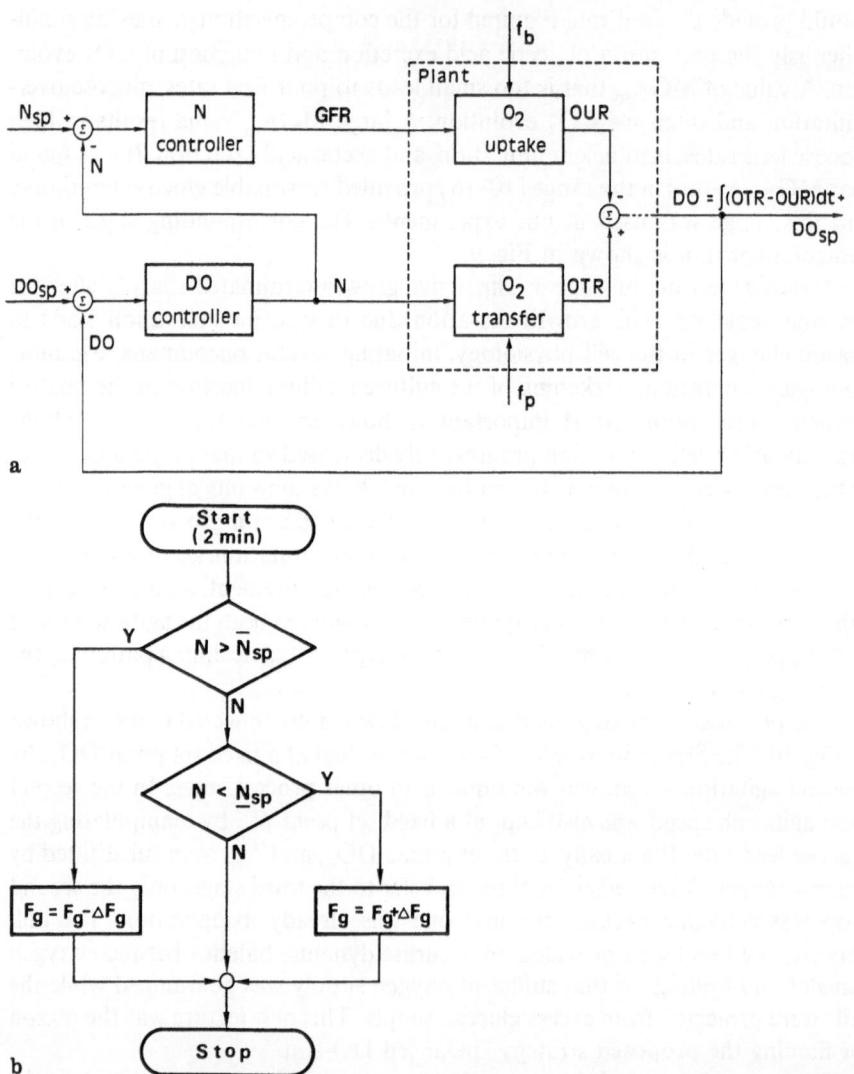

Fig. 10. Scheme of the balanced DO-stat control strategy used in Stage 3. (a) two interconnected loops for control of the oxygen balance by manipulating its uptake rate OUR and transfer rate OTR; (b) incremental algorithm for control of the agitation speed N by the glucose feed rate GFR. The algorithm for control of the DO by the agitation speed N is shown in Fig. 8

equation can be transformed to

$$\text{GFR} = \frac{1}{R_{o/g}} \times \text{OUR}. \tag{3}$$

It is simple to conclude that the glucose feed rate should be gradually decreased to maintain a constant OUR, which will result in a decreasing GUR* time-profile. This mechanism is exploited by the balanced DO-stat. Under constant DO, the oxygen transfer rate OTR was equal to the oxygen uptake rate OUR. In its turn, OUR can be kept stable if the agitation speed is controlled at a certain value by manipulation of the glucose feed rate, which provides the desired effect.

The proposed control scheme was well-suited for on-line computer applications. It worked satisfactorily with simple incremental algorithms, which proved to be robust and do not require any mathematical models. This scheme resulted in smooth and reliable control of phenylalanine cultivation. As shown in Fig. 11. DO was maintained within a 2% range during the control period. OUR and CER also fluctuated within rather narrow limits. The OUR stability is an intrinsic feature of the balanced DO-stat strategy, which can be easily drawn out from the oxygen balance equation.

*Glucose feeding in stage 4: During the final process stage, cell metabolic power became very low, and the phenylalanine synthesis almost completely stopped. The $R_{o/g}$ changed its increasing trend and began to rapidly decrease. Hence, the balanced DO-stat strategy cannot perform adequately any more. Therefore, after detection of the transfer to the last process stage, the GFR was fixed constant at its last value before the detection. The feeding with this rate continued until the detection of intensive acetic acid excretion, manifested by high $R_{o/g}$ values (rule 4.1). This indicated the right time for termination of the process.

3.2.3 Application of the Proposed Complex of Techniques to a Real Process

The described complex of techniques was applied to diagnose and control the phenylalanine production process. The results from the work of the higher level of the control system are shown in Fig. 12, which depicts the time-courses of the on-line calculated degrees of certainty for every process stage.

The knowledge-based procedure provided accurate detection of the process stage transfers. Since the first two transfers had abrupt character, the static form of the rules did not hamper the recognition and they were detected very reliably. The utilization of flat fuzzy membership functions for the segmentation of the structural variables provided robust performance. Even strong noises in the process variables did not disturb the recognition procedure. However, the last transfer, which was gradual, confused the system and its detection was somewhat delayed. The reason for this was the principal limitation of the static facts to interpret reliably alterations in the cultivation system with more complicated dynamics. For better diagnosis, the corresponding rule should be enriched by dynamic facts. Except for this problem, the performance of the knowledge-based diagnosis procedure was satisfactory, and its decisions were similar to the decisions of an experienced operator. Based on the stage diagnosis, the second group of rules was capable of timely detection of the local events.

Time-courses of the main process variables in this cultivation are shown in Fig. 13. In total, 3.8 g tyrosine and 1620 g glucose were fed during the process;

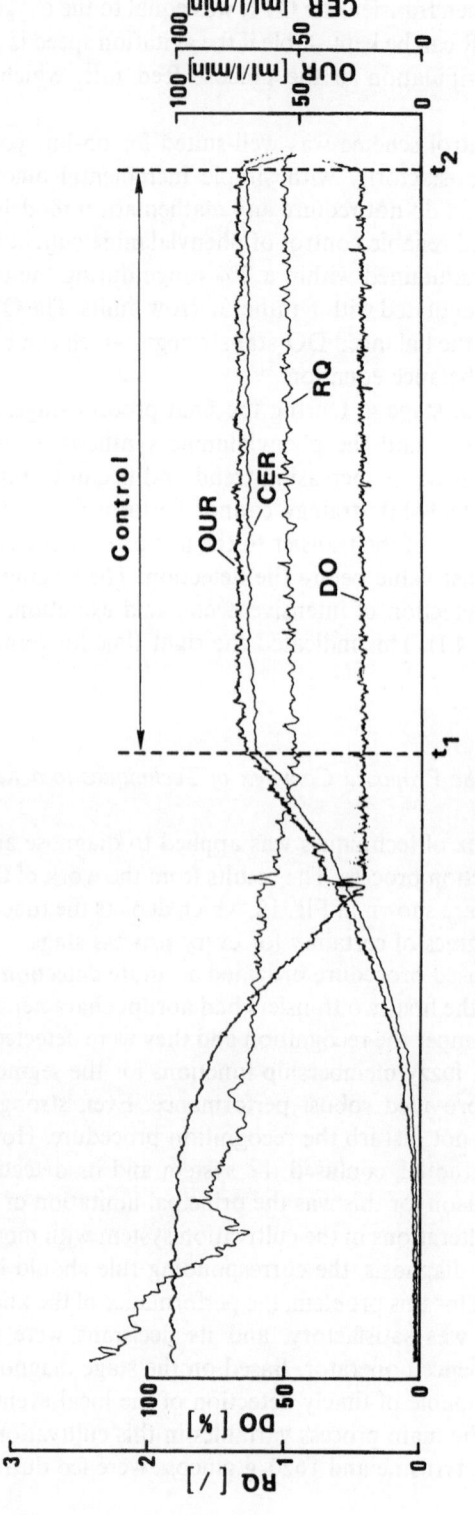

Fig. 11. Time courses of DO, OUR and CER in an experiment by a balanced DO-stat in the period $[t_1-t_2]$

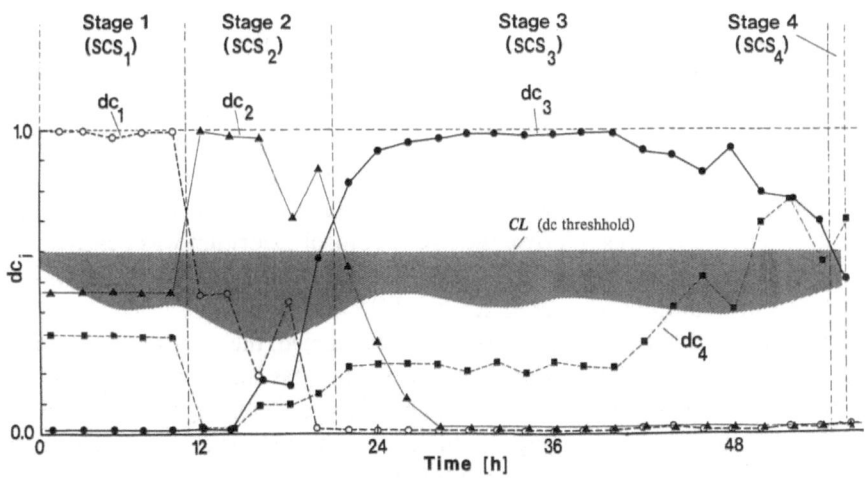

Fig. 12. Results from the recognition procedure: time-courses of the degrees of certainty dc_1–dc_4 used for detection of the stage transfers. The first transfer (*Stage 1*→*Stage 2*) and the second transfer (*Stage 2*→*Stage 3*) were adequately detected. However, the response of the system to the last transfer (*Stage 3*→*Stage 4*) was somewhat delayed. The confidence level CL, set at 0.6 in this experiment, indicates the dc value above which the decision was considered as valid

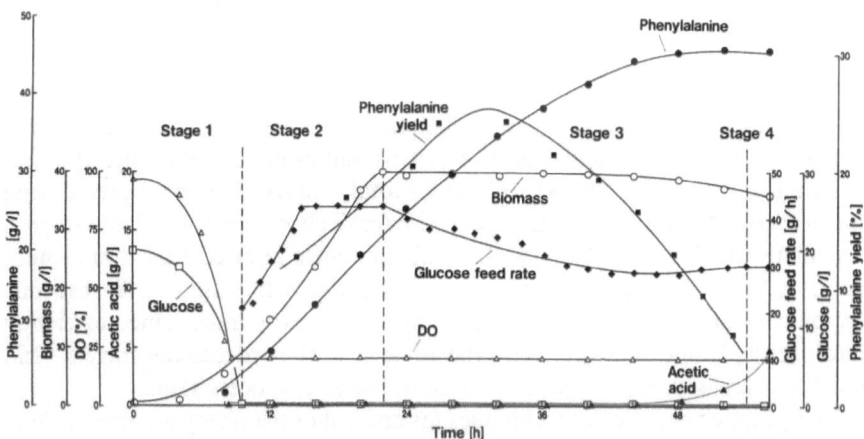

Fig. 13. Typical time-courses of some variables in the phenylalanine production process: (●) phenylalanine concentration, (○) biomass concentration, (▲) acetic acid concentration, (◆) glucose feed rate, (□) glucose concentration, (■) phenylalanine yield, (△) DO

initial glucose concentration was $20 \, \mathrm{g \, L^{-1}}$. During the first stage, glucose was not fed. The first loop for the control of DO by agitation was activated at the 9th hour, immediately after DO entered the set range [20–22]% (rule 1.2).

Feeding started at the 10th hour under command from the higher level after consumption of the initial amount of glucose (rule 1.1). The glucose feed rate was gradually increased by the MGA control scheme, with MGA being controlled in the interval [12–16] seconds. During this period agitation speed increased together with the glucose feed rate to compensate for the change in the OUR. Around the 15th hour agitation speed entered a previously set range, selected to provide a proper OTR [19] which, according to Eq. (3), acts as a gain. Too high an agitation speed will result in an increased OTR and a shift of the GUR* profile into the supracritical region. In this experiment, agitation speed range was set at [1200–1230] rpm, which was empirically found to provide a suitable OTR. After entering this interval, the MGA control algorithm was deactivated to prevent further increase in the glucose feed rate and agitation speed, which might result in acetic acid excretion (rule 2.2).

Around the 22nd hour, the tyrosine was consumed, the cell growth stopped and the system detected the transfer to the 3rd process stage, and the balanced DO-stat control strategy was activated. This control algorithm worked throughout the 3rd process stage, until around the 53 h when the higher level reported a transfer to the 4th stage. Soon after, increasing $R_{a/g}$ values signaled for intensive acetic acid excretion and the process was terminated (rule 4.1).

A complete prevention of the acetic acid excretion throughout the period of phenylalanine synthesis was achieved in this experiment. The obtained GUR* time-profile was similar to the GUR^*_{crit} as shown in Fig. 4. A negligible low concentration of acetic acid (about $0.1 \, g \, L^{-1}$ was found around the 14th hour during the second cultivation stage, when GUR* entered the region of supracritical feed rates. It was quickly consumed late, and the acetic acid concentration was maintained at zero until the last cultivation stage, when around the 50th hour its uncontrollable excretion begun.

Cells grew free of inhibition and their concentration increased to about $40 \, g \, L^{-1}$ at the moment of tyrosine depletion. This amount of biomass, determined by the total tyrosine fed (3.8 g), was sufficient to metabolize the predetermined amount of glucose (1620 g). In total, $640 \, L \, CO_2$ was released during the process. Phenylalanine was produced intensively during stage 2 and stage 3. Calculations of the momentary phenylalanine yield showed that a maximum value of 25% was reached several hours after the growth cessation. At that time the glucose uptake rate was still high, and glucose was transformed efficiently into phenylalanine. However, with the advance of the process, the current yield quickly decreased, reaching a low value in the final process stage.

The applied complex of techniques for control of the phenylalanine production process resulted in a remarkably improved efficiency. The final phenylalanine concentration reached $46 \, g \, L^{-1}$, the yield was nearly 18%, and the productivity $- 0.9 \, g \, L^{-1} \, h^{-1}$. These results are about 50% higher than the data obtained without application of the described techniques.

4 Conclusion

The applied control approach enhanced the phenylalanine synthesis through improvement of the balance between the main process products. The problem of acetic acid excretion was completely eliminated by combined control of the oxygen supply and the glucose feeding. The total biomass synthesis was controlled by the amount of tyrosine fed, selected to provide the metabolic potential sufficient for effective transformation of the predetermined amount of glucose. Excessive carbon dioxide evolution was prevented by the glucose feed rate, preventing an unnecessarily strong glucose limitation. Apart from the improvement in the process efficiency, the control system provided an "intelligent" and fully automatic mode of operation. The knowledge-based module was capable of diagnosing the major alteration in the cell population, as well as detecting some other events arising during the process. Based on this, the control module could dynamically adapt to the current process situation and to function adequately.

It should be stated that there are still some possibilities for additional improvement of the phenylalanine production process, mainly in two directions. First, there is the sophistication of the temperature regime for full utilization of the genetic potential of the thermo-controlled promoter system. Second, there is the release of the cell population from the phenylalanine inhibition, which appears to have a negative effect on the overall process performance. Further study of the corresponding physiological factors and mechanisms will additionally contribute to the enhancement of the efficiency of the phenylalanine production process.

5 Nomenclature

AFR   ammonia feed rate ($g\,L^{-1}$)
CER   carbon dioxide evolution rate ($g\,L^{-1}$)
CF_i certainty factor of the ith rule
CL   confidence level
DO   dissolved oxygen (%)
DO_{sp} dissolved oxygen set point (%)
ΔDO increment in the DO (%/min)
dc   degree of certainty
GFR   glucose feed rate ($g\,L^{-1}$)
GUR   glucose uptake rate ($g\,L^{-1}$)
GUR*   specific glucose uptake rate (h^{-1})
GUR^*_{crit} critical specific glucose uptake rate (h^{-1})
MGA   marker of glucose accumulation (s)
MGA_{sp} MGA set point (s)
N agitation speed (rpm)
OTR   oxygen transfer rate ($g\,h^{-1}$)
OUR   oxygen uptake rate ($g\,h^{-1}$)
$R_{a/g}$ AFR to GUR ratio ($g\,g^{-1}$)
$R_{o/g}$ OUR to GUR ratio ($g\,g^{-1}$)
RQ   respiratory quotient (dimensionless)

SGR   specific growth rate (h^{-1})
w_{ij} weight coefficient
$Y_{c/g}$ carbon dioxide yield $(g\,g^{-1})$
$\mu_{ij}(x_j)$ membership function

6 References

1. Kickert WJM, Bertrand JWM, Praagman J (1978) IEEE Trans Syst Man Cybern SMC-8: 805.
2. Waterman D (1985) In: A Guide to Expert Systems. Addison Wesley–Reading, MA.
3. Halme A (1988) Proc 4th Int Sump Computer Applications in Fermentation Technology, Cambridge, UK, p 159.
4. Xu E, Xu G (1989) Proc IFAC Workshop AI in Real-Time Control, Shenyang, PRC, p 97.
5. Smuts W, McLeod I (1989) Proc IFAC Workshop on AI in Real-Time Control, Shenyang, PRC, p 31.
6. Halme A (1989) Proc IFAC Workshop on AI in Real-Time Control, Shenyan, PRC, p 11.
7. Lubbert A (1989) Proc IFAC Workshop on Expert Systems in biotechnology, Helsinki, p 11.
8. Aarts R, Suviranta A, Rauman-Aalto P, Linko P (1990) Food Biotechnol 4: 301.
9. Iserman R (1987) Proc 3rd IFAC Congress, Munich, FRG, p 6.
10. Rod MG, Suski GJ (1988) Proc IFAC Workshop on AI IN Real-Time Control, Clyne Castle, UK.
11. Rod MG, Li H, Su S (1988) Proc IFAC Workshop on AI IN Real-Time Control, Shenyang, PRC.
12. Cooney C, O'connr G, Sanchez-Riera F (1988) Proc 8th Int Biotechnol Symp, Paris, France, p 563.
13. Chen Q, Wang S, Wang J (1988) Proc 4th Int Symp Computer Applications in Fermentation Technology, Cambridge, UK, p 253.
14. Linko P (1988) Ann New York Acad Sci 542: 83.
15. Konstantinov KB, Kishimoto M, Yoshida T (1988) Annual Reports of ICBiotech, Osaka University 11: 419.
16. Konstantinov KB, Yoshida T (1989) Biotechnol Bioeng 33: 1145.
17. Konstantinov KB, Yoshida T (1990) J Ferment Bioeng 70: 48.
18. Konstantinov KB, Nishio N, Yoshida T (1990) J Ferment Bioeng 70: 253.
19. Konstantinov KB, Kishimoto M, Seki T, Yoshida T (1990) Biotechnol Bioeng 36: 750.
20. Konstantinov KB, Yoshida T (1990) J Ferment Bioeng 70: 421.
21. Konstantinov KB, Nishio N, Yoshida T (1991) J Ferment Bioeng 71: 350.
22. Konstantinov KB, Yoshida T (1991) IEEE Trans Syst Man Cybern 21: 908.
23. Konstantinov KB, Yoshida T (1992) J Biotechnol 24: 33.
24. Konstantinov KB, Yoshida T (1991) submitted to IFAC Workshop on Software Structures Integrating Artificial Intelligence and Knowledge-Based System in process Control, Norway.
25. Sugimoto S, Yabuta M, Kato N, Seki T, Yoshida T, Taguchi H (1987) J Biotechnol 5: 237.
26. Field R (1971) Biochem J 124: 581.
27. DeBoor C (1979) In: A practical Guide to Splines, Springer, Berlin Heidelberg New York.
28. Weiner N, Dracoszy P (1961) J Pharmcol Exp Ther 132: 299.
29. Landwall P, Holme T (1977) J Gen Microbiol 103: 345.
30. Landwall P, Holme T (1977) J Gen Microbiol 103: 353.
31. Mori H, Yano T, Kobayashi T, Shimizu S (1979) J Chem Eng Japan 12: 313.
32. Yano T, Mori H, Kobayashi T, Shimizu S (1980) J Ferment Technol 58: 259.
33, Kashket E (1981) J Bacteriol 146: 377.
34. Re aske D, Adler J (1981) J Becteriol 145: 1196.
35. Smirnova G, Oktyabrskii O (1982) Microgiologiya (Russian) 54: 252.
36. Shimizu N, Fukuzono S, Fujimori K, Nishimura N, Odawara Y (1988) J Ferment Technol 66: 187.
37. Andersen K, von Meyenburg K (1980) J Bacteriol 144: 114.
38. Sonnleitner B, Kappeli O (1986) Biotechnol Bioeng 28: 927.
39. Meyer H-P, Leist C, Fiechter A (1984) J Biotechnol 1: 355.

4.6 Modulator Sorption in Gradient Elution Chromatography

Ajoy Velayudhan[1] and Michael R. Ladisch[1,2]

[1]Laboratory of Renewable Resources Engineering
[2]Department of Agricultural Engineering,
 Purdue University, 1295 Potter Center, West Lafayette, IN 47907-1295, USA

Contents

Gradient elution chromatography is widely used to separate both small molecules and macromolecules. The mobile phase additive (modulator) used to modify adsorbate retention is usually considered to be either unretained or linearly retained. In both cases, the shape of the gradient does not change as it moves down the column. However, the high mobile phase concentrations at which such a modulator is commonly used makes it likely to adsorb according to its nonlinear sorption isotherm. Here the quantitative consequences of such nonlinear modulator sorption are reviewed. Nonlinear sorption deforms the shape of the gradient during its passage through the column; ultimately a shock (or shock layer) could be formed. The condition for shock formation is discussed, and numerical simulations using representative parameters illustrate the magnitude of gradient deformation and the consequences for separation.

1 Introduction

Gradient elution, invented by Tiselius and his co-workers [1], employs a mobile phase whose composition at the column inlet varies with time. This is in contrast to isocratic elution, where the mobile phase composition remains constant at every point in the column throughout the separation. In fact, one of the reasons gradient elution was introduced was to hasten isocratic separations, by reducing tailing and by moving widely separated compounds closer together. It is also well-suited to the separation of macromolecules, whose retention tends to vary strongly with mobile phase composition [2], and is consequently a widely used technique in biotechnology.

The mobile phase composition is usually altered by changing the concentration of a small molecule, e.g., a salt such as sodium chloride in ion-exchange and hydrophobic interaction chromatography, and an organic modifier such as

Bioproducts and Bioprocesses 2
Editors: Yoshida, Tanner

acetonitrile in reversed-phase chromatography. These mobile phase additives will for convenience be referred to by the general term "mobile phase modulators."

When adsorbate concentrations are low enough to lie within the linear regions of their sorption isotherms, a reasonably complete understanding of isocratic elution exists. An explicit analytical solution for the effluent sample peak that accounts for axial dispersion, boundary-layer mass transfer, pore diffusion, and finite sorption–desorption kinetics was derived by Rasmuson [3]. While the added complexity of the temporal and spatial dependence of the modulator concentration—which in turn causes the adsorbates' retention factors to change with time and position within the column—has precluded analogously complete solutions in gradient elution, a good deal of information about the characteristics of the sample peaks emerging from the column is available. Under the assumption that the modulator does not itself bind to the chromatographic stationary phase, good estimates of the retention times (centers of mass) and band widths of the peaks, as well as of peak resolution (which is a measure of how well adjacent peaks are separated), are available. Comprehensive reviews of these results can be found in Snyder [4, 5] and Jandera and Churacek [6]. Analogous results are summarized in Yamamoto et al. [7], for the case where the distribution constant for the modulator is a constant (i.e., unaffected by its own concentration or those of the samples).

The present chapter focuses on the consequences of allowing the modulator to adsorb according to a nonlinear binding isotherm. This phenomenon, termed "solvent demixing", has been qualitatively described in the literature [4]; here we examine its quantitative implications. It will be seen that, when the input gradient is sufficiently steep, solvent demixing can cause the gradient shape to become significantly deformed, and even form shock layers, as it moves down the column. The nature of such gradient deformation and its implications for separation are discussed with the help of illustrative simulations.

2 Theoretical Development—Literature Survey

A considerable amount of work on gradient elution has been reported in the literature; this section summarizes the important theoretical developments.

A useful classification is in terms of how modulator sorption is treated. The vast majority of published work either considers the modulator to be an unadsorbed component (i.e., one that travels at the mobile phase velocity), or assumes it to be linearly adsorbed. Analytical results based on these possibilities are examined in turn in this section. Throughout, the adsorbates' concentrations are assumed to lie in the linear regions of their own sorption isotherms. Finally, numerical results adsorbates at high concentrations are briefly described.

2.1 Unretained Modulator

Most analytical results have been obtained under the assumption of an un-retained modulator. The bandspreading suffered by the modulator is usually neglected, since it is less important than that of the sample components. In addition, the modulator is usually a small molecule, and its bandspreading is small (i.e., the efficiency of the column with respect to the modulator—as measured by its plate count, for instance—will be high). Then, if the shape of the gradient at the inlet is given by

$$C_M(x, t)|_{x=0} = f(t) \tag{1}$$

the location and shape of the gradient for all further times and positions is given by

$$C_M(x, t) = f\left(t - \frac{x}{v}\right), \tag{2}$$

where v is the chromatographic or mobile-phase velocity, t is time and x is distance into the column.

The interaction between the modulator and the sample components must be specified next. A large number of such relationships, varying from the theoretical to the empirical, have been put forward and are summarized in Jandera and Churacek [6]. Here we restrict attention to two very commonly used expressions. The first is

$$\ln K_i = \alpha_i - \beta_i \ln C_M, \tag{3}$$

where K_i is the distribution coefficient for the ith adsorbate, C_M the mobile-phase modulator concentration, and α_i and β_i are constants. This expression is frequently used in ion exchange chromatography [4, 6] although theoretical justification can only be furnished under the following assumptions. Activities must be replaced by concentrations; the ion-exchange process is regarded solely as a displacement of counterions, and effects such as Donnan exclusion are neglected. In spite of these rather restrictive assumptions, this simple expression often describes experimental data reasonably well.

The other form that is frequently used is

$$\ln K_i = \alpha_i^* - \beta_i^* C_M \tag{4}$$

where α_i^* and β_i^* are constants. This is usually applied to reversed-phase chromatography; again, theoretical validation of this expression is only available under many simplifying assumptions. "Linear solvent strength" (LSS) conditions are said to exist [4] when the following two conditions are satisfied: the gradient shape at the inlet as given by Eq. (2) must be linear, and Eq. (4) must be applicable to all the adsorbates.

Combining Eq. (2) with either Eq. (3) or (4) specifies the distribution coefficient of each component as a function of time and position. These results can then be substituted into the usual mass balance equations describing transport and sorption in a packed bed for each sample component (since the sample components are assumed to obey linear isotherms, each component can be treated independently). However, these equations are extremely complex, and analytical solutions for the effluent concentrations are not available. The method of moments has, however, been applied to simplified descriptions, the most common being the lumping of all bandspreading terms into a single lumped dispersion coefficient, D_i^*, so that the mass balances become

$$(1 + \phi K_i) \frac{\partial C_i}{\partial t} + v \frac{\partial C_i}{\partial x} = D_i^* \frac{\partial^2 C_i}{\partial x^2}, \tag{5}$$

where K_i is a function of x and t through Eqs. (1)–(3) or (4) and ϕ is the overall volumetric phase ratio. If Eq. (5) is made dimensionless through the usual definitions, $\tau = vt/L$ and $\xi = x/L$, there follows

$$(1 + \phi K_i) \frac{\partial C_i}{\partial \tau} + \frac{\partial C_i}{\partial \xi} = \frac{1}{Pe_i^*} \frac{\partial^2 C_i}{\partial \xi^2}, \tag{6}$$

where $Pe_i^* = vL/D_i^*$ is the lumped Peclet number.

Then, the method of moments [8] can be used to extract the retention time (the first moment) and the band width (as the second central moment). Even the second moment can be difficult to calculate analytically, but the first moment is usually easy to obtain [9]. First moments can also be calculated under the assumption of ideal chromatography, i.e., when $D_i^* = 0$ in Eq. (5), by the method of characteristics. This has been done by several authors for linear input gradients [4, 5, 10, 11] and is summarized by Jandera and Churacek [6]. It is found that the results from the moment calculation agree with those from the method of characteristics, which shows that—as in isocratic elution—the retention time is essentially independent of non-equilibrium terms. Such a calculation has not been done, to our knowledge, for the general model which rigorously accounts for the various nonequilibrium phenomena.

The calculation of the band width by the method of moments has also been carried out by several workers. Here the modulator concentration is usually considered to be essentially constant across the width of the adsorbate peak(s) [12, 13]. Poppe et al. [9], generalized this by considering a linear variation of local adsorbate retention factor (or, equivalently, distribution coefficient) across the peak. This allowed them to estimate band compression under the assumption of a slowly varying gradient for which the linear variation of retention factor described above would hold. Band compression arises due to the difference in modulator concentration across an adsorbate peak, which causes the trailing portion of the peak to move faster than the leading portion, thus compressing the peak. An approximate calculation for this peak compression effect was given by Snyder et al. [4, 14].

From these retention time and band width calculations, the resolution can be estimated, as in isocratic elution. Such results for the retention mechanisms given by Eq. (3) and (4) are summarized in Snyder [4], and Jandera and Churacek [6].

2.2 Linearly Sorbed Modulator

Here the distribution coefficient of the modulator, K_M, is assumed to be constant. Then Eq. (2) must be replaced by [15, 16]

$$C_M(x, t) = f\left(t - \frac{x}{v}\{1 + \phi K_M\}\right). \tag{7}$$

Drake [10] and Freiling [11] have discussed the consequences of Eq. (7) qualitatively. Yamamoto et al. [7, 17, 18] have made extensive use of Eq. (7) in a continuous-flow plate theory to estimate peak retention times and band widths as well as resolution.

McCoy and co-workers have applied their version of the moment method [19–21] to gradient elution [22]. Again, Eq. (7) applies, and the gradient is assumed not to vary appreciably over the width of the sample peak so that band compression effects are neglected. Resolution is also estimated. A similar calculation was carried out by Gibbs and Lightfoot [23], who also discuss more general issues such as the geometrical configurations that give rise to efficient column separations. The work of Poppe et al. [9] is also applicable to Eq. (7), although they restrict it to a completely unretained modulator (i.e., one whose movement is described by Eq. (2)).

2.3 Finite Sample Concentration

When the sample components are at concentrations high enough to lie in the nonlinear regions of their own isotherms, interference occurs between the sample components as well as between the sample components and the modulator. This makes the problem extremely complex, and analytical results cannot be expected. However, it has been pointed out [7, 24] that the linear-isotherm assumption often remains an acceptable approximation. This is because of the linearizing effect of the gradient: the modulator decreases the retention factor of the sample components. For a well-chosen gradient shape, the adsorbates could reach the linear portions of their respective sorption isotherms before having travelled very far into the column, thus restoring isotherm linearity. It is only when the column loading is extremely high that the adsorbates interfere with each other over a significant fraction of the column.

Snyder et al. [25–27], have carried out Craig simulations of gradient elution at high sample concentrations. The issue is of practical interest, since gradient elution could be an extremely efficient technique for preparative separations. An experimental comparison at laboratory scale of gradient elution with isocratic elution and displacement chromatography was presented by Liao et al. [28].

Selectivity reversal becomes possible under gradient elution (i.e., the elution order of the sample components changes at different modulator levels); Snyder et al. [29], and Antia and Horvath [30] have examined this phenomenon numerically.

3 Modulator Sorption

3.1 Motivation

As we have described in the previous section, it is usually assumed that the modulator is either unretained or linearly retained (i.e., zero, or nonzero but constant, distribution coefficient). These assumptions simplify the analysis considerably since the gradient shape remains undeformed in both cases, and is merely delayed when a constant distribution coefficient is assumed. There is, however, some reason to believe that, at least when the modulator achieves its effect by itself competing for binding sites on the stationary phase, the modulator undergoes nonlinear sorption. The feed components tend to be much more retentive than the modulator itself; this is particularly true of macromolecular adsorbates. Thus the modulator can only compete effectively by being at much higher concentrations, when it is likely to have by-passed the linear region of its own sorption isotherm. Then the modulator will suffer self-interference [31], and different modulator concentrations will move at different speeds [32]. This will cause the shape of the gradient to change as it moves down the column, which could alter the separation significantly.

Modulator sorption is likely to be most significant for normal and reversed phase chromatography, where the increasing concentrations of modulator at the inlet are likely to give rise to increased levels of adsorbed modulator. This is in contrast to ion exchange chromatography, where electroneutrality requires that the stationary phase be saturated with buffer counterions even before the gradient is fed into the column. If the modulator salt is the same as the buffer salt, there will be no increase in modulator sorption as the gradient progresses, and consequently no gradient deformation results. Such deformation becomes possible when the feed components are themselves at concentrations high enough to displace large amounts of initially-bound modulator. Then, both the feed components and the incoming gradient of modulator are free to compete for the freed binding sites, and this could result in gradient deformation. This possibility could be important in preparative ion exchange, and needs to be further examined. However, we restrict ourselves here to linear chromatography, so that by definition the feed concentration in ion exchange is not large enough to compete effectively with the bound modulator, and no gradient deformation (as a result of sorption) occurs. The other possibility for gradient deformation arises when the buffer salt is different from the modulator salt; complex interference effects among the two salt counterions and the feed components are then expected.

Normal-phase chromatography, in which the modulator can be extremely retentive, is likely to generate considerable sorption-induced gradient deformation and indeed unusual experimental results have been found that are extremely sensitive to modulator level [33, 34]. In what follows, however, we restrict our attention to the more widely-used mode of reversed-phase chromatography.

The following subsections discuss the specification of the modulator sorption isotherm and possible consequences of gradient deformation on separations.

3.2 Modulator Sorption Isotherm

In what follows, we assume a binary mobile phase, consisting of the modulator and the carrier. In order to quantify the extent of gradient deformation due to modulator sorption, it is necessary to specify its (absolute) sorption isotherm. Unfortunately this is usually difficult because both the modulator and the carrier are found at high concentrations and only the excess isotherm can be determined experimentally for liquid mixtures [35]. Thus the individual isotherm of the modulator can only be extracted from the excess isotherm data if some kind of model is used, e.g., that the carrier does not adsorb at all.

Reversed-phase chromatography was examined as a representative example for which considerable data can be found in the literature. Excess isotherm data on octadecyl silica (C-18) columns, with water as the carrier and acetonitrile as the modulator, is available in the literature [36, 37]. Assuming that water does not adsorb at all resulted in an individual acetonitrile isotherm that went through a maximum, which is physically unlikely [36]. Therefore the layer model [38] was used by Tani and Suzuki [37] to calculate the individual isotherms of both water and acetonitrile.

3.3 Gradient Deformation and Shock Formation

We now consider the formation of the modulator shock. Under the assumption of ideal chromatography, the velocity at any point in a continuous transition (proportionate pattern) is related to the concentration at that point through the well-known relation [31, 32]

$$v_{\text{continuous}} = \frac{v}{1 + \phi(dq/dC)}, \tag{8}$$

where v is the mobile phase velocity, ϕ is the phase ratio, and dq/dC is the slope of the isotherm at the given concentration C. For a concave-down (Type I) modulator isotherm, such as that depicted in Fig. 1, the slope decreases with increasing concentration. It therefore follows from Eq. (8) that higher concentrations move faster. This would eventually give rise to the unphysical situation of overlapping concentrations; shocks, or discontinuous transitions, were introduced to avoid this possibility [31]. The analog of Eq. (8) for a shock is [31, 32]

$$v_{\text{shock}} = \frac{v}{1 + \phi(\Delta q/\Delta C)}, \tag{9}$$

where the shock spans the concentrations C_1 and C_2 that differ by ΔC; the corresponding q_1 and q_2 differ by Δq.

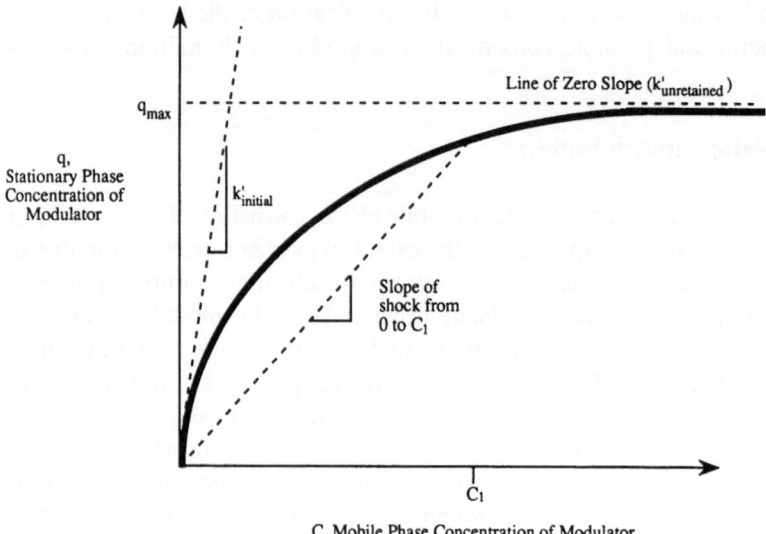

Fig. 1. Representative concave-down monotonic modulator sorption isotherm. The limiting slopes at very low and very high mobile phase concentration are shown, as is the slope of a representative shock

In order to analyze when shocks could form in gradient elution, it is necessary to know the modulator isotherm. In recent work, we fitted the individual isotherm data for acetonitrile from Tani and Suzuki [37] to both Langmuir and BET (Brunauer–Emmett–Teller) functional forms (this was a purely formal process with no mechanisitic significance), and provided an analytical solution via the method of characteristics to the gradient deformation problem for the Langmuir form [39]. Specifically it was shown that the gradient deformation could eventually lead to the formation of a shock. Ideal chromatography was assumed for the modulator; in the presence of non-equilibrium and dispersive terms the shocks would become attenuated into shock layers [40]. An explicit condition for shock incipience within the column when the inlet gradient was linear was formulated as

$$\frac{(1 + b_M C_{M,0})^3}{2\phi a_M b_M \alpha t_0} < 1, \tag{10}$$

where a_M and b_M are the Langmuir parameters and $C_{M,0}$ the initial concentration of the modulator, α the slope of the inlet gradient, ϕ the volumetric phase ratio, and t_0 the retention time of an unretained adsorbate.

From Fig. 1, which depicts a representative concave down sorption isotherm, it can be seen that the initial slope (which is used in the linearly sorbed model) is higher than any other. Consequently, from Eq. (8), the linearly sorbed model generates the slowest-moving modulator front. On the other hand, an unretained modulator would be equivalent to reaching saturation on the isotherm (in both cases, $dq/dC = 0$) and the resulting modulator front will move at v, the maximum possible velocity. Thus the shock, which begins at zero modulator concentration, must move at a velocity in between these two extremes, which consequently serve as useful upper and lower bounds on the actual speed of the nonlinearly sorbed modulator. In fact the retention factor k' corresponding to the linearly sorbed modulator is a dimensionless measure of the "distance" between the two bounds and can therefore be used as a simple yardstick to decide whether the complicated theory of nonlinear modulator sorption is necessary. If $k' = 0.1$, there is a 10% difference in time of emergence between the unretained and the linearly retained descriptions of the modulator. Since accounting for nonlinear sorption will give a front that lies in between these two limiting fronts, the details of nonlinear sorption can be neglected as long as the consequences of accepting an error of up to 10% in the modulator front are acceptable. However, for separations involving macromolecules, a 10% difference in modulator concentration could cause considerable change in adsorbate retention. Then the nonlinear sorption of the modulator must be dealt with.

In order to compare the various models of modulator sorption, the method-of-characteristics calculation [39] was used to evaluate modulator effluent profiles for various linear gradients at the column inlet. The parameters used are listed in Table 1. The results of this calculation (which assumes ideal chromatography) are compared in Fig. 2 with those derived from regarding modulator sorption as unretained or linearly retained. It can be seen that there is considerable difference between the three descriptions of the modulator. In the absence of a shock, the effluent history of the nonlinearly sorbed model begins at the same point as the linearly sorbed model, and moves towards the unretained model as the mobile-phase concentration increases, in agreement with the previous discussion.

For the parameters given in Table 1, the value of k' for the linearly retained modulator is 1.23, giving a difference of more than 100% between the unretained and the linearly sorbed descriptions. Such a large difference, which is reflected in

Table 1. Modulator and column parameters used in the simulations

Phase ratio $\phi = 0.67$ (dimensionless)
Isotherm parameter $a = 1.85$ (dimensionless)
Isotherm parameter $b = 0.26$ M^{-1}
Initial modulator concentration $C_{M,0} = 0$ M
$t_0 = 1.44$ min
The inlet gradient slope α: 0.6, 1.1, 1.5 M/min
Retention factors of adsorbates given by $k' = k'_0 e^{-SC_M}$

Adsorbate	k'_0	S	N^1 (plate count)
1	3.6	1.2	250
2	10.0	1.0	250

[1] The plate counts of the adsorbates need not be equal.

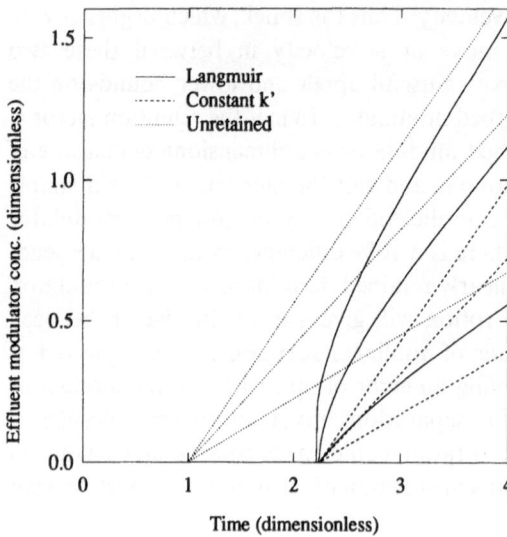

Fig. 2. Effluent modulator concentrations assuming ideal chromatography for the cases where the modulator is unretained, linearly sorbed and nonlinearly sorbed. Three inlet gradient slopes (0.6, 1.1 and 1.5 M/min) are shown for each case. The modulator concentration is made dimensionless by multiplication with the isotherm parameter b ($= 0.26$ M^{-1}). Similarly the time is divided by t_0 ($= 1.44$ min). Relevant data are in Table 1

the widely different retention times of the three models in Fig. 2, requires that we account for nonlinear modulator sorption. On the other hand, the assumption of an unretained modulator has been shown to describe experiment reasonably accurately for a large class of data [4], which implies that for some columns the retention factor of the modulator is low. There is a need for experimental determination of the sorption of the various modulators used in chromatography (particularly in reversed-phase chromatography) on the various sorbents in common use.

Equation (10), and its consequences as described above, was obtained under the assumption of ideal chromatography. In order to ensure the accuracy of this theoretical calculation, and to assess the effects of modulator band spreading, plate simulation for different plate counts were carried out for the Parameters in Table 1. The results are shown in Fig. 3, where the ideal-chromatography result is repeated for convenience. It can be seen that the ideal results are in essential agreement with the simulations. Further, the differences between the simulations and the ideal result are small, even at relatively low plate count. (The modulator, being typically a small molecule, would have a plate count in the tens of thousands for the usual lab-scale HPLC column, and perhaps a plate count of several hundred to a thousand for the usual pilot-plant scale preparative columns). Thus the assumption of ideal chromatography for the modulator is likely to be quite reasonable for most practical situations.

3.4 Implications for Separations

It is clear that such gradient deformation could strongly affect a separation, particularly when a shock layer is formed. We have carried out numerical

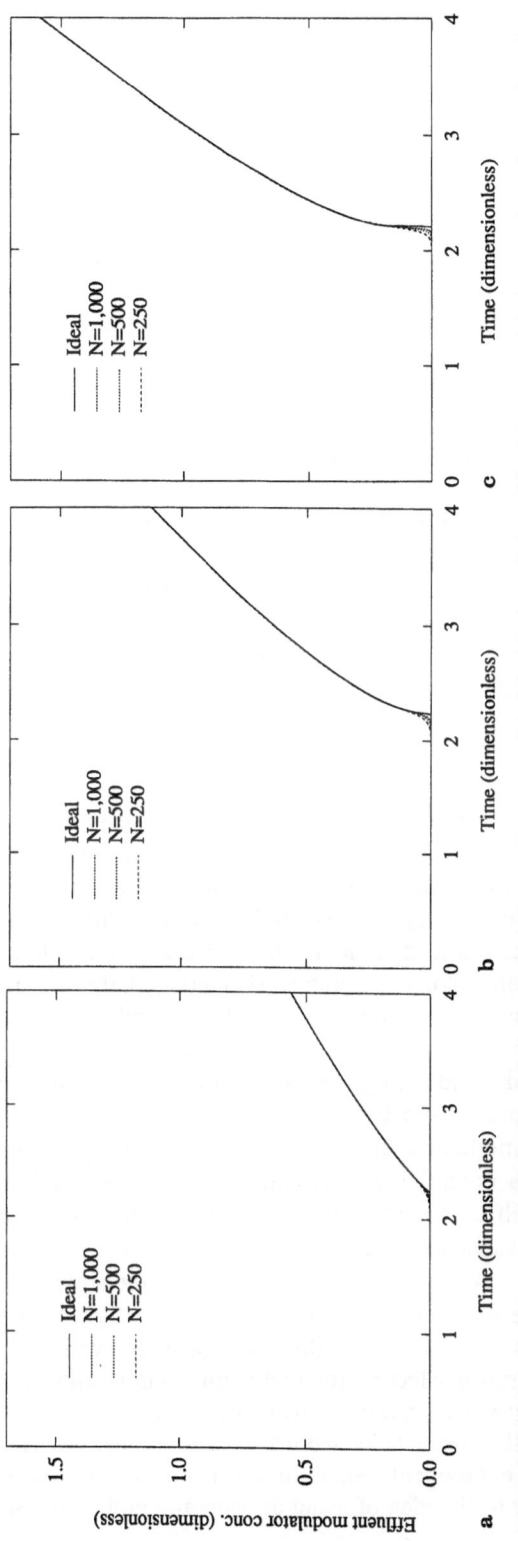

Fig. 3. Effluent modulator concentrations from plate simulations for nonlinear sorption. The ideal result from Fig. 2 is repeated for comparison. Parts **a**, **b** and **c** show inlet gradient slopes of 0.6, 1.1 and 1.5 M/min, respectively. All other data as in Fig. 2

simulations [41] on representative adsorbates under conditions where a modulator shock layer was formed for reversed-phase chromatography on a C-18 column with an acetonitrile–water mobile phase. It was found that, in the vicinity of the modulator shock layer, the adsorbate peak was considerably altered. Peaks lying just behind the shock layer were concentrated by as much as 100%. Peaks that straddled the shock layer exhibited shoulders, because of the considerably different retention and band spreading characteristics of the same component on either side of the shock layer. Peaks that eluted ahead of the shock layer (this would only occur in practice with weakly-retained adsorbates) were broader than might have been expected, because they were eluting isocratically and therefore were not subjected to gradient-induced band compression over some portion of the column. Thus, modulator sorption can affect a separation in several ways, and should be accounted for, particularly in scale-up.

As has been shown earlier, ideal chromatography is a good assumption for the modulator. It has therefore been used in simulations to illustrate the effects described in the previous paragraph. The modulator concentration at any point in the column at any time can be determined by the method of characteristics. This is then used in the plate simulations of the adsorbates, the appropriate value of the modulator concentration being used in each plate at the appropriate time. This extends the ideal assumption of Yamamoto et al. [7, 42] to nonlinear modulator sorption, and substantially simplifies the calculations.

An important consequence of modulator shock formation, particularly for preparative separations, is the possibility of sample concentration. This occurs because of a difference in modulator concentration across the shock layer that is large enough to cause the retention of the adsorbate to be much lower upstream of the shock relative to its value downstream. Downstream, the modulator concentration is lower than it is upstream, resulting in less interference with the adsorbates which are consequently more strongly retained. Thus, if part of the adsorbate peak were to move ahead of the shock layer, it would then find itself in a region where its retention factor was significantly increased, thus holding it up and allowing the shock layer to overtake it again. The modulator's shock layer therefore acts as a barrier for an adsorbate whose retention varies appropriately across it and can be used to advantage to recover this component in concentrated form. Figure 4 shows a representative simulation; the separation produced by an unretained modulator under otherwise identical operating conditions is also shown. The potential disadvantage is that more than one feed component can find itself compressed in this fashion: here concentration is achieved at the coat of a substantial loss in resolution as these two components are likely to emerge in a narrow mixed band just behind the shock. In Fig. 5, where the adsorbates are the same as in Fig. 4, but the inlet gradient slope is increased, the adsorbates are so close together that recovery might be difficult. Nonetheless, the concentration effect on the first component is considerable. At even higher inlet gradient slopes, mixing is to be expected.

On the other hand, the shock layer would serve as an extremely efficient means of separating two components which found themselves on either side of the shock. This is similar to the idea of using an isocratic mobile phase that is

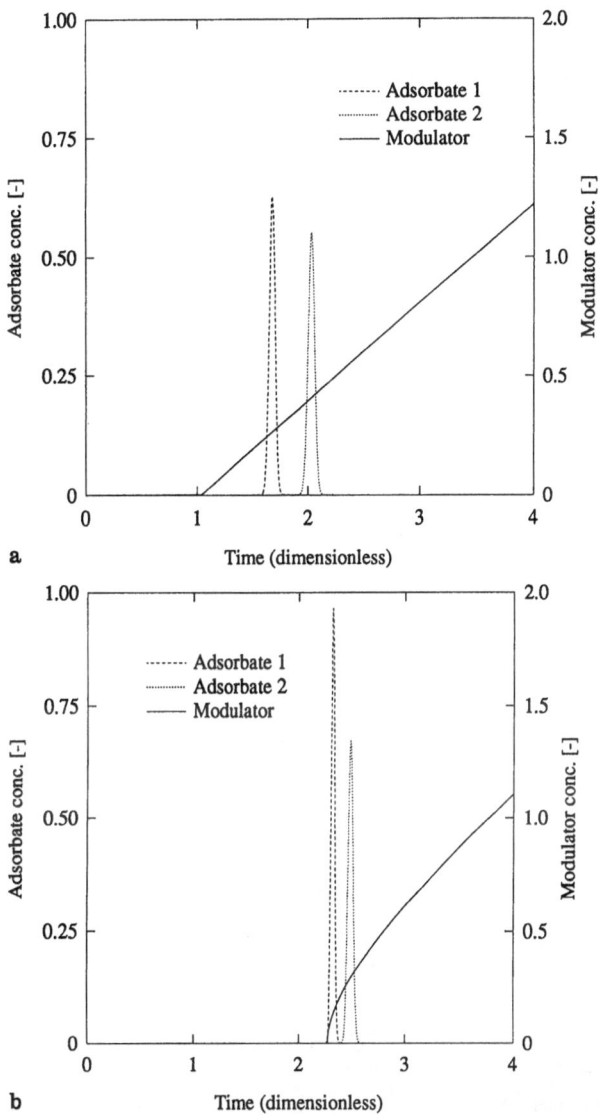

Fig. 4. Representative 2-component gradient elution separation. The slope of the gradient at the inlet is 1.1 M/min, and is started after sample injection is completed (0.04 dimensionless time). Adsorbate parameters as in Table 1. Figure 4(a) shows the separation when the modulator is unretained, and 4(b) shows the same separation for a nonlinearly-sorbed modulator. Adsorbate concentration made dimensionless through dividing by its input concentration, assuming a rectangular pulse injection. Modulator concentration made dimensionless as in Fig. 2

intermediate in binding strength to the two components to be separated [32, 43]. In this context it could be of interest, again with particular reference to preparative separations, to choose strongly-retained modulators (that remain less retentive than the feed components) so as to increase the shock strength and enhance the concentration effect.

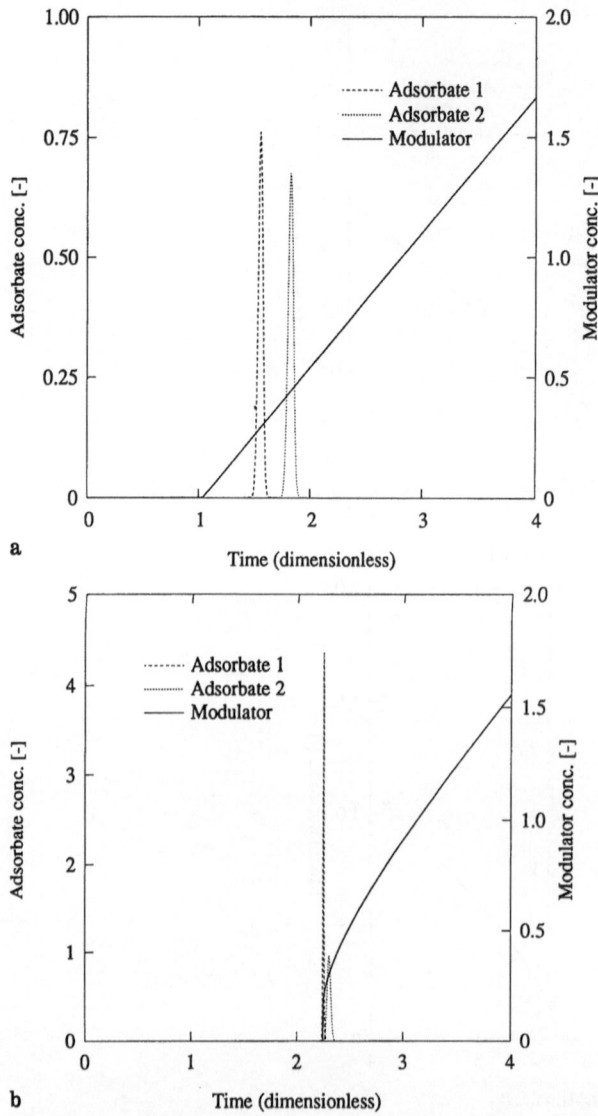

Fig. 5. Representative 2-component gradient elution separation for (**a**) unretained modulator and (**b**) sorbed modulator. Inlet gradient slope is 1.5 M/min; all other data as in Fig. 4

An additional complication could arise if an adsorbate were very highly concentrated. We have assumed that all adsorbates were in the linear regions of their own sorption isotherms, but the local concentration in the vicinity of the modulator shock layer could become high enough (as in Fig. 5b) for it to begin to compete effectively with the modulator for binding sites. This could result in the modulator shock being attenuated or even suppressed, and extremely complex profiles (for the adsorbates as well as the modulator) could ensue. These possibilities need to be investigated further.

4 Conclusions and Outlook

The effects of nonlinear modulator sorption on gradient elution separations seem important enough to examine further. A fundamental aspect is that of determining sorption isotherms for commonly used mobile phase mixtures on the appropriate sorbents. In this context, it is interesting that such sorption effects could be quite significant in reversed-phase chromatography, where the modulator sorption is usually neglected, while they may be less important in ion-exchange chromatography, where the modulator is known to bind to the stationary phase.

The primary need now is for experimental results, first to see whether such modulator shocks are formed and if they are consonant with the simple theory discussed above, and then to see what consequences they have in practical separations, and how best to take advantage of them. Such experimental work has been initiated in our laboratory.

Acknowledgments. This work was supported by the National Science Foundation through grants BCS8912150 and EET8613167. We thank Professor Fred Regnier and Dr. Paul Westgate for their helpful comments during the preparation of this manuscript.

5 Nomenclature

a_M	Langmuir isotherm parameters of modulator, dimensionless
b_M	Langmuir isotherm parameter of modulator $[M^{-1}]$
C_i	mobile phase concentration of the ith adsorbate [M]
C_M	mobile phase concentration of the modulator [M]
$\dfrac{dq}{dC}$	slope of the modulator isotherm at a mobile phase concentration C, dimensionless
D_i^*	lumped diffusivity of the ith adsorbate (Eq. (5)) $[m^2\ s^{-1}]$
K_i	distribution coefficient of the ith adsorbate, dimensionless
k_i'	retention factor of the ith adsorbate, dimensionless
L	column length, m
Pe_i^*	lumped Peclet number of the ith adsorbate (Eq. (6)), dimensionless
t	time elapsed since onset of chromatographic run [s]
t_0	column hold-up time [s]
v	chromatographic (mobile phase) velocity $[m\ s^{-1}]$
$v_{continuous}$	concentration velocity in a continuous transition $[m\ s^{-1}]$
v_{shock}	front velocity for a discontinuous transition (shock) $[m\ s^{-1}]$

Greek Symbols

α	inlet slope of (linear) gradient $[M\ s^{-1}]$
$\alpha_i,\ \alpha_i^*$	constants in Eqs. (3) and (4)

β_i, β_i^* constants in Eqs. (3) and (4)
ϕ volumetric phase ratio, dimensionless
ξ scaled distance ($=x/L$), dimensionless
τ scaled time ($=vt/L$), dimensionless

6 References

1. Alm RS, Williams RJP, Tiselius A (1952) Acta Chem Scand 6: 826.
2. Jennissen HP (1976) Biochemistry 15: 5683.
3. Rasmuson A (1982) Chem Eng Sci 37: 787.
4. Snyder LR (1980) In: Horvath C (ed) High performance liquid chromatography: Advances and perspectives, vol 1. Academic, New York, p 108.
5. Snyder LR, Stadalius MA (1986) In: Horvath C (ed) High performance liquid chromatography: Advances and perspectives vol 4. Academic, New York, p 195.
6. Jandera P, Churacek J (1985) Gradient elution in column liquid chromatography: Theory and practice. Elsevier, New York.
7. Yamamoto S, Nakanishi K, Matsumo R (1988) Ion-exchange chromatography of proteins. Marcel Dekker, New York.
8. Aris R (1956) Proc Roy Soc (London) A235: 67.
9. Poppe H, Paanakker J, Bronckhorst M (1981) J Chromatogr 204: 77.
10. Drake B (1955) Arkiv Kemi 8: 1.
11. Freiling EC (1955) J Am Chem Soc 77: 2067.
12. Snyder LR, Saunders DR (1965) J Chromatogr Sci 7: 195.
13. Liteanu G, Gocan S (1974) Gradient elution chromatography. Halstead (Wiley), New York.
14. Snyder LR, Dolan JW, Grant JR (1979) J Chromatogr 165: 3.
15. Wilson JN (1940) J Am Chem Soc 62: 1583.
16. Weiss J (1943) J Chem Soc 297.
17. Yamamoto S, Nakanishi K, Matsumo R, Kamikubo T (1983) Biotechnol Bioeng 25: 1373.
18. Yamamoto S, Nakanishi K, Matsumo R, Kamikubo T (1983) Biotechnol Bioeng 25: 1465.
19. McCoy BJ (1979) Sep Sci Tech 14: 515.
20. Duarte PE, McCoy BJ (1982) Sep Sci Tech 17: 879.
21. McCoy BJ (1984) J Chromatogr 291: 339.
22. Kang K, McCoy BJ (1989) Biotechnol Bioeng 33: 786.
23. Gibbs SJ, Lightfoot EN (1986) Ind Eng Chem Fundam 25: 490.
24. Frey DD (1990) Biotechnol Bioeng 35: 1055.
25. Snyder LR, Dolan JW, Cox JB (1989) J Chromatogr 483: 63.
26. Snyder LR, Dolan JW, Cox JB (1989) J Chromatogr 484: 409.
27. Snyder LR, Dolan JW, Cox JB (1991) J Chromatogr 540: 21.
28. Liao AW, El Rassi Z, LeMaster DM, Horvath C (1987) Chromatographia 24: 881.
29. Snyder LR, Dolan JW, Cox JB (1989) J Chromatogr 484: 437.
30. Antia FD, Horvath C (1989) J Chromatogr 484: 1.
31. Helfferich F, Klein G (1970) Theory of multicomponent chromatography. Marcel Dekker, New York.
32. DeVault D (1943) J Am Chem Soc 65: 532.
33. Boehme W, Engelhardt H (1977) J Chromatogr 133: 67.
34. Paanakker JE, Kraak JC, Poppe H (1978) J Chromatogr 149: 111.
35. Adamson AW (1967) Physical chemistry of surfaces. Interscience, New York.
36. Slaats EH, Markovski W, Fekete J, Poppe H (1981) J Chromatogr 207: 299.
37. Tani K, Suzuki Y (1989) J Chromatogr Sci 27: 698.
38. Everett DH (1986) Pure Appl Chem 58: 967.
39. Velayudhan A, Ladisch MR (1992) Chem Eng Sci 47: 233.
40. Rhee HK, Amundson NR (1971) Chem Eng Sci 26: 1571.
41. Velayudhan A, Ladisch MR (1991) Anal Chem 64: 2028.
42. Yamamoto S, Nomura M, Sano Y (1990) J Chromatogr 512: 89.
43. Morbidelli M, Storti G, Carra M, Niederjaufner G, Pontoglio A (1985) Chem Eng Sci 40: 1155.

5 Cell Culture

5.1 Strategies for Improving Productivity in Plant Cell, Tissue, and Organ Culture in Bioreactors

M.L. Shuler

School of Chemical Engineering, Cornell University, Ithaca, New York 14853, USA

Contents

The purpose of this paper is to highlight questions that need to be addressed in designing bioreactor systems that will maximize volumetric productivity. Since animal cell bioreactor systems have been more intensely studied, a comparison of animal cell with plant cells is made. Individual strategies that often are successful are: (1) strain selection, (2) production medium development, (3) cell immobilization, (4) product secretion, (5) in situ product removal, and (6) elicitation. It is critical to recognize that these strategies are interactive and their integration into a coherent process strategy is problematic.

Of particular note is the problem of mass transfer at both the whole reactor level and at the microscopic level. In particular, mass transfer limitations within aggregates, immobilized cell preparations, and root clumps can be important. Such limitations can be adverse (e.g. oxygen starvation) or stimulatory by providing a growth environment (e.g. concentration of endogenously generate growth factors) that triggers higher levels of differentiation, organization, and secondary metabolite production. Because aggregate size, etc. is constrained by the plant cell's tolerance to shear, reactor scale-up must include considerations of changes at the "catalyst" level.

1 Why Plant Cell Culture?

Plant cell tissue culture has two primary applications: micropropagation of elite plants and production of chemicals (e.g. medicinals, flavors, pigments, fragrances, and pesticides). In both cases bioreactor designs and operating strategies that can direct cellular physiology towards more productive states are required. The focus of this article will be on the production of chemicals.

Plants represent a vast source of untapped chemicals—the chemical diversity of plants is far greater than from microbes. In 1985, of the 3500 new chemical structures reported, 2519 were from higher plants [1]. Only 5000 of the estimated 250 000 to 300 000 plants have been studied extensively for production of useful medicinals. Nonetheless, there are 121 clinically useful prescription drugs and about 25% of the US pharmaceuticals are derived from plants.

Because of regulatory and economic constraints, it has been difficult for plant cell culture to succeed commercially; almost all of the plant chemicals used commercially derive from field grown material. Based on reported production

Bioproducts and Bioprocesses 2
Editors: Yoshida, Tanner
© Springer-Verlag Berlin Heidelberg 1993

rates for secondary metabolites from plant cell tissue culture (g product per litre per day), it has been estimated that plant cell products must have a wholesale value of at least $150/kg and for many products more than $1000/kg [2]. More importantly, a product must have a sufficient market (ca. 50 million dollars/yr) to make the development of new technology attractive.

This author believes that the most likely candidates for commercial production by plant cell tissue culture are those products that come from slow growing species and for chemicals not yet established as commercial products. If the product can be extracted from an annual plant, it will be difficult for plant cell culture to be more cost effective. If a product is already made commercially from whole plant material, it will be difficult for plant cell culture to displace whole plant material.

The most likely products for development in the US are pharmaceuticals from slow growing woody species. However, because plant cell tissue culture is not an established technology, firms that have found potentially interesting chemicals from such plant species have not sought to proceed with further development if de novo synthesis appeared infeasible. Thus, the establishment of plant cell tissue culture as a technically and economically attractive process could have a great impact on drug development. Plant cell tissue culture would be an important enabling technology for many compounds [2]. With improved methods for drug screening (e.g. receptor-based assays) the development of this enabling technology becomes all the more important.

An important example is taxol. Taxol is extracted from the bark of the Pacific Yew tree (*Taxus brevifolia*). This tree is one of the slowest growing trees and is relatively rare; currently attempts are being made to have it listed as an endangered species. Taxol, discovered by a NCI (National Cancer Institute) screen, has proved to be one of the best anti-cancer drugs discovered in the last decade. It has recently been approved as drug for treatment of refractory ovarian cancer. The NCI has to cut 2000 trees to obtain about 1 kg. To treat one patient for ovarian cancer requires three to six 60–100 yr old trees. In spite of serious efforts, the de novo synthesis of taxol has not been accomplished. Only three conceivable routes exist for producing taxol: plant cell tissue culture, a semi-synthetic route using extraction of a precursor from the leaves of the more common yew and the subsequent multi-step conversion to taxol, and attempts to extract taxol directly from needles of yews. The more complete understanding of how bioreactor conditions alter physiology and metabolism in plant cell culture is essential if this technology is to be commercially useful.

2 Comparison to Animal Cell Culture

Since many readers will be more familiar with animal cell culture, I review here similarities and differences between the two techniques. Table 1 summarizes some of the key differences between the nature of plant cells in culture and insect and mammalian cells, while Table 2 summarizes the culturing requirements with respect to some key environmental parameters.

Table 1. Comparison of key culture characteristics among plant, mammalian, and insect cultures

Characteristic	Plant	Insect	Mammalian
1. Is transformation necessary to establish continuous cell lines?	No	No	Yes
2. Contact inhibition?	3-D Growth	Monolayer typically	Monolayer for normal cells
3. Cryopreservation	Difficult	Fairly routine	Fairly routine
4. Is differentiation of cells irreversible?	No	Yes	Yes
5. Aggregate formation in suspension?	Yes	Unusual	Unusual
6. Large central vacuole present?	Yes	No	No
7. Cell wall?	Yes	No	No
8. Intercell connections	Plasmodesmata	Gap-junctions	Gap-junctions
9. Size	15–150 μm	10–20 μm	ca. 10 μm
10. τ_d	24–100 h	14–26 h	10–30 h
11. Typical attainable cell mass in suspension	10–50 g dw/L	< 1.0 g dw/L	< 0.5 g dw/L

Table 2. Comparison of key environmental parameters for plant, insect, and mammalian cells

Parameter	Plant	Insect	Mammalian
1. Temperature optimum	25–30°C	27–30°C	37°C
2. pH optimum	5.5–5.8	6.0–6.5	7.0–7.5
3. CO_2 incubator needed?	No	No	Typical because of bicarbonate buffer
4. Shear tolerance	Moderate	Low	Low
5. O_2 requirements (m mol O_2/g dw h)	0.4	0.4	0.2

One factor that is particularly important to note is that plant cell cultures are "totipotent". It is possible to produce unorganized, dedifferentiated callus or suspension cultures from fully differentiated whole plant material; it is also possible to regenerate whole plants from this dedifferentiated plant material. With animals terminally differentiated cells (e.g. hepatocytes) cannot be used to regenerate other cell types. The ability of plant cells to be continually able to respond to developmental clues is an important factor to consider in bioreactor strategy.

Related factors are the possibilities of cell-to-cell communication. Plant cells often grow in large aggregates and diffusional limitations may result in concentration gradients of key nutrients or metabolic by-products. Such gradients may result in microenvironments that foster cellular response to developmental cues. Further, the cells in a plant tissue are interconnected with plasmodesmata which are channels that interconnect the cytoplasm of plant cells in a tissue; the plasmodesmata allow the more rapid exchange of low molecular weight metabolites (<800 kDa) than would be possible by simple diffusion.

Plant cells have a large central vacuole and typically store products of interest within such vacuoles, while animal cells typically secrete products of interest into the extracellular compartment. Extracellular product release usually simplifies recovery and purification; mechanisms to increase product release in plant cells are under investigation and are discussed later in this article.

Another key point of differentiation between plant and animal cell culture is the level of cell density that can be attained. The high cell number densities that can be achieved by plant cells increases product concentration and possibly volumetric productivities. It also complicates the delivery of oxygen. Plant cells contain large amounts of water so in many cases more than 50% of the bioreactor volume is cellular volume even at dry weight cell densities of 20 to 30 g/L. This large amount of cells alters the fluid dynamics in a reactor. Although the specific O_2 requirement for plant cells is low, the volumetric requirements are relatively high due to the high cell density and constraints on mixing due to shear sensitivity. Because plant cells have a cell wall, they are less sensitive to shear than animal cells. Nonetheless, their large size makes them more vulner-

Table 3. Comparison of media for plant, insect, and mammalian cells

Nutrient	Plant	Insect	Mammalian
1. Chemically-defined?	Yes	Possible	Not usually
2. Autoclavable?	Yes	No (filter sterilize)	No (filter sterilize)
3. Typical cost	Low to moderate	High	High
4. Carbon/energy source	Usually not photosynthetic; typically sucrose @ 30 g/L	Complex mixture of sugars and amino acids	Glucose, glutamine and other amino acids; less complex than for insect cells
5. Nitrogen sources	NO_3^- and NH_4^+	Amino acids	Amino acids

able to turbulent shear since their cell size is on the order of the size of the smallest eddies.

Table 3 summarizes key aspects of nutrition and medium formulation. The media used for plant cell culture are typically chemically-defined, autoclavable, and much less expensive than media for animal cells. A more detailed comparative analysis of media for plant and animal cell culture is available elsewhere [3]. The requirements with respect to carbon/energy and nitrogen sources are much simpler for plant cells, and their metabolism yields fewer problems with build-up of metabolic by-products (alanine, lactate, and ammonia) that limits animal cell culture. For plant cells lactate and ammonia are not waste products. Lactate build-up can be a problem for insect cells, but ammonia can be converted to uric acid. With mammalian cells both lactate and ammonia accumulation are important.

These differences between plant and animal cells presents unique considerations in developing appropriate bioreactor strategies.

3 Bioreactor Considerations with Dedifferentiated Cultures

The major problem in commercializing plant cell culture has been low product concentrations and volumetric productivities. Fortunately, strategies to improve product levels to levels that exceed that found in the natural plant have been worked out for many compounds [2]. Strain selection is important but must be combined with an effective cultivation strategy. The primary challenge is to induce specific pathways for secondary metabolism. These pathways may be associated with particular organs and tissues and the use of undifferentiated cultures presents special challenges.

Three levels of cell-to-cell association in a bioreactor are possible:

1) suspensions of fine aggregates
2) pseudo-tissues
3) tissue or tissue-like.

In the first case (suspensions) concentration gradients are minimized and metabolic by-products are diluted to low levels. A pseudo-tissue is formed by forcing cells from a fine suspension together (e.g. by immobilization in a gel or entrapment between membranes). In such a case concentration gradients of nutrients and metabolic by-products will exist. A tissue or tissue-like system can be formed by allowing cells to grow in place in a matrix (e.g. a foam matrix) or using large natural aggregates (e.g. >2 mm). In a tissue-like system the cells will be interconnected by plasmodesmata allowing an additional level of interaction. Thus, the choice of reactor (suspension or immobilized cell) can alter cell physiology significantly by altering the degree of cell-to-cell contact.

Hallsby and Shuler [4] have shown the importance of concentration gradients. Tobacco cells were entrapped between two stainless steel screens (11 mm apart). In one case nutrient flow was parallel to both sides of the screens while in the second case the system was operated with a perfusion mode where nutrient was forced up through the bottom screen and out through the top screen. In the parallel flow mode concentration gradients were present while in the perfusion mode the convective flow of nutrient would eliminate all concentration gradients.

When operated for extended periods (about 30 days) three distinct cell regions were formed. Of particular note was the formation of primitive, but organized nodular structures. These nodules were metabolically highly active and formed lipids and polysaccharides in much greater amounts than callus or suspension cultured cells. The nodules exhibited radial alignment of cells and were much denser in terms of dry weight content than the corresponding callus culture (80 vs 30 g/L). Unlike callus, nodule growth was hormone independent.

The hypothesis that these structures were a result of imposition of concentration gradients is supported by (1) the total absence of the formation of cell layers and nodules in an identical reactor with cells from the same inoculum when that reactor was operated in the perfusion mode, and (2) the absence of nodules from any suspension cultures over a wide variety of nutrient conditions.

The importance of culture techniques and the resulting degree of cellular association is further indicated by dramatic increases or decreases in specific productivity of suspended cells due to immobilization. Some examples of positive increases include:

1. Greater than 100-fold increase in capsaicin (hot pepper flavor) from *Capsicum frutescens* in a foam matrix [5] and from *C. annum* in a gel matrix [6];
2. A 20-fold increase in thiophenes from *Tagetes patula* using natural large aggregates [7], and;
3. A 13-fold increase in methylxanthines from *Coffe arabica* in a gel matrix [8].
4. A 2.5-fold increase in specific productivity for shikonin from *Lithospermum erythrorhizon* in calcium alginate [9].

Examples of decreases include:

1. The complete, but reversible, suppression of indole alkaloid synthesis in *Catharanthus roseus* immobilized in dialysis tubes [10];
2. A significant reduction in accumulation of tryptamine (3X) and catharanthine (4X) in *C. roseus* surface-immobilized on a fiberglass support [11]; and,
3. A 2.5-fold decrease in cryptoshanshinone in *Salvia miltiorrhiza* which was gel immobilized [12].

Different forms of immobilization are likely to lead to different degrees of cellular association. For example, in gels or between membranes, immobilization leads to pseudo-tissue unless cells are seeded at a low level and are allowed to grow in place. In foam particles, where cells must enter the foam compartment and grow, and in natural aggregates, tissue-like structure is attained and cells should be highly interconnected by plasmodesmata.

Bringi and Shuler [13] described a conceptual framework to understand better how the type of immobilization could alter the level of complexity in an

Table 4. Expected effects of different types of cell immobilization

| | Type of immobilization* | | | |
	Fine suspension	Gel immobilization	Foam immobilization	Natural aggregates
1. Physical	−	+	+	+
2. Matrix	−	+	±	−
3. Mass transfer gradients	−	+	+	+
4. Plasmodesmata	−	−	+	+

* A " − " indicates that the effect is not present or is likely negligible. A " + " indicates that the effect *may* be important and must be considered.

immobilized system. Table 4 summarizes these types of interactions. In some cases the physical support provided by the matrix is important (e.g. protection of evolving structures from shear). In other cases the matrix itself may modify the local environment (e.g. nutrient partitioning). As described in the membrane immobilization case, concentration gradients of nutrients or metabolic by-products may be important. Finally, cell-to-cell interaction due to plas-modesmata can lead to enhanced coordination of metabolism between cells.

As an example of the effect of the immobilizing matrix consider the case of ajmalicine production from *C. roseus*. Calcium alginate [14] is stimulatory, while κ-carrageenan (M. Asada, unpublished data), entrapment in dialysis tubing and self-immobilization on cheesecloth [10] and on fiberglass [11] are inhibitory to ajmalicine formation. However, agar biofilm formation at certain thicknesses has been reported to enhance indole alkaloid formation [15] over suspension cultures, although the analytical technique used by Kargi et al. [15] is not specific for ajmalicine and a high background level of phenolics could have led to an apparent increase in alkaloids. A non-woven polyester has been used for immobilization, but the maximum alkaloid level reported is low (13 mg/g dw) [16]. The stimulatory effect of calcium alginate over other matrices is not yet understood. However, Bramble et al. [17] have presented evidence that calcium, used to increase the stability of the beads, was correlated with a decrease in intracellular phosphate levels and with an increase in alkaloid production in *Coffea arabica*. Other systems have shown increased alkaloid production in response to decreases in phosphate concentrations in the medium [18].

In addition to the significant impact immobilization can have on secondary-metabolite formation, three other strategies have been found to be effective. Plant cells can be "fooled" into inducing pathways for secondary metabolites when challenged with "elicitors". Many of the secondary metabolites have anti-fungal, anti-bacterial or anti-insecticed activity. When a cell "senses" the presence of attack (e.g. fungal cell wall components, plant cell wall components, or some stress inducing agents), it activates many pathways associated with secondary metabolism. See DiCosmo and Misawa [20] for a discussion of elicitation.

A second strategy that has been effective in increasing production in several cases is the use of in situ extraction of product. Either resins or immiscible solvents have been used; in situ extraction may be effective due to protecting the desired product from further reaction, or relieving feedback inhibition. Several successful examples of this strategy exist [21–25].

I will elaborate by considering the specific example of in situ adsorption applied to ajmalicine production in the presence of the neutral resin, Amberlite XAD-7. Although the exact mechanism of product enhancement is unknown, sufficient information exists to offer a reasonable hypothesis. Our data demonstrates that the increased ajmalicine production is preferential over other alkaloids, and the ratio of ajmalicine to serpentine is greatly increased [25, 14]. The lower pK_a of ajmalicine [6] when compared to serpentine [10] allows its preferential passage across the plasma membrane in the neutral form. In situ extraction from the medium provides a sink for ajmalicine and may increase its transport out of the cell. Normally ajmalicine is found in the vacuole. Blom et al. [26] have provided evidence consistent with the ion-trap hypothesis: ajmalicine in the neutral form diffuses across the tonoplast membrane and becomes trapped when protonated since the tonoplast pH is lower than the cytoplasmic pH. The neutral form of ajmalicine can be converted to serpentine by vacuolar peroxidase further trapping ajmalicine. Thus by experimentally providing an external "sink" for ajmalicine through in situ adsorption, its exit across the plasma membrane is enhanced relative to vacuolar uptake since the concentration gradient into the tonoplast and to the extracellular compartment would become more nearly equal. Since the surface area available for transport is increased significantly if transport through the plasma membrane is allowed, the cytosolic concentration of ajmalicine may decrease, relieving potential feedback inhibition on its formation (see Ref. [27] for evidence of potential feedback effects with precursors to ajmalicine). Such a mechanism would be consistent with the observed synergistic interaction of in situ adsorption with elicitation and use of production media [14].

In almost all cases the medium which supports cell growth is suboptimal for product formation. Thus, the formation of a "production medium" and a two-step process is almost inevitable. Typically production media have elevated sucrose levels, reduced nitrogen and/or phosphate levels, altered hormone levels and types, and, in a few cases, alteration of trace metals. Examples of some production media for various products are: shikonin [28]; anthocyanins [29, 30]; and indole alkaloids [14, 18, 19, 31, 32].

Asada and Shuler [14] combined alginate immobilization, elicitation, medium development, and in situ adsorption to increase extracellular ajmalicine production by almost fifty fold from C. roseus. Kim and Chang [33] have reported a 65-fold increase in maximum shikonin productivity (mg/L day) from Lithospermum erythrorhizon when combining elicitation with solvent (n-hexadecane) extraction. The same authors [9] have reported a 25-fold improvement in shikonin specific productivity by combining immobilization and solvent extraction.

Although these studies do not provide the basis for a general recommendation on a single bioreactor approach to enhance productivity, it is clear that the choice of suspension versus immobilized cells is critical. Further, the type of immobilization strategy adopted may lead to radically different results. Combining immobilization with strategies such as solvent extraction and elicitation have led to unexpected synergistic effects in some cases.

However, these strategies to enhance productivity require that the suspension cultures have at least a basal-level of production. In cases where callus and suspension cultures show negligible activity the use of organized tissues may be successful.

4 Organ Culture

Many products of commercial interest are not formed in suspension or callus culture but are formed in organized tissues. For example, suspension cultures of C. roseus will not form vindoline which is necessary for the formation of the important anti-cancer compounds, vinblastine and vincristine. However, shoots from plantlets regenerated from these same suspension cultures will synthesize vindoline demonstrating that the suspensions had the genetic information necessary for synthesis. However, the level of intracellular organization necessary for vindoline synthesis is not attained in dedifferentiated cultures as shown in a series of papers from the Plant Biotechnology Institute in Canada [34].

The primary forms of organized tissue are embryo, shoot, and root. The use of these tissues is reviewed in detail in Payne et al. [2]. Most attention has been focused on root cultures [35]. The use of *Agrobacterium rhizogenes* to induce "hairy roots" has resulted in dramatic improvements in growth rate resulting in root cultures with growth rates equal to or exceeding growth rates for suspension cultures of the same plant species. In other cases (particularly for species resistant to *A. rhizogenes* infection) the correct and timely application of hormones can result in increases in growth rate. As described elsewhere in this volume [36], significant progress has been made on genetic manipulation of roots for production of secondary metabolites. Thus, high yields and good growth rates make these cultures quite attractive.

The engineering issues involved in using root cultures have not been fully addressed. Root cultures usually form large clumps (effective diameters of 1 to 10 cm or more). The mass transfer of nutrients into these roots present significant problems. Probably the intrinsic kinetics of root metabolism in submerged culture have never been adequately measured. Prince et al. [37] have analyzed the probable contribution of diffusion and intraparticle convection to supply of oxygen to a root clump. For a root clump of *Allium cepa* (onion) with a hydraulic permeability of ca. 4×10^{-4} cm^2 in a sphere of 1.8 cm radius an external fluid velocity of 1 cm/s is necessary to insure supply of oxygen to all parts of the clump. Clearly, in submerged culture, special reactor designs to insure high fluid velocity past a root clump will be necessary.

An alternative design is to supply roots with a nutrient mist [38]. The mist provides a thin liquid film around the root surface and penetrates the interior of the clump. Because the diffusivity of oxygen and its concentration in the gas phase are much higher than in the liquid phase, the supply of oxygen is greatly facilitated.

With either of these reactor approaches it should be recognized that roots pass through several developmental stages. The method of culture will effect root physiology greatly. Very little work has been reported on how root metabolism will respond to changes in culture conditions.

5 Summary

Plants contain a wealth of potentially useful chemical structures. For chemicals from rare or slow-growing species plant cell tissue culture offers a potentially attractive means for large scale production. Plant cell culture differs from the requirements for animal cell cultures. Plant cells grow readily to high cell density in a chemically-defined medium. Bioreactors can be operated to promote growth as dispersed suspensions, pseudo-tissues, or tissue-like structures. These choices can profoundly alter metabolism and response to other product enhancement strategies (i.e. medium manipulations, use of elicitors and in situ extraction). In some cases, organ cultures may be necessary to achieve good product formation. Reactor issues with organ cultures are just now being explored; mass transfer within root clumps is an example of such an issue.

Acknowledgement. The work described in this paper is a result, in part, of support from NSF (EET-8801492). Many of the ideas described in this paper were developed with graduate students and visitors working in my laboratory: Dennis Kubek, Om Sahai, Greg Payne, Masanori Asada, Anders Hallsby, Chris Prince, Tom Hirasuna, Carolyn Lee, and "Bobby" Bringi.

References

1. Abelson PH (1990) Science 247: 513.
2. Payne GF, Bringi V, Prince CL, Shuler, ML (1991) Plant cell and tissue culture in liquid systems. Hanser, New York.
3. Spier RE, Fowler MW (1985) In: Moo-Young M (ed) Comprehensive Biotechnology vol 1 Pergamon, Elmsford, New York, p 301.
4. Hallsby GA, Shuler ML (1986) Biotechnol Bioeng Symp Series 17: 741.
5. Lindsey K, Yeoman MM (1984) Planta 162: 495.
6. Ravishankar GA, Sarma KS, Venkataraman LV, Kadyan AK (1988) Current Sci 57: 381.
7. Hulst AC, Meyer MMT, Brefeler H, Tramper J (1989) Appl Microbiol Biotechnol 30: 18.
8. Haldimann D, Brodelius P (1987) Phytochem 26; 1431.
9. Kim DJ, Chang HN (1990) Biotechnol Bioeng 36: 460.

10. Payne GF, Payne NN, Shuler ML (1988) Biotechnol Bioeng 31: 905.
11. Facchini PJ, DiCosmo F (1990) Appl microbiol biotechnol 33: 36.
12. Miyasaka H, Nasu M, Yamamoto T, Endo Y, Yoneda K (1986) Phytochem 25: 1621.
13. Bringi V, Shuler ML (1990) In: deBont, JAM, Visser J, Mattiasson B, Tramper J (eds) Physiology of immobilized cells. Elsevier, Amsterdam, The Netherlands, p 161.
14. Asada M, Shuler ML (1989) Appl Microbiol Biotechnol 30: 475.
15. Kargi F, Ganapathi B, Maricick K (1990) Biotechnol Prog 6: 243.
16. Rho D, Bedand C, Archambault J (1990) Appl Microbiol Biotechnol 33: 59.
17. Bramble JL, Graves DJ (1991) Biotechnol Bioeng 37: 859.
18. Berlin J, Forche E, Wray V, Hammer J, Hosel W (1983) Z Naturforsch 38c: 346.
19. Knobloch KH, Berlin J (1983) Plant cell, tissue, and organ culture 2: 333.
20. DiCosmo F, Misawa M (1985) Trends Biotechnol 3: 318.
21. Becker H, Reichling JH, Bisson W, Herold S (1984) Proc 3rd Eur Congr Biotechnology 1: 209.
22. Berlin J, Witte L, Schubert W, Wray V (1984) Phytochem 23: 1277.
23. Rhodes MJC, Hilton M, Parr AJ, Hamill JD, Robins RJ (1986) Biotechnol Lett 8: 415.
24. Robins RJ, Rhodes MJC (1986) Appl Microbiol Biotechnol 24: 35.
25. Payne GF, Payne NN, Shuler ML (1988) Biotechnol Lett 10: 187.
26. Blom TJM, Sierra M, Vliet TB, Franke-van Dijk MEI, Koning P, Iren F, Verpoorte R, Libbenga KR (1991) Planta 183: 170.
27. Madyastma KM, Coscia CJ (1978) Recent Adv Phytochem 13: 85.
28. Tabata M, Fujita Y (1985) In: Zaitlin M (ed.) Biotechnology in plant science. Academic, New York, p 207.
29. Yamakawa T, Kato S, Ishida K, Kodama T, Minoda Y (1983) Agr Biol Chem 47: 2185.
30. Hirasuna TJ, Shuler ML, Lackney VK, Spanswick RM (1991) Plant Sci.
31. Zenk MH, El-Shagi H, Arens H, Stockigt J, Weiler EW, Deus B (1977) In: Barz W, Reinhard E, Zenk MH (eds) Plant tissue culture and its biotechnological applications. Springer, Berlin Heidelberg New York, p 27.
32. Morris P (1986) Planta Medica 2: 121.
33. Kim DJ, Chang HN (1990) Biotechnol Lett 12: 443.
34. DeLuca V, Balsevich J, Tyler RT, Kurz WGW (1987) Plant Cell Rep 6: 458.
35. Flores HE, Hog MW, Pickard JJ (1987) Trends Biotechnol 5: 64.
36. Hashimoto T, Yamada Y (1991) This volume, p 247.
37. Prince CL, Bringi V, Shuler ML (1991) Biotechnol Prog 7: 195.
38. Weathers PJ, DiIorio A, Cheetham RD (1989) in Proceedings of the BIOTECH USA Conference, San Francisco, CA.

5.2 Genetic Engineering of Medicinal Plants

Takashi Hashimoto and Yasuyuki Yamada

Department of Agricultural Chemistry, Faculty of Agriculture, Kyoto University, Kyoto 606-01, Japan

Contents

Over the next several years, we anticipate the application of molecular biology for improving yields of useful secondary products both in whole plants and in plant tissue and cell cultures. Possible strategies for yield improvement in medicinal plants are proposed. We here describe our recent work on improving alkaloid composition by a genetic engineering approach. A cDNA of hyoscyamine 6β-hydroxylase, which catalyzes the two consecutive oxidation reactions in the biosynthetic pathway from hyoscyamine to scopolamine in several solanaceous plants, was cloned from *Hyoscyamus niger* and placed under the control of the cauliflower mosaic virus 35S promoter in a plant expression vector. When the gene was introduced and expressed constitutively in hairy roots and regenerated plants of normally hyoscyamine-rich *Atropa belladonna*, scopolamine synthesis was markedly enhanced in the transgenic tissues.

1 Introduction

Despite our efforts in the last two decades to improve yields of useful secondary products in plant cell cultures, the compounds that are now, or about to be, produced commercially are very limited in number. Past experiences have taught us that selection of stable, high-producing cell lines and optimization of culture conditions are both important. We know that elicitation of plant cells can induce or hasten the timing of biosynthesis of some compounds. Organized cultures, especially root cultures, have been found to be useful in some cases when undifferentiated cell cultures fail to synthesize compounds of interest.

Bioproducts and Bioprocesses 2
Editors: Yoshida, Tanner
© Springer-Verlag Berlin Heidelberg 1993

Most of these approaches, however, are empirical and underlying cellular mechanisms are usually left unexplored.

Our research group at Kyoto University has been studying the biosynthesis of tropane alkaloids with the goals to understand the in vivo regulation of the pathway and the molecular mechanisms of cell- and tissue-specific expression. Recently we have succeeded in the molecular cloning of several alkaloid synthesis genes. It seems likely that such advances in biochemistry and molecular biology will soon be seen in the biosynthesis of other useful secondary products as well. Accordingly, we should see the application of molecular biology for improving yields of secondary products both in whole plants and in plant tissue and cell cultures over the next several years. We here list several such strategies that may soon become possible and give one example from the biosynthesis of tropane alkaloids to demonstrate the usefulness of genetic engineering when applied to product formation in medicinal plants.

2 General Strategies for Yield Improvement

2.1 Overproduction of Simple Precursors of the Secondary Compounds

The biosynthesis of amino acids in plants is primarily controlled by feedback inhibition of one or a few regulated enzymes in each biosynthetic pathway by the end product amino acids from that pathway. The regulated enzymes usually catalyze the first reaction in the pathway common to a group of amino acids or the reaction after a branch point to the biosynthesis of a particular amino acid. Molecular cloning of such regulatory enzymes has been reported from several higher plants, e.g. see Ref. [1]. Site directed mutagenesis or random mutagenesis, followed by functional screening, may enable us to engineer mutant enzymes that are no longer feedback-inhibited by the end product amino acids. Transgenic plant cells expressing such mutant enzymes would overproduce the amino acids whose synthesis is controlled by the wild-type enzyme. Because the biosynthesis rate of secondary products usually is not controlled by the supply of amino acid precursors, overproduction of precursor amino acids in normally secondary-product, low-producing cells or tissues would not cause product formation. It is expected, however, that in high producing cells increased supply of primary precursors may improve the end product formation still further.

2.2 Increasing the Concentration of Rate-limiting Enzymes

Most enzymes in a given pathway of secondary metabolism are coordinately regulated and there are no clear rate-limiting enzymes as found in primary metabolism. However, the catalytic activities of individual enzymes in a pathway often vary considerably, which may result in modest accumulation of some intermediates unless metabolic channeling or compartmentation occurs. By fortifying the enzymes of limiting catalytic activities through genetic engineering, normally accumulating intermediates may be efficiently converted to the final product of the pathway. If the introduced enzymes can have access to the intermediates during their translocation from synthesis tissues to storage tissues, the conversion would be still more efficient.

2.3 Creating a New Branch from a Preexisting Pathway

A major limitation for modifying an existing biosynthetic pathway by introducing a foreign enzyme is the substrate specificity of the introduced enzyme. It must act on an intermediate in the pathway of interest. We can look for modification enzymes in the degradation system in secondary product-utilizing microorganisms or the xenobiotic detoxifying microsomal system in animals. However, a more promising approach might be to search in other plant species for similar enzymes that have different substrate specificities. A well known example is the generation of a new petunia flower color by transformation of a petunia mutant with a maize dihydroflavonol 4-reductase gene[2]. The experiment was based on the different substrate specificities of the reductases from petunia and maize on dihydrokaempferol substrate.

2.4 Reducing the Rate of an Existing Side Reaction

The theoretical basis of antisense mRNA technology (although not proven) is that an RNA strand complementary to the target mRNA is expressed as the product of transgene, and a hybrid duplex forms between the two RNAs, inhibiting translation of the target mRNA or promoting its degradation. The technology has been used successfully to suppress color formation in petunia [3], and should be useful for inhibiting undesirable side reactions that may limit the formation of desired compounds. Recently, similar gene-silencing effects are reported by sense transgenes in several plant systems, Ref. [4] and references therein.

2.5 Manipulation of Regulatory Genes

Co-ordinate tissue-specific, developmentally regulated, or signal-responsive expression of structural genes for each group of secondary metabolites is believed to be controlled by one or more regulatory genes. The products of the regulatory genes may be DNA-binding proteins that function as transcriptional activators or repressors. Several such genes that control flavonoid biosynthesis have been cloned [5–7]. Although regulatory genes for other groups of secondary products are yet to be identified, we can expect rapid progress in this field. When regulatory genes controlling the genes for enzymes catalyzing the biosynthesis of useful secondary products are isolated, an obvious experiment will be to express such genes constitutively in cell culture or in every cell of a whole plant. The expression of these regulatory genes in undifferentiated cells may allow us to decouple biochemical differentiation from morphological differentiation, a decoupling which, at present, has only been achieved in a few cases. Whether we can solve the long-standing problem of differentiation in cell culture will depend on future advances in the studies on the molecular mechanisms of tissue-specific expression.

3 Genetic Engineering for Increased Scopolamine Formation

3.1 Utility of Scopolamine

Hyoscyamine and scopolamine, the 6, 7-epoxide of hyoscyamine, are two major tropane alkaloids that accumulate in plants of the family Solanaceae. Although both of these alkaloids are important anticholinergic agents that act on the parasympathetic nerve system, they exert different effects on the central nerve system; hyoscyamine excites the nerve system, whereas scopolamine suppresses it. Ratios of hyoscyamine contents to scopolamine contents vary significantly between plant species, but high scopolamine contents are found only in a few species. These differences result in higher commercial demand for scopolamine than the demand for hyoscyamine (and its racemic form atropine). Since several tropane alkaloid-producing species (e.g. many *Datura*, *Scopolia* and *Atropa* species) accumulate hyoscyamine as a major alkaloid with scopolamine present in minor quantities, it is important to increase scopolamine content in these species.

3.2 Biosynthesis of Scopolamine

The epoxide oxygen of scopolamine is derived from molecular oxygen; this addition of oxygen is catalyzed by a 2-oxoglutarate-dependent dioxygenase, hyoscyamine 6β-hydroxylase (H6H; EC 1.14.11.11) (Fig. 1). H6H incorporates one atom of dioxygen into the 6β-position of [S]-hyoscyamine as a hydroxyl group while simultaneously using the other dioxygen atom for oxidative decarboxylation of 2-oxoglutarate. This oxidative reaction requires ferrous ion and ascorbate, and probably involves a ferryl enzyme as a highly reactive reaction intermediate [8–10]. In plants, the product of the enzyme reaction, 6β-hydroxyhyoscyamine (Hyos-OH) is further converted to scopolamine by 7β-dehydrogenation, thus retaining the 6β-hydroxyl oxygen in scopolamine [11]. Low levels of enzyme activity catalyzing this epoxidation reaction in the presence of co-factors of 2-oxoglutarate-dependent dioxygenases were found in cultured roots of *Hyoscyamus niger*, and co-purified with H6H during partial purification [12]. Also. a monoclonal antibody raised against purified H6H

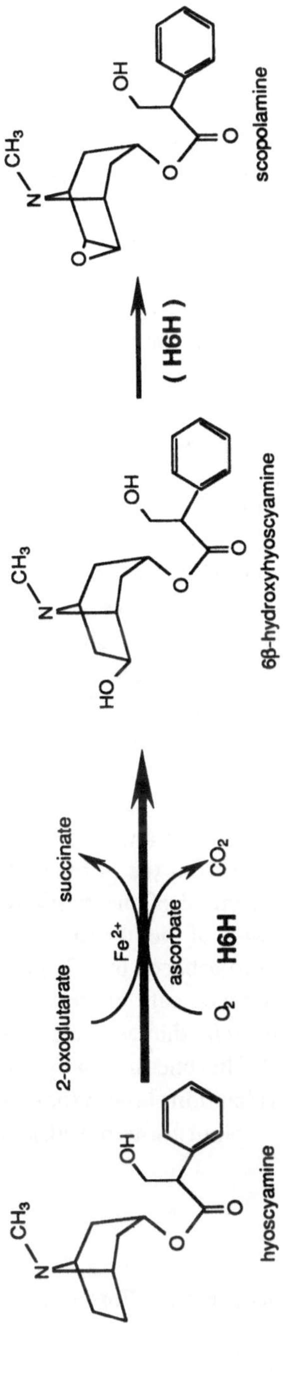

Fig. 1

inhibited not only the hydroxylase activity but also the epoxidase activity to a similar extent (our unpublished result). On the basis of these results, it seems likely that the low epoxidase activity may also be catalyzed by H6H.

Tropane alkaloids are mainly synthesized in the root and transported to the aerial parts. Our recent immunohistochemical studies using an H6H-specific antibody and immunogold-silver enhancement detected H6H only in the pericycle, the outermost cell layer of young vascular cylinder in the root. in several scopolamine-producing plants [13]. This pericycle-specific localization of scopolamine biosynthesis provides an anatomical explanation for the root-specific biosynthesis of tropane alkaloids and may be important for translocation of alkaloids from the root to the aerial parts through the vascular cylinder.

3.3 Molecular Cloning of H6H cDNA from *H. niger*

H6H was purified to homogeneity from cultured roots of *H. niger* by a series of chromatographic steps [14]. After the homogeneous H6H protein was digested with lysylendopeptidase, the internal peptides of H6H were seperated by HPLC and analyzed by a protein sequencer. Amino acid sequences of 12 peptides were determined, three of which were used to design oligonucleotide probes. After screening a cDNA library made from mRNA from cultured *H. niger* roots, several hybridization positive clones were obtained [15]. All the cDNA clones encoded a polypeptide of MW 38999 and only differed in the length of their 5'-untranslated regions and in the polyadenylation sites. All the 12 internal peptide fragments determined in the purified H6H were found in the amino acid sequence deduced from the cDNA and the predicted amino acid composition matched well with the composition determined from the purified H6H.

The H6H cDNA was placed under the control of the cauliflower mosaic virus (CaMV) 35S promoter with the terminator of the nopaline synthase gene in a plant expression vector and introduced into tobacco by a Ti plasmid-based binary transformation system. Transgenic tobacco calli expressing the cDNA had H6H enzyme activity and a 38-kDa protein that was recognized by an H6H-specific antibody on Western bolts [16]. This unequivocally demonstrated that the cDNA encodes H6H. Northern hybridization showed that H6H mRNA is abundant in cultured roots, and present in plant roots but absent in leaves, stems and cultured cells of *H. niger*.

3.4 Expression of *H. niger* H6H Enhances Scopolamine Formation in Transgenic *A. belladonna*

H6H is a promising candidate that, when expressed at high levels in hyoscyamine-accumulating plants, would result in increased oxidation of hyoscyamine, and higher levels of scopolamine. We used two methods to introduce the *H. niger* H6H gene (placed in the above mentioned plant expression vector) into

Atropa belladonna, a typical hyoscyamine-rich plant. First, the *Agrobacterium rhizogenes*-based binary vector system was used for transformation of *A. belladonna* leaf explants. Resulting hairy root clones were selected for kanamycin resistance, which would be expected from the neomycin phosphotransferase II gene placed next to the H6H gene in vector. The presence of the *H. niger* H6H gene in several kanamycin-resistant root clones was confirmed by the use of polymerase chain reaction. These hairy root clones were further screened with Western blotting using an H6H-specific antibody for the clones with high expression of H6H protein. A hairy root clone thus obtained showed 3- to 10-fold higher H6H enzyme activity and similarly higher contents of Hyos-OH and scopolamine than wild-type hairy root clones [17].

Next, a transgenic *A. belladonna* plant having the *H. niger* H6H gene was obtained using the *A. tumefaciens*-based transformation method. In the transgenic plant, the introduced H6H gene (driven by the CaMV 35 S promoter) was expressed in almost all tissues, including the leaf and the stem where the H6H protein is normally absent, and the alkaloid composition in the leaf constituted almost exclusively of scopolamine. It seems likely that the H6H enzyme expressed in the leaf and the stem has access to the hyoscyamine during translocation of the alkaloid from the root to the leaf [18].

4 Conclusion

The ability of plant cells to synthesize a vast array of secondary product in a temporal and cell-specific manner far exceeds the biosynthetic diversity of the other kingdoms. Using some of the strategies proposed above, we will soon be able to manipulate such native expression pattern of biosynthesis genes to satisfy our biotechnological demands. Our initial success to create scopolamine-rich transgenic *A. belladonna* plants is very encouraging for the future advances in genetic engineering of medicinal plants. A good combination of basic research and technological applications will continue to support biotechnology of secondary products.

5 Abbreviations

H6H = hyoscyamine 6β-hydroxylase
Hyos-OH = 6β-hydroxyhyoscyamine
HPLC = high-performance liquid chromatography
CaMV = cauliflower mosaic virus

6 References

1. Kaneko T, Hashimoto T, Kumpaisal R, Yamada Y (1990) J Biol Chem 265: 17451.
2. Meyer P, Heidmann I, Forkmann G, Saedler H (1987) Nature 330: 677.

3. Mol JNM, van der Krol AR, van Tunen AJ, van Blokland R, de Lange P, Stuitje AR (1990) FEBS Lett 268: 427.
4. Goring DR, Thomson L, Rothstein SJ (1991) Proc Natl Acad Sci USA 88: 1770.
5. Paz-Ares J, Ghosal D, Wienand U, Peterson P, Saedler H (1987) EMBO J 6: 3553.
6. McCarty DR, Carson CR, Stinard PS, Robertson DS (1989) Plant Cell 1: 523.
7. Ludwig SR, Wessler SR (1990) Cell 62: 849.
8. Hashimoto T, Yamada Y (1986) Plant Physiol 81: 619.
9. Hashimoto T, Yamada Y (1987) Eur J Biochem 164: 277.
10. Yamada Y, Hashimoto T (1989) Proc Japan Acad Sci Ser B 65: 156.
11. Hashimoto T, Yamada Y (1989) Agric Biol Chem 53: 863.
12. Hashimoto T, Kohno J, Yamada Y (1989) Phytochemistry 28: 1077.
13. Hashimoto T, Hayashi A, Amano Y, Kohno J, Iwanari H, Usuda S, Yamada Y (1991) J Biol Chem 266: 4648.
14. Yamada Y, Okabe S, Hashimoto T (1990) Proc Japan Acad Sci Ser B 66: 73.
15. Matsuda J, Okabe S, Hashimoto T, Yamada Y (1991) J Biol Chem 266: 9460.
16. Yun D-J, Hashimoto T, Yamada Y (1993) Biosci Biotech Biochem 57: 502.
17. Hashimoto T, Yun D-J, Yamada Y (1993) Phytochemistry 32: 713.
18. Yun D-J, Hashimoto T, Yamada Y (1992) Proc Natl Acad Sci USA 89: 11799.

5.3 Engineering Growth Factor/Receptor Processes: Effects of EGF Receptor Trafficking Dynamics on Cell Proliferation Responses To EGF

D.A. Lauffenburger[1], C. Starburck[2], and H.S. Wiley[3]

[1](to whom correspondence should be addressed) Departments of Chemical Engineering and Cell & Structural Biology, University of Illinois at Urbana-Champaign, Urbana, IL 61801, USA
[2]Department of Chemical Engineering, University of Pennsylvania, Philadelphia, PA 19104, USA
[3]Division of Cell Biology and Immunology, University of Utah, Salt Lake City, UT 84132, USA

Contents

Approaches of biochemical engineering, cell biology, and molecular biology have been combined in an effort to increase quantitative understanding of how binding and trafficking aspects of growth factor/receptor interactions influence mammalian cell proliferation, with an aim to exploit this understanding for improved design of cell culture bioreactor systems. Epidermal growth factor (EGF) and its receptor (EGFR) have served as the example of focus for our investigations to date. A mathematical model for EGF/EGFR binding and trafficking dynamics has been developed and experimentally validated, using site-directed EGFR mutations to elucidate subtle features of the trafficking process. This model predicts that changes in EGF/EGFR trafficking dynamics, achieved by alterations in key EGFR cytoplasmic tail domains, can lead to dramatic modification of the cell proliferation response to EGF. We suggest that serum growth factor requirements may be significantly reduced for cell lines incorporating engineered growth factor receptors.

1 Introduction

A central issue in modern biotechnology is the control of mammalian cell proliferation by hormones and growth factors. The bioprocessing industry is attempting to produce a wide range of peptide growth factors for mammalian cells, for applications in human and animal health care such as wound healing and immune response stimulation. Roughly 60% of sales in the bioprocessing industry over the next decade are projected to be in human therapeutics, and over 3/4 of the current or planned therapeutic products are peptide growth factors [1]. At the same time, use of mammalian cell themselves in bioreactors to produce pharmaceutical agents is gaining favor, due to their specialized capability to synthesize proteins with appropriate conformation and activity [2, 3]. Proliferation of these cells require growth factors or complex mixtures of

Bioproducts and Bioprocesses 2
Editors: Yoshida, Tanner
© Springer-Verlag Berlin Heidelberg 1993

regulatory proteins. Finally, growth factor interactions with their cellular targets are critically implicated in many issues arising in the investigation of cancer [4]. Clearly, an improved understanding of the mechanisms by which growth factors regulate mammalian cell proliferation would have tremendous beneficial consequences for both production and application aspects of growth factor-based biotechnology.

There has been a dramatic accumulation of information concerning molecular details of growth factors [5], their interactions with cell receptors [6], and the cell proliferation cycle [7]. Very little has been accomplished, however, by way of incorporating this information into a quantitative, integrated framework relating molecular, physical and cellular properties (e.g. rates of ligand secretion, receptor/ligand binding and trafficking processes, signal transduction) to the resulting cell response. Peptide growth factors are produced by a stimulatory cell and are bound by receptors on target cells, leading to generation of an intracellular mitogenic signal [8]. This resultant signal will be affected by the availability of ligand and the number of activated receptors [9]. Modifications of ligand structure could alter signal transduction by changing the conformation of ligand/receptor complex [10, 11], and could also change the rate of ligand depletion by uptake and degradation since this is mediated by the same receptor [9]. Traverse of several intracellular compartments by the ligand/receptor complex could generate different signals due to localization of particular receptor substrates [12]. The physical environments of intracellular compartments differ and the stability of ligand/receptor complexes within these compartments can influence subsequent cellular routing [13, 14]. Thus, changes in growth factor and receptor structures could have an effect on signal duration and/or specificity, either of which may change the target cell response.

Predicting the effect of any given alteration in ligand or receptor structure is thus a complex, dynamic problem. Understanding the relationship of ligand and receptor structure to mitogenic signal magnitude and duration should aid in accelerated development of growth factor-based technologies in both the bioprocess and health care industries. Quantitative information relating cell proliferation to ligand and receptor biochemical and biophysical properties will provide a more informed basis for rational design of mammalian cell bioreactor growth media and culture regimens, or of therapeutic protocols. There is a clear role for efforts at the interface of biochemical engineering with cell biology to characterize key molecular and cellular properties in a formulation ready to be exploited for design purposes.

An excellent candidate for quantitative investigation of cell proliferation dependence on growth factor/receptor properties is epidermal growth factor (EGF), a acidic polypeptide with a molecular weight of 6045. It is known that binding of EGF to its cell surface receptor (EGFR) activates its intrinsic protein tyrosine kinase activity [15] and initiates a cascade of intracellular reactions leading to chromosomal replication. For fibroblastic cells, EGF is believed to be a key regulator of the rate of progress through G1 phase of the cell cycle into S phase [7] and thus to essentially determine cell proliferation rate [16]. EGF provides a good model system for quantitative analysis not only because of its relevance as a possible wound-healing agent [17] and as a medium requirement for cell growth in culture [18], but also because of the tremendous amount of pertinent molecular information which has been gathered [19, 20]. Both EGF and EGFR have been cloned and sequenced, and can be experimentally manipulated at the physical/chemical level by the use of site-directed mutagenesis and cellular expression systems [10, 21, 22]. Detailed, quantitative models of the binding of EGF to EGFR have been developed over the last 10 years by a number of different investigators [23–28] and the biochemistry of the receptor is well understood [8, 19, 20]. We have successfully exploited the favorable features of the EGFR system over the past several years and have derived models that accurately describe the trafficking pattern of the EGF/EGFR complex [9, 23, 27, 28]. In this paper we summarize our most recent work in development and experimental validation of a mathematical model for EGF/EGFR trafficking and application of this model to experiments on EGF stimulation of cell proliferation.

We have accomplished the following major tasks to date in our collaborative research on growth factor/receptor phenomena, primarily in association with Drs. G. N. Gill and M. G. Rosenfeld (University of California at San Diego). First, we have developed a mathematical model for the dynamics of EGF/EGFR trafficking in fibroblast cell lines, based on fundamental mechanistic hypotheses. Second, we have established experimental protocols to validate the various rate expressions in this mathematical model and have measured each of the model parameters in separate assays. Third, we have used site-directed EGFR mutations to demonstrate that changes in receptor primary sequence can result in changes in key model parameter values. Fourth, we have made predictions from the receptor trafficking model for influence of specific receptor sequences on cell proliferation response to EGF, based on the associated parameter value changes. Each of these accomplishments will now be briefly described.

2 EGF/EGFR Trafficking Model

A schematic illustration of our receptor trafficking model is given in Fig 1. The key steps in EGF/EGFR binding and trafficking are as follows:

a) new receptor synthesis, with rate k_s;
b) reversible binding of ligand (concentration L) with surface free receptors (number density R_s) to form surface binary complexes (number density C_s), with association and dissociation rate constants k_f and k_r, respectively;
c) internalization of free receptors and of complexes with endocytic rate constants k_t and k_e, respectively;
d) recycling of intracellular receptors (number densities R_i, C_i) back to the cell surface with recycling rate constant k_x;

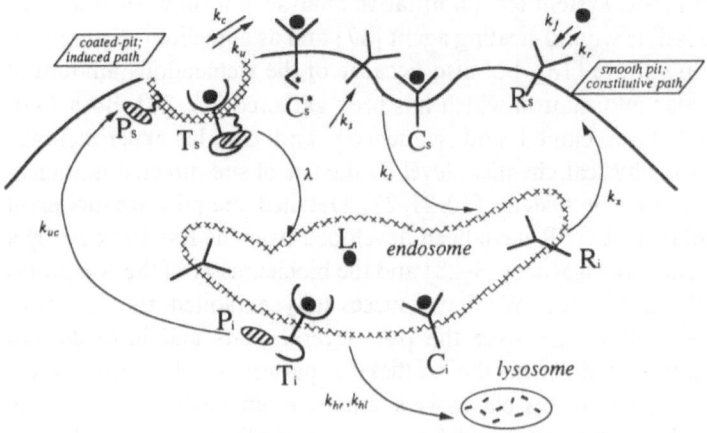

Fig. 1. Schematic illustration of EGF/EGFR binding and trafficking model, based on work in references [27], [28], [29], [37]. k_f and k_r are ligand/receptor association and dissociation rate constants. k_c and k_u are rate constants for ligand/receptor complex coupling with and uncoupling from coated-pit endonexin molecules. k_x is the recycling rate constant and k_{hr} and k_{hl} are the degradation rate constants for receptor and ligand, respectively. k_s is the receptor synthesis rate. k_t is the constitutive internalization rate constant and λ is the coated-pit internalization rate; hence, $k_e = [(\lambda T_s + k_t C_s)/C_s]$ is the net specific internalization rate constant which includes both constitutive and ligand-induced endocytic pathways. R_s, C_s, P_s, and T_s are surface numbers of free receptors, bound receptors, free endonexins, and ternary complexes. R_i, C_i, T_i, and P_i are intracellular numbers of these same quantities. C_s^* represents active signalling ligand/receptor complexes, which we assume to be identical to the surface complexes C_s unless the receptor tyrosine kinase is inactive. Return of coated-pit endonexins to the surface after internalization is assumed to be rapid and thus is not explicitly included in the model

e) degradation of intracellular receptors with rate constant k_{hr} and of intracellular ligand (essentially present at number density C_i as part of the binary complex) with hydrolytic rate constant k_{hl}.

An important complicating feature is that the endocytic rate constant, k_e, for receptor/ligand binary complexes is not in fact constant but rather can vary with complex number density, C_s (see Fig. 2). Our hypothesis accounting for this observation, validated by a number of different kinetic experiments, is that these complexes can interact in a reversible fashion with components (which we term "endonexins") leading to receptor aggregation in clathrin-coated pits [27–29]. This aggregation provides for a saturable "high affinity/low capacity" ligand-induced internalization pathway in contrast to the nonsaturable "low affinity/high capacity" constitutive internalization pathway in which receptors and/or complexes are merely found transiently in coated pits by random surface diffusion. Hence, k_e is actually a function of underlying parameters: the constitutive internalization rate constant k_t, the coated pit turnover rate constant λ, the total number of endonexins P_T, and the affinity of the binary complex/endonexin ternary complex, K_{CP} (equal to the ratio of coupling to uncoupling rate constants, k_c/k_u). Although the identities of endonexins have not yet been

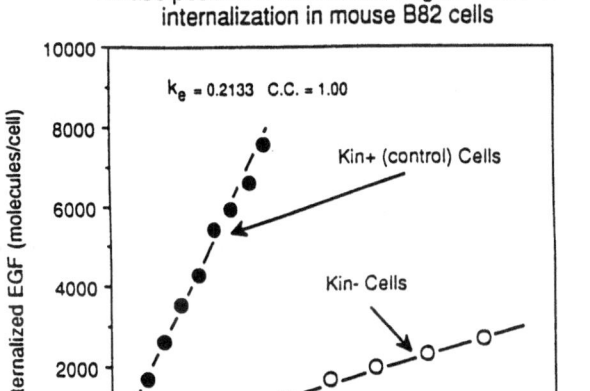

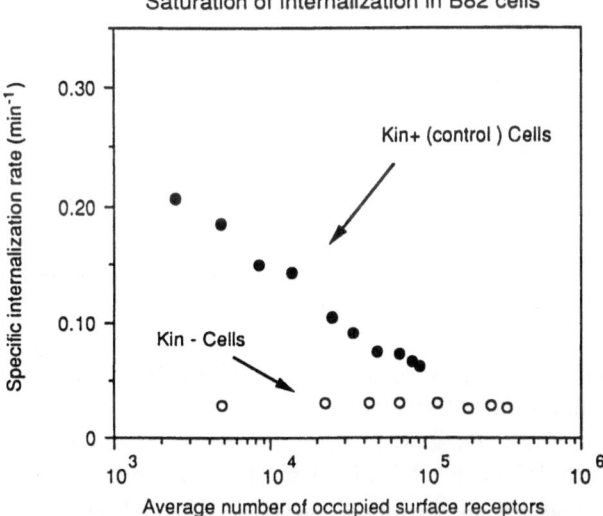

Fig. 2. (Bottom) Experimental plots of the specific internalization rate constant, k_e, versus number of surface complexes, C_s, for wild-type ($Kin +$) EGFR and kinase-inactive ($Kin -$) mutant EGFR [50]. k_e decreases as C_s increases for the wild-type receptor, while it is independent of C_s for the kinase-inactive receptor. The reason is that the kinase-inactive receptors do not interact with the coated-pit endonexins, and therefore are internalized only through the non-saturable constitutive pathway. **(Top)** k_e is obtained from the slope of a plot of internalized EGF versus the integral of surface-bound EGF for any given EGF concentration, following the procedure of Wiley and Cunningham [34]. The lesser slope of this plot for the Kin-receptor cells is indicative of the slower internalization rate constant

documented, plausible candidates exist in the so-called "adaptors" [30] and the coated pit assembly proteins [31].

We should also point out that while some investigators have proposed receptor dimerization as a key event in EGFR operation [32, 33], work to date by ourselves and others [25, 26, 29, 34] find that this process, even if it occurs to a significant extent, is not a rate-limiting event in EGFR binding and trafficking dynamics. Hence, in the interest of limiting model complexity to crucial essentials we have for now chosen to neglect dimerization in our present analysis. If further data suggest otherwise, we could consider its additional incorporation.

3 Experimental Assays for Model Validation and Parameter Measurement

We have designed experimental procedures in which individual steps of the trafficking model can be measured, so that these may in turn be validated and associated parameters evaluated (e.g. [26–29, 34, 50]). This strategy is in contrast to one in which all parameters are estimated simultaneously from overall kinetic data [25]. We believe that our approach is preferable for two reasons: first, in the alternative approach individual steps cannot be validated; second, we can in many cases offer explicit formulas for parameter determination.

Our experiments have primarily utilized the B82 mouse fibroblastic L cell line, which lacks endogenous EGFR [35]. Transfection of these cells with human EGFR permits studies of wild-type (WT) receptor behavior as well as behavior of site-directed receptor mutants. An especially useful feature of this cell line is that nonspecific ligand binding—generally a bane of quantitative receptor/ligand trafficking studies—can be quantified very simply by using nontransfected cells.

Cell surface aspects of the model have been validated by EGF binding experiments on normal human fibroblasts conducted at 4°C, at which internalization is negligible [26]. Values for k_f, k_r, k_c, k_u, R_T (total receptor number), and P_T were obtained from ^{125}I-labelled EGF association and dissociation kinetic experiments. The values of k_f, R_T, and P_T carry over to physiological temperature, 37°C; the latter two because they are absolute quantities, and the receptor/ligand association rate constant because its value turns out from experimental measurement to be diffusion-limited in this case (though this must not be generally assumed to be true [36]). The values of all these parameters (k_f, k_r, k_c, k_u, R_T, P_T) were reevaluated for B82 cells transfected with human EGFR with experiments at 4°C and 37°C [27, 29, 37]. k_r and k_u were found from ^{125}I-labelled EGF dissociation kinetic experiments [38] by preventing internalization at that temperature with phenylarsene oxide [34]. k_c is determined from the half-time for transition between the induced pathway value for k_e at low surface bound receptor number and the constitutive pathway value for k_e at high surface bound receptor number [27].

Values for k_e and k_t were evaluated for B82 cells using short-term (< 5 min) ^{125}I-labelled EGF and anti-EGFR antibody internalization experiments, respectively [27, 28, 37]. For such short periods, recycling and degradation can be largely neglected (though we have derived analytical correction expressions for these effects [28]). The slope of a straight-line plot of $C_i(t)$ versus the integral of $C_s(t)$ over t yields k_e [23, 29]. Data for k_e as a function of C_s (obtained from experiments using a range of EGF concentrations), as shown in Fig. 2, yield K_{cp}, and thus k_c as well as P_T, using an expression derived by Starbuck et al. [27]

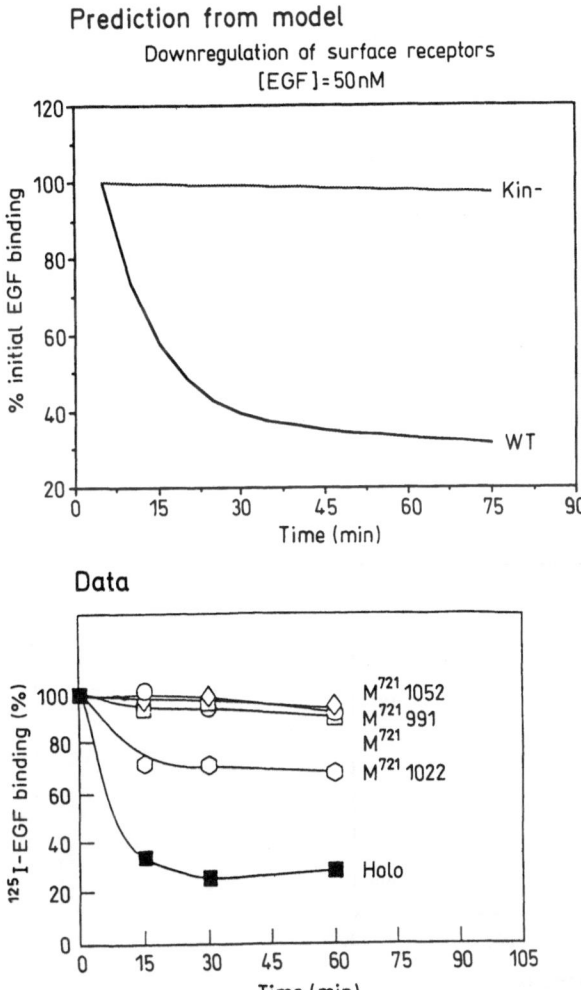

Fig. 3. Example comparison of a priori predictions for downregulation of surface EGFR upon incubation in EGF, with all parameters determined from separate experiments as explained in the text. (**Top**) The number surface receptors is predicted to decrease upon ligand incubation for wild-type receptors, but not for kinase-active receptors [50]. The reason is that the kinase-inactive receptor apparently have a negligible affinity for the coated-pit endonexins, and thus are not internalized by the ligand-induced pathway. (**Bottom**) Experimental data [22] show the predicted trends

predicting the saturation of the induced, "high-affinity/low capacity" pathway. This type of figure is termed a "Satin plot", for saturation of induced internalization [29].

k_x has been determined, also for B82 cells, from pulse/chase experiments in which $C_s(t)$ and $C_i(t)$ are measured; a short initial pulse of labelled EGF is provided followed by a chase with high concentrations of unlabelled EGF [28].

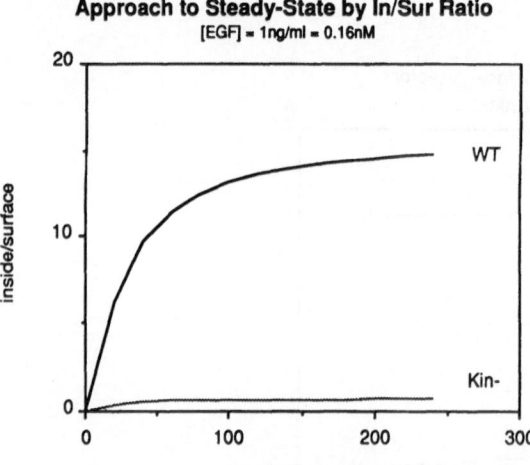

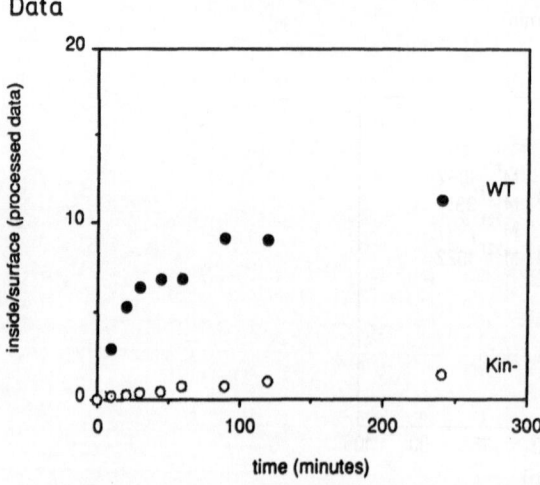

Fig. 4. Example comparison of a priori predictions for ratios of internalized: surface EGFR upon incubation in EGF, with all parameters determined from separate experiments as explained in the text. (**Top**) The ratio of internalized to surface receptors is predicted to be much greater for wild-type EGFR than for kinase-inactive EGFR [50], because the former can make use of the induced internalization pathway while the latter cannot. (**Bottom**) Experimental data [50] show the predicted trends

The decrease in C_i during the chase period is then due primarily to receptor recycling, and the slope of a plot of ln C_i versus t gives k_x (as before, an explicit correction accounting for incomplete dissociation and subsequent re-endocytosis has been derived). k_{hr} was determined simultaneously with the new receptor synthesis rate k_s from ^{35}S-methionine pulse/chase experiments in which transient receptor turnover is followed both in the presence and absence of EGF [28]. Finally, k_{hl} has been determined by measuring the appearance rate of iodotyrosine degradation products in the medium [28].

In this manner, all of the model terms have been validated individually and their associated parameters evaluated independently, in the B82 mouse fibroblast L cell line. Completely a priori predictions have been made for quantities not used in the parameter determination procedures, such as ligand depletion from the medium, rate and extent of surface receptor downregulation,

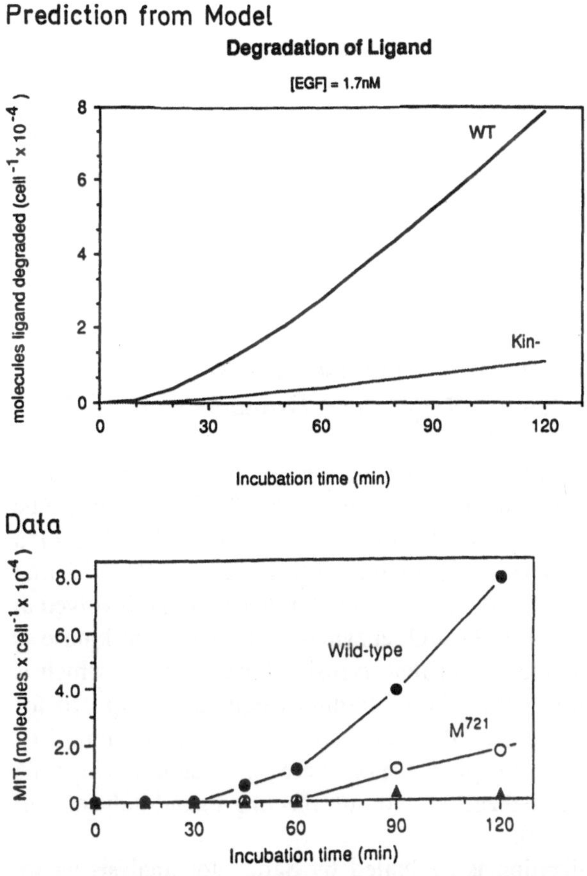

Fig. 5. Predictions of ligand degradation during incubation. (**Top**) EGF degradation mediated by kinase-inactive receptors is predicted to be substantially less than that mediated by wild-type receptors [50]. The reason is that the kinase-inactive receptors do not utilize the ligand-induced internalization pathway. (**Bottom**) Experimental data [28] show the predicted behavior

and transient ratios of internalized to surface complexes; each of these a priori predictions exhibits excellent match with corresponding experimental data [27]. Example comparisons are shown in Figs. 3–5 for receptor downregulation, ratios of internalized to surface receptors, and ligand degradation, respectively. In each figure, predictions and data are shown for both the wild-type EGF receptor and for a mutant receptor in which the tyrosine kinase activity has been suppressed by replacing the 721 lysine residue by a methionine residue (see below).

Our models of ligand-receptor dynamics predict that alterations in particular system parameters can have major effect on cellular behavior. Validity of our model predictions can be experimentally tested by systematically varying intrinsic ligand-receptor rate constants. Two approaches have to date been experimentally determined to be feasible: (a) modification of endocytic rate constants by site-directed mutagenesis of cytoplasmic regulatory domains in the EGFR; and, (b) modification of association and dissociation rate constants by site-directed mutagenesis of residues in the binding domain of EGF [11]. These particular approaches have been chosen because the physical basis of the alterations in rate constants is relatively well established. In addition, these modifications in ligand and receptor do not interfere with intrinsic signal transduction. Structural features of both EGF and its receptor required for signal generation are well established [8, 19, 20] and are not altered in our experimental system. We have so far focused on studies using site-directed EGFR mutations, and plan investigations using EGF mutants in the near future.

4 Modification of Endocytic Rate Constants by Site-directed Mutagenesis of Cytoplasmic Regulatory Domains in EGFR

Work over the last several years has established the role of the C' terminus of EGFR in mediating internalization and downregulation [22, 28, 37, 40–43]. The cytoplasmic domain of EGFR can be classified into three parts (see Fig. 6): a perimembrane region (residues 645–688), a kinase region (residues 689–957) and a regulatory region (residues 958–1186). The perimembrane region is involved in heterologous regulation of EGFR through enzymes such as protein kinase C [37]. The kinase region contains the protein tyrosine kinase activity which is essential for receptor function [44]. The C' terminus domain is required for receptor association with coated pits which mediate rapid internalization [29]. Importantly, removal of the receptor region between residues 973–1186 abrogates ligand-induced internalization without affecting ligand-induced mitogenesis or gene induction [22].

Ligand-induced internalization is evaluated by Satin plot analysis as we have previously described [29]. By analyzing a series of receptor mutants, we have localized the specific domain required for ligand-induced internalization. This region maps between residues 973–1100 [22]. Receptors with progressive

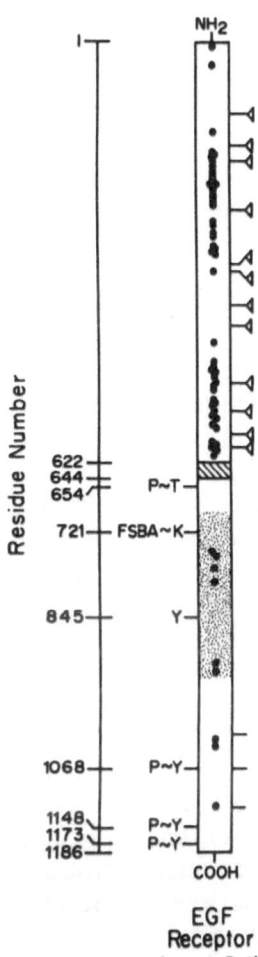

Fig. 6. Structural features of the EGF receptor (from [8]). The *top region* is the extracellular domain, the *bottom region* is the cytoplasmic domain, the *hatched region* is the transmembrane domain. *Triangles* represent glycosylation sites, *filled circles* represent cisteine residues. Residue 721 is the key lysine necessary for tyrosine kinase activity. Tyrosine and threonine residues known to be phosphorylated are indicated

truncations within this region display progressive reduction in k_e, but all possess the ability to stimulate mitogenesis and display identical k_f and k_r values. In the Δ973 receptor, the final 213 residues of EGFR are deleted [22, 42]. This receptor possesses normal ligand-induced kinase activity, and is thus capable of generating a mitogenic response to EGF, but exhibits a low constant value of k_e (roughly equal to k_t over the entire range of C_s) similar to those for receptors lacking kinase activity [29].

Internalization can be restored to truncated EGFR by transferring specific sequences to the C' terminus. Fusing sequences 992–1010 onto a Δ958 receptor yields the t958/992–1010 construct which displays rapid ligand-induced receptor internalization and degradation. Although the Δ973 receptor displays very little internalization, it shows strong biological activity [42]. In contrast, the t958/992–1010 receptor has low activity. This inverse correlation between receptor internalization and mitogenic signalling supports the hypothesis that EGF-stimulated mitogenic signalling occurs principally at the cell surface [22, 42, 45] rather than requiring internalization for action upon intracellular kinase substrates as has been previously suggested [13, 46]. A clear implication is that ligand-induced receptor internalization and degradation—or downregulation— is essentially an attenuating process for the mitogenic signal.

5 Predictions of Proliferation Responses from Trafficking Model

Our experiments indicate that ligand-induced internalization leads to receptor degradation, signal attenuation and ligand depletion from the medium. This has important consequences for growth factor regulation of cell proliferation. Qualitatively, one can predict that prevention of ligand-induced internalization should result in an enhancement of cell sensitivity to ligand. We have explored this intriguing possibility using our experimentally-validated mathematical model for the EGFR system described above. Setting the model parameter values at those measured for B82 cells transfected with either WT or Δ973 EGFR, the level of signalling EGF/EGFR complexes at the cell surface, C_s^*, can be predicted as shown in Fig. 7a. This comparison leads to a very important inference: given identical EGF concentrations, C_s is much greater for Δ973 receptors than for WT receptors. This increase should lead to enhanced cell proliferation if this signalling step is rate-limiting. That is, increased C_s levels should allow faster progress through G0/G1 phase of the cell cycle until this control becomes no longer the limiting factor for proliferation. Indeed, quantitative verification of this relationship has been obtained for human fibroblasts [9] (see Fig. 7b). (It should be noted that contact inhibition, which can serve as another limiting factor at high cell densities, may itself be at least partially related to modulation of EGF/EGFR interactions by cell density [47].) If our model is accurate, then, cell lines expressing growth factor receptors with reduced ligand-induced endocytic rates should show a decreased requirement for serum growth factors. Experimental results consistent with this prediction have been obtained by Wells et al. for a cell line, NR6 (an NIH3T3 cell mutant lacking endogenous EGFR) transfected with the Δ973 receptor or the wild-type receptor (see Fig. 8b) [42]. The Δ973-transfected cells showed an EGF dependence reduced by a factor of about 10 compared with that for the wild-type-transfected cells, as predicted. A quantitative comparison is made in Fig. 8a, using the correlation found between the estimated number of signalling complexes and the cell growth rate for this cell type [51] from the data of Wells et al.

Hence, it appears that alterations in receptor trafficking can, in fact, dramatically alter dependence of cell proliferation on growth factor levels. This result has clear implications for cell culture applications in both bioprocessing and tissue engineering.

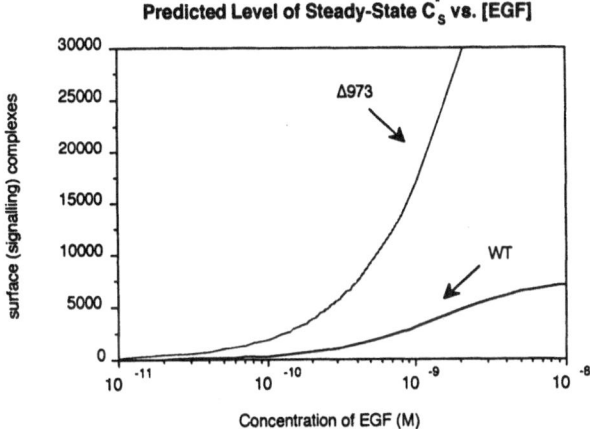

Predicted Level of Steady-State C_s^* vs. [EGF]

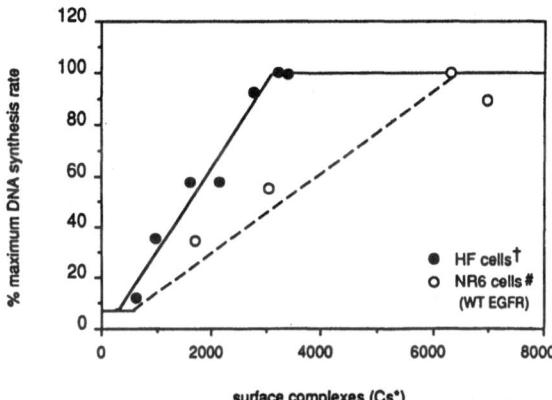

Relationship Between Occupancy and Steady-State Response

[†] data from Knauer, Wiley, and Cunningham *J Biol. Chem.* **259**: 5623-5631 (1984).

[#] data from Wells *et al., Science* **247**: 962-964 (1990).

Fig. 7. (Top) Model predictions of the number of signalling EGF/EGFR complexes, C_s^*, at steady-state versus EGF concentration using the parameter values as in Figs. 3–5 [51]. For the truncated receptor mutant, Δ973, interaction affinity with the coated-pit endonexins is assumed to be negligible, so that $k_e = k_t$ as with the kinase-inactive receptor mutant. This leads to the substantial increase in C_s^* for the Δ receptor compared to the wild-type receptor, since it is not down-regulated in the presence of ligand. **(Bottom)** Plots of DNA synthesis rate versus number of surface complexes, as estimated for human fibroblasts (*filled circles*) [9] and for NR6 cells (*open circles*) [50] using data from Wells et al, for wild-type EGFR transfects [42]

6 Experimental Test Systems

Two cell types have been used in these initial experiments: B82 and NR6. B82 cells are EGFR-negative derivatives of mouse L cells [35]. They are extremely useful for analysis of receptor dynamics because they can be transfected efficiently with various receptor genes, they express large numbers of receptors, they grow very rapidly, and their endocytic system is insensitive to hormonal

Prediction from Model

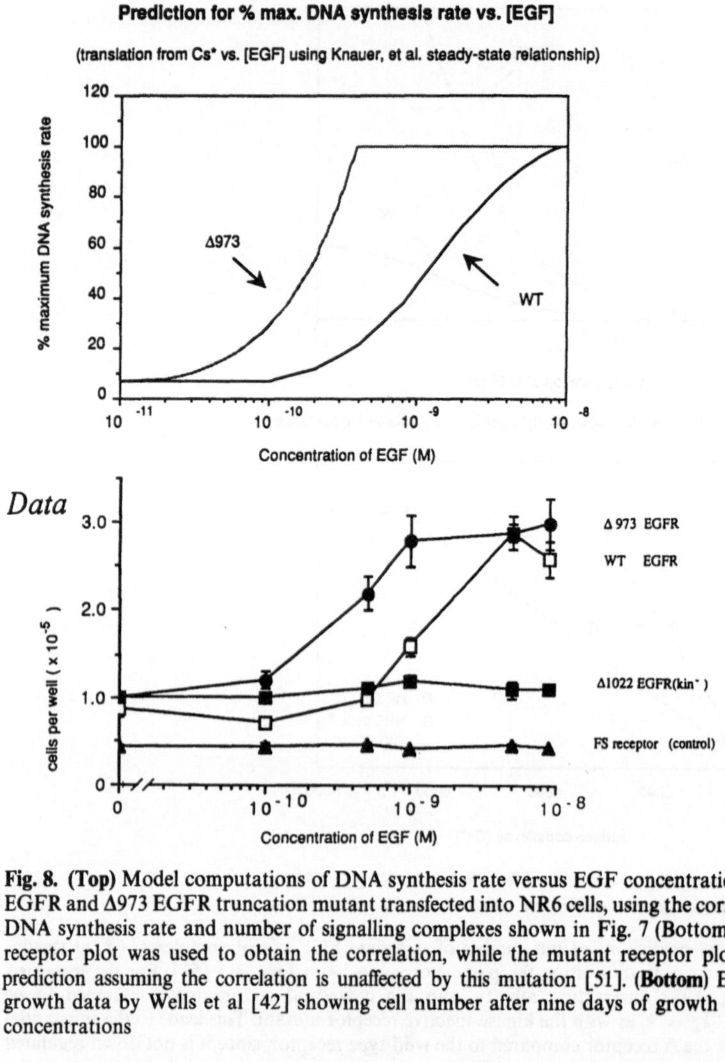

Prediction for % max. DNA synthesis rate vs. [EGF]

(translation from Cs* vs. [EGF] using Knauer, et al. steady-state relationship)

Concentration of EGF (M)

Data

Concentration of EGF (M)

Fig. 8. (Top) Model computations of DNA synthesis rate versus EGF concentration, for wild-type EGFR and Δ973 EGFR truncation mutant transfected into NR6 cells, using the correlation between DNA synthesis rate and number of signalling complexes shown in Fig. 7 (Bottom). The wild-type receptor plot was used to obtain the correlation, while the mutant receptor plot is an a priori prediction assuming the correlation is unaffected by this mutation [51]. **(Bottom)** Experimental cell growth data by Wells et al [42] showing cell number after nine days of growth at various EGF concentrations

stimulation. This minimizes perturbations by growth factors on the systems we are attempting to analyze. We currently have access to over 120 EGFR mutants expressed in B82 cells. The NR6 cells are EGFR-negative derivatives of 3T3 cells [48]. They grow more slowly, require fibronectin supplements, and express transfected receptors at lower levels than B82 cells. However, they exhibit a strong mitogenic response to activated EGFR, unlike B82 cells [42]. It is most efficient to conduct receptor dynamics studies with B82 cells and then apply resulting insights to a more technologically useful cell line such as Chinese hamster ovary (CHO) cells, because with our present pX plasmid vector (see [22]) expression levels in CHO cells are quite low, making quantitative studies of ligand/receptor trafficking very difficult to carry out. We are aware that improved expression vectors for production cell lines are being developed by industrial investigators so that results from our work may before long be exploited in such systems.

Acknowledgements. A grant from the NSF Biotechnology Program to D.A. Lauffenburger and H.S. Wiley is gratefully acknowledged.

7 References

1. Shamel RE, Chow JJ (1989) Chem Eng Prog December p 33.
2. Hu W-S, Dodge TC (1985) Biotech Prog 1: 209.
3. Mizrahi A (1986) Bio/Tech 4: 123.
4. Guroff G (1987) Oncogenes, genes, and growth factors. John Wiley, New York.
5. Pimentel E (1987) Hormones, growth factors, and oncogenes. CRC, Boca Raton, FL.
6. Sporn MB, Roberts AB (1990) Growth factors and their receptors. Springer Berlin Heidelberg New York.
7. Pardee AB (1989) Science 246: 603.
8. Carpenter G (1987) Annu Rev Biochem 56: 881.
9. Knauer DJ, Wiley HS, Cunningham DD (1984) J Biol Chem 259: 5623.
10. Engler DA, Matsunami RK, Campion SR, Stringer CD, Stevens A, Niyogi SK (1988) J Biol Chem 263: 12384.
11. Matsunami RK, Campion SR, Niyogi SK, Stevens A (1990) FEBS Lett 264: 105.
12. Willingham MC, Haigler HT, Fitzgerald DJ, Gallo MG, Rutherford AV, Pastan IH (1983) Exp Cell Res 146: 163.
13. Basu SK, Goldstein JL, Anderson RG, Brown MS (1981) Cell 24: 493.
14. Davis CG, Goldstein JL, Sudhof TC, Anderson RG, Russell DW and Brown MS (1987) Nature 326: 760.
15. Ushiro H, Cohen S (1980) J Biol Chem 255: 8363–8365.
16. Baserga R (1985) The Biology of cell reproduction. Harvard University Press.
17. Cohen S, Carpenter G (1975) Proc Natl Acad Sci USA 72: 1317.
18. Cherington PV, Pardee AB (1980) J Cell Physiol 105: 25.
19. Gill GN, Bertics PJ, Santon JB (1987) Molec Cell Endocrinol 51: 169.
20. Carpenter G, Wahl MI (1990) In reference 6, p. 69.
21. Lin CR, Chen WS, Kruiger W, Stolarsky LS, Weber W, Evans RM, Verma IM, Gill GN, Rosenfeld MG (1984) Science 224: 843.
22. Chen WS, Lazar CS, Lund KA, Welsh JB, Chang CP, Walton GM, Der CJ, Wiley HS, Gill GN and Rosenfeld MG (1981) Cell 59: 33.
23. Wiley HS, Cunningham DD (1981) Cell 25: 433.
24. Gex-Fabry M, DeLisi C (1986) Am J Physiol 250: R1123.
25. Waters CM, Oberg KC, Carpenter G, Overholser KA (1990) Biochemistry 29: 3563.
26. Mayo KH, Nunez M, Burke C, Starbuck C, Lauffenburger DA, Savage CR (1989) J Biol Chem 264: 17838.
27. Starbuck C, Wiley HS, Lauffenburger DA (1990) Chem Eng Sci 45: 2367.
28. Wiley HS, Herbst JJ, Walsh BJ, Lauffenburger DA, Rosenfeld MG, Gill GN (1991) J Biol Chem 266: 11083.
29. Lund KA, Opresko LK, Starbuck C, Walsh BJ, Wiley HS (1990) J Biol Chem 265: 15713.
30. Pearse BMF (1988) EMBO J 7: 3331.
31. Kirchhausen T, Nathanson KL, Matsui W, Vaisberg A, Chow EP, Burne C, Keen JH, Davis AE (1989) Proc Natl Acad Sci USA 86: 2612.
32. Yarden Y, Schlessinger J (1987) Biochemistry 26: 1443.
33. Yarden Y, Schlessinger J (1987) Biochemistry 26: 1434.
34. Wiley HS, Cunningham DD (1982) J Biol Chem 257: 4222.
35. Davies RL, Grosse VA, Kucherlapati R, Bothwell M (1980) Proc Natl Acad Sci USA 77: 4188.
36. Lauffenburger DA, DeLisi C (1983) Int Rev Cytol 84: 269.

270   D. A. Lauffenburger et al.

37. Lund KA, Lazar CS, Chen WS, Walsh BJ, Welsh JB, Herbst JJ, Walton GM, Rosenfeld MG, Gill GN, Wiley HS (1990) J Biol Chem 265: 20517.
38. Wiley HS, Walsh BJ, Lund KA (1989) J Biol Chem 264: 18912.
39. Opresko LK, Wiley HS (1987) J Biol Chem 262: 4116.
40. Livneh E, Benveniste M, Prywes R, Felder S, Kam Z, Schlessinger J (1986) J Cell Biol 103: 327.
41. Livneh E, Reiss N, Berent E, Ullrich A, Schlessinger J (1987) EMBO J 6: 2669.
42. Wells A, Welsh JB, Lazar CS, Wiley HS, Gill GN, Rosenfeld MG (1990) Science 247: 962.
43. Heisermann GJ, Wiley HS, Walsh BJ, Ingraham HA, Fiol CJ, Gill GN (1990) J Biol Chem 265: 12820.
44. Chen WS, Lazar CS, Poenie M, Tysien RY, Gill GN, Rosenfeld MG (1987) Nature 328: 820.
45. Masui H, Wells A, Lazar CS, Rosenfeld MG, Gill GN (1991) Cancer Res 51: 6170.
46. Lauffenburger DA, Linderman JJ, Berkowitz L (1987) Ann NY Acad Sci 506: 147.
47. Lichtner RB, Schirrmacher V (1990) J Cell Physiol 144: 303.
48. Schneider CA, Lim RW, Terwilliger E, Herschman HR (1986) Proc Natl Acad Sci U.S.A. 83: 333.
49. Muramatsu T (1990) Cell surface and differentiation Chapman and Hall, New York.
50. Starbuck C (1991) PhD Thesis, University of Pennsylvania.
51. Starbuck C, Lauffenburger DA (1992) Biotech Progr 8: 132.

5.4 The Effect of Light Irradiation on Secondary Metabolite Production by *Coffea arabica* Cells

Shintaro Furusaki[1], Minoru Seki[1], Hiroyuki Kurata[1], and Tsutomu Furuya[2]

[1]Department of Chemical Engineering, The University of Tokyo, Tokyo 113, Japan
[2]School of Pharmaceutical Sciences, Kitasato University, Tokyo 108, Japan

Contents

The effect of light irradiation on secondary metabolite production by *Coffea arabica* cells was investigated for the purpose of developing a photo-bioreactor suitable for plant cells. The light irradiation significantly enhanced the caffeine production by the *Coffea arabica* cells. With increasing the light irradiation rate, the specific caffeine productivity became greater but the cell growth was suppressed. In addition, the caffeine production was activated to a considerably high level even at a low rate of light irradiation of ca. 0.30×10^{18} quanta/s/L-medium. Consequently, the maximum production was achieved at a moderate irradiation rate of 1.9×10^{18} quanta/s/L-medium in the 20-day batch culture. Semicontinuous cultures were conducted using light for 56 days. The transfer of photocultured cells into darkness was effective for recovering the decreased growth activity caused by light irradiation.

1 Introduction

Plant cell cultures have been investigated for the production of valuable secondary metabolites such as pharmaceuticals, pigments and flavors. Since plant tissue cells are known to be capable of growing in vitro, the use of plant tissue cultures for production of useful bioproducts has become popular in bioengineering research. The use of plant tissue cultures has the following merits: (1) production is not affected by seasons and climate; (2) the area for plants to produce materials using cell culture is much smaller than that necessary for production in the field; (3) the duration of production is shorter in the case of the tissue culture production; (4) some cultured plant cells are able to accumulate the useful products in higher concentrations than intact-plant cells. Thus, plant cell cultures have apparently greater productivity than cultivation of whole plants for the production of biochemicals. However, only a few examples [1, 2] of the industrial application of plant cell cultures have been reported because of their high costs.

In order to enhance the productivity, various kinds of stresses are imposed upon the plant cells. Some examples of stresses are given as follows: (1) limitation of the nitrogen source in terms of the concentration and the species; (2) limitation of the phosphorus source; (3) control of hormones; (4) addition of elicitors, natural or artificial [3, 4]; (5) shear stress; (6) immobilization [5–7]; and (7) light irradiation [8].

The effect of the light irradiation is considered to be easiest for the control of secondary metabolite production from the reactor design viewpoint. In many cases, the intensity, wavelength and cycle of light irradiation are known to affect the secondary-metabolite productivity [9–11], and it is known that these effects can be controlled by appropriate reactor design. However, little research has been conducted on analyzing quantitative relations between light irradiation and secondary metabolite synthesis or cell-growth for plant cell culture.

Thus, the purpose of this study was to investigate quantitatively the behavior of the response of a plant cell culture to light irradiation and seek a way to increase the productivity of secondary metabolites by appropriate reactor design and its operation. The system under investigation was the production of caffeine by *Coffea arabica* cells. Many researchers have reported using this system, because the cells secrete caffeine into the medium and it is easy to analyze the characteristics of the metabolite production. For example, Frischknecht and Baumann [8] studied the qualitative effects of light irradiation for production of purine alkaloids by *Coffea arabica*. According to their study, some stressors such as high light intensity and high NaCl concentration stimulated secondary-metabolite production.

2 Materials and Methods

2.1 Cells and Media

The suspension cells of *Coffea arabica* were obtained from seed-derived callus tissue. Two kinds of media were prepared. They were the DK medium and the B2K medium, which contained different growth regulators, 1 mg/L 2, 4-D (2, 4-dichlorophenoxyacetic acid) and 2 mg/L IBA (3-indolebutyric acid), respectively. The other components in these media were based on the MS medium, supplemented with sucrose, 10 mg/L thiamine hydrochloride, 5.0 mg/L nicotinic acid, 10 mg/L pyridoxine hydrochloride, 100 mg/L myo-inositol, 2.0 mg/L glycine and 0.1 mg/L kinetin. The pH in the media was adjusted to 5.8.

2.2 Culture Conditions

Preculture was carried out for 3 weeks in 300-ml flasks (100 mL liquid volume) with a reciprocating shaker at 78 strokes/min in the darkness. The temperature of the incubator was controlled 27°C. After the preculture, the cell suspension was filtered and cells were washed several times with the medium used for main culture. Main culture was carried out at 27°C in 500-mL flasks (100 mL liquid volume), which was shaken with a rotary shaker at 78 rpm. Light sources were two 15-w fluorescent lamps inside the incubator.

2.3 Analysis

About 2 mL of cell suspensions were taken directly from the media for analysis. They were centrifuged for 12 min at $11000 \times g$ and then the supernatants were analyzed. Caffeine concentration was measured using a computer controlled HPLC system equipped with an ODS column (TSK-gel ODS-80TM; TOSOH, Tokyo). The elution was carried out isocratically with a methanol : water (40 : 60) solvent. Detection was carried out at 271 nm. Sucrose concentration was determined according to the phenol–sulfuric acid method.

The cell suspension in the sample tube was permitted to stand for 30 min. The relative cell volume (packed cell volume) was measured, this value was converted into dry cell weight. The chlorophyll-a was extracted from the cells with acetone and its content was determined by measuring the absorbance of the extract at 665 nm.

The rate of light irradiation was measured with ferrioxalate chemical actinometry [12]. An aqueous solution of ferrioxalate (6 mM, 100 mL) was added to the same 500-mL flask as those used for the main cultures. The flask was exposed to light for a certain period of time (5–10 min) under the culture conditions in the incubator and the amount of photo-reacted ferrioxalate was measured using o-phenanthroline. The rate of light irradiation in the solution of the actinometry was approximately calculated from the measured amount of reacted ferrioxalate, considering the spectrum profile of the fluorescent lamp, the quantum yield of the ferrioxalate reaction and the fraction of the light that was transmitted through the actinometer without being absorbed.

3 Results

3.1 Effect of the Light-Irradiation Period

The effect of the light-irradiation period on caffeine production by *Coffea arabica* cells was investigated in the batch culture using the DK medium. The results are shown in Fig. 1. The cells were continuously exposed to light for a certain period from the beginning of cultivation. Caffeine production by the cells irradiated for the initial 3 days was little enhanced. However, the cells irradiated for the initial 6 days showed an increase in caffeine production. Caffeine production of the cells irradiated for more than 11 days was greatly increased. The cells irradiated for 11 days maintained the increased production even after being transferred into darkness and the caffeine production was more than that of the cells irradiated for 22 days. It has been suggested that the caffeine

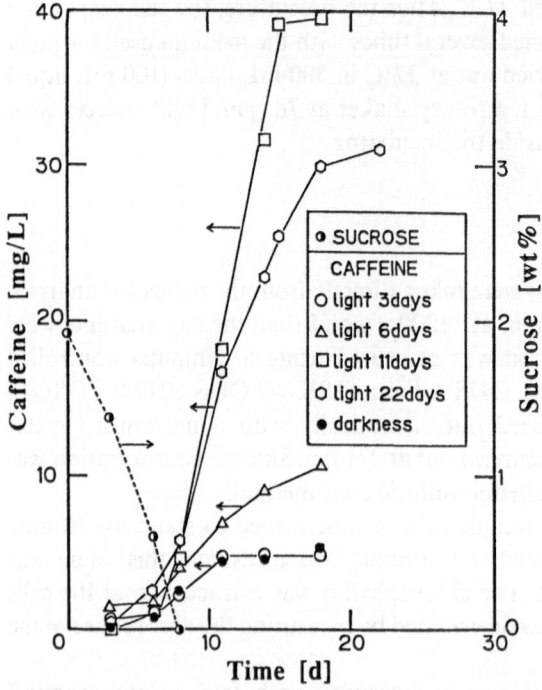

Fig. 1. The effect of light irradiation time period on caffeine production in the batch culture using the DK medium. The light irradiation rate was 3.2×10^{18} quanta/s/L-medium. The initial dry cell weight was 0.76 g. The initial sucrose concentration was 2 wt%. Caffeine concentration in the medium: (●) darkness; (○) light for initial 3 days; (△) light for initial 6 days; (□) light for initial 11 days; (◯) continuous light. Sucrose concentration in the medium in the continuous light (◐)

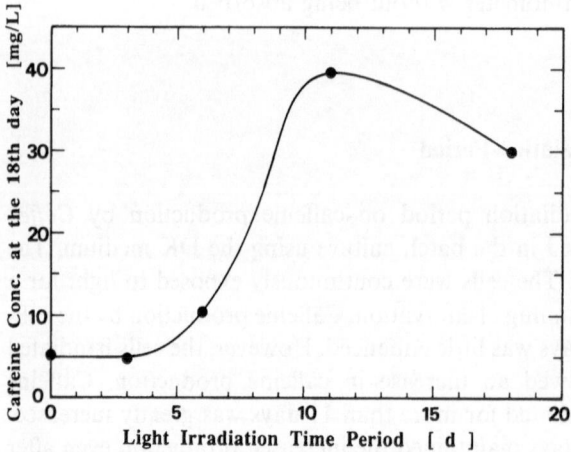

Fig. 2. The effect of light irradiation time period on caffeine production

production was suppressed by long-term continuous light and there is an optimal light irradiation period to obtain the maximum production of caffeine (Fig. 2). All the sucrose uptake rates of the cultivated cells, in spite of the variations in irradiation period, were nearly equivalent. Increased productivity of caffeine was found after sucrose was completely consumed.

3.2 Effect of a Growth Regulator

The light effect on caffeine production was investigated in the batch culture using the B2K medium. The results are shown in Fig. 3. Caffeine production was increased 7.7 - fold by light and was even more greatly enhanced than that in the DK medium. The production was increased before sucrose was exhausted, i.e. while the cells were still growing.

3.3 Effect of the Light-Irradiation Rate

The effect of the irradiation rate on caffeine production and cell growth was investigated in the batch culture using the B2K medium. The irradiation rate

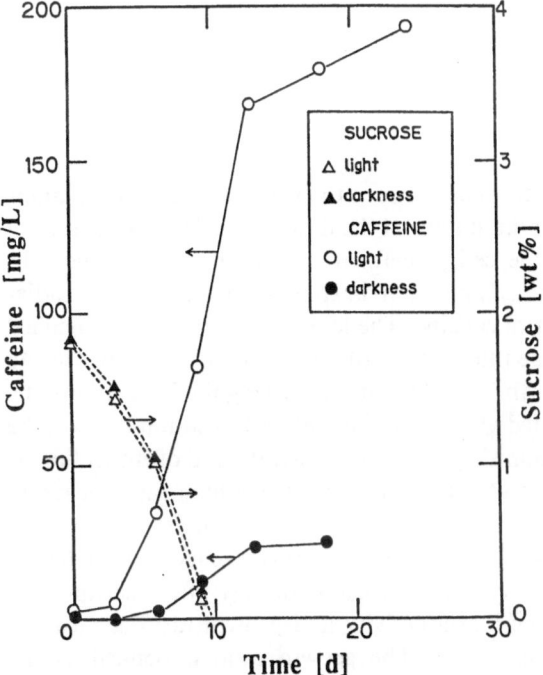

Fig. 3. The effect of light irradiation on caffeine production in the batch culture using the B2K medium. Caffeine concentration in the medium: (●) in darkness; (○) in light. Sucrose concentration in the medium: (▲) in darkness; (△) in light

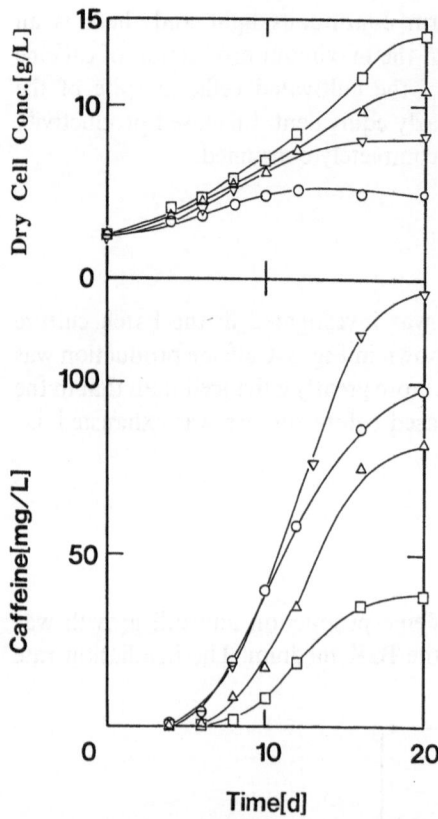

Fig. 4. The effect of the light irradiation rate on caffeine production and cell growth. The initial sucrose concentration was 2 wt%. Light irradiation rate per unit volume of the medium: (○) 3.2×10^{18} quanta/s/L; (▽) 1.9×10^{18} quanta/s/L; (△) 0.29×10^{18} quanta/s/L; (□) 0.13×10^{18} quanta/s/L

was controlled by varying the distance between the flask and fluorescent lamps or by painting a certain area of the surface of the flask black. The time course of the caffeine production and the cell growth are shown in Fig. 4. The cells irradiated at 0.29×10^{18} quanta/s/L-medium, in spite of a lower rate of irradiation, produced a large amount of caffeine. The level of production approached that of the cells irradiated at the rate of 3.2×10^{18} quanta/s/L-medium. The lag time in caffeine production became shorter with increasing the irradiation rate. The growth of the cells irradiated at 3.2×10^{18} quanta/s/L-medium began to be suppressed after 8 days and diminished down to zero at the end of cultivation. It was shown that the growth was greatly suppressed at the high rate of irradiation. The time course of the caffeine production rate per 1 g of dry cells (specific productivity) is shown in Fig. 5. It showed the convex curve which had a maximum at ca. the 10–12th day, and decreased rapidly after reaching the maximum. The maximum specific productivity and growth ratio against the irradiation rate can be seen in Fig. 6. The growth ratio is defined as the comparative value for the cell weight after 20 days of cultivation divided by an initial cell weight. The growth ratio decreased with increasing irradiation rate. The maximum specific productivity increased rapidly in the narrow range of a

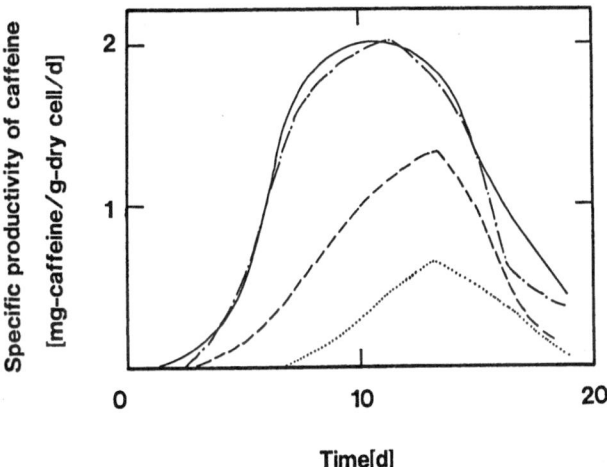

Fig. 5. The time course of the specific productivity of *C. arabica* cells cultured at the various rates of light irradiation. Light irradiation rate per unit volume of the medium: (———) 3.2×10^{18} quanta/s/L; (—.—) 1.9×10^{18} quanta/s/L; (————) 0.29×10^{18} quanta/s/L; (.....) 0.13×10^{18} quanta/s/L

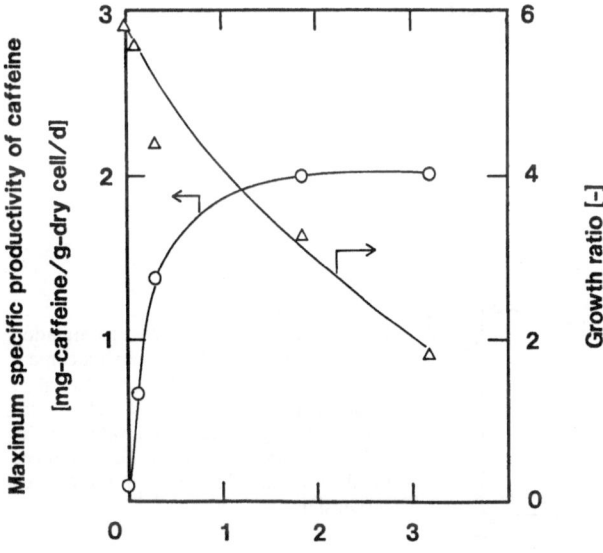

Irradiation Rate $[10^{18}$ quanta/s/L-medium]

Fig. 6. The effect of the irradiation rate on the maximum specific productivity of caffeine and the growth ratio of *C. arabica* cells. (○) Maximum specific productivity of caffeine; (△) growth ratio

low irradiation rate below 0.3×10^{18} quanta/s/L-medium, and was almost steadily maintained at a high value in the wide range of the irradiation rate over 1.0×10^{18} quanta/s/L-medium. Consequently, the irradiation rate for caffeine production had a maximum value at about 2×10^{18} quanta/s/L-medium as

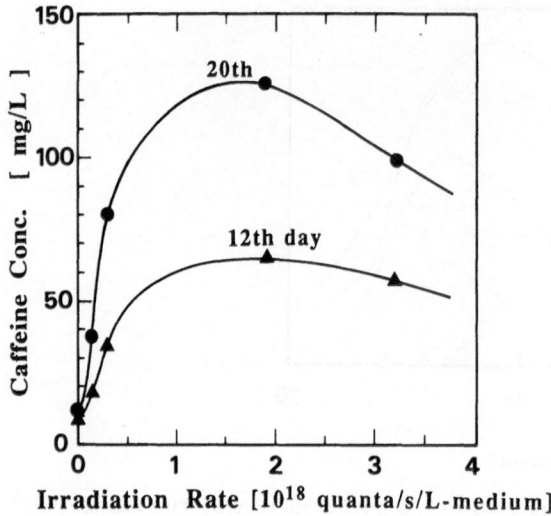

Fig. 7. The effect of the light irradiation rate on the caffeine production: the 12th day (▲) and the 20th day (●) in the batch culture

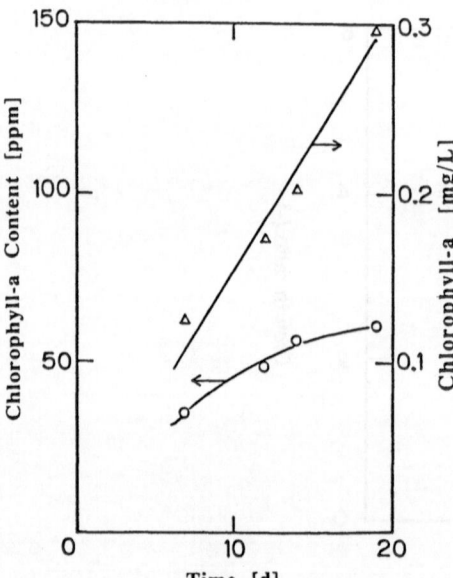

Fig. 8. The time course of the accumulation of chlorophyll a: (○) Chlorophyll-a content in the *Coffea arabica* cell. (△) Amount of chlorophyll-a per unit volume of the cell suspension; light irradiation rate was 1.9 × 10¹⁸ quanta/s/L-medium. The initial cell weight was 8.2 g-fresh cell per 120 mL-cell suspension

shown in Fig. 7. The color of the cells irradiated at 1.9×10^{18} quanta/s/L-medium turned green gradually because of chloroplast formation (Fig. 8).

3.4 Semicontinuous Cultures

Semicontinuous cultivation of *Coffea arabica* cells was carried out in the B2K medium under light. One tenth of the medium was replaced by the fresh medium with 4 wt% sucrose every 3 days. The effect of long-term light irradiation on caffeine production and cell growth was investigated. The time course of the caffeine production and the cell growth is shown in Figs. 9 and 10. Two types of photo-cultivation were carried out. One was the cultivation under continuous light for 51 days (Fig. 9), the other was the cultivation in which the cells were transferred into the darkness after the initial 11 days of light irradiation (Fig. 10). The productivity of the cells cultivated under continuous light reached a

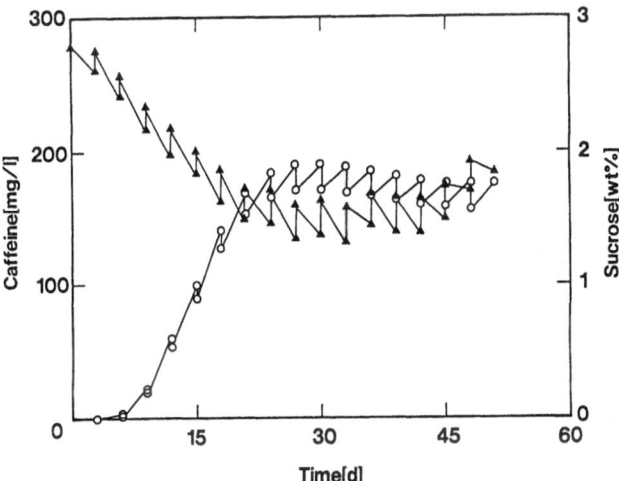

Fig. 9. The semicontinuous culture of *C. arabica* cells. The cells were cultivated under continuous light for 51 days. The irradiation rate was 3.2×10^{18} quanta/s/L-medium. The initial dry cell weight was 0.76 g. (○) Caffeine; (▲) sucrose

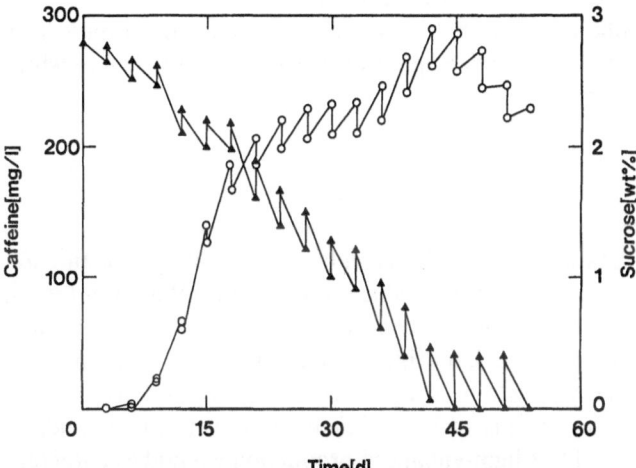

Fig. 10. The semicontinuous culture of *C. arabica* cells. The cells were transferred into darkness after the initial 11 days of continuous light. The irradiation rate was 3.2×10^{18} quanta/s/L-medium. The initial dry cell weight was 0.76 g. (○) Caffeine; (▲) sucrose

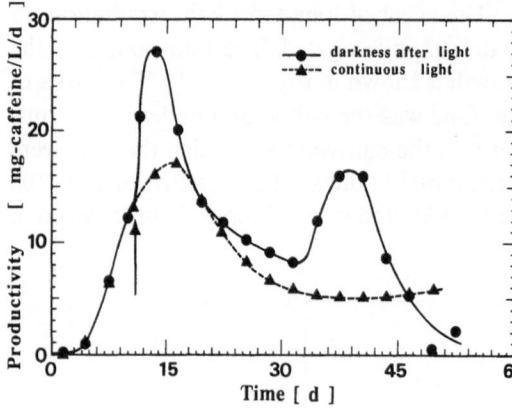

Fig. 11. The time course of the productivity (production rate per unit volume) in the semicontinuous culture. (●) Darkness after 11 days light irradiation. (▲) Continuous light

maximum at ca. the 15th day. After the maximum, the production decreased gradually and reached a steady state after 20 days of cultivation. The sucrose consumption began to be suppressed after 20 days and diminished nearly to zero after 45 days. The color of the cells turned dark green and lysis of the cells was observed near the end of the cultivation. It was found that the growth was greatly inhibited by the long-term continuous light. On the other hand, the caffeine production by the cells irradiated for 11 days increased more than that of the cells cultured under continuous light, after they were transferred into darkness. The caffeine production gradually decreased after reaching a maximum on approx. the 13th day. The production was, however, increased again after 35 days of cultivation. This reactivation was not seen in the culture under 51-day's continuous light. It was suggested that cultivation in darkness would lead to a recovery in the decreased activity of the cell. The sucrose consumption was little suppressed, differing from that by the cells irradiated for 51 days. The green cells, colored by the initial 11-day light irradiation, gradually turned white after being transferred into darkness.

4 Discussion

It is important to investigate the kinetics of secondary metabolite production for the development of a photo-bioreactor suitable for plant cell culture. Photosynthesis or photochemical reactions generally occur in proportion to the amount of light absorbed. On the other hand, the light irradiation is considered to affect the secondary-metabolite production as a promoting signal like hormones. A low rate of light irradiation seems to be sufficient to activate the secondary metabolism. The kinetics of this light-enhanced production would be characterized by several factors, i.e. irradiation period, periodic light irradiation, light irradiation rate, light wavelength, etc.

4.1 Effect of the Irradiation Period and Its Rate

Considering that irradiation for at least 6 days is required to enhance caffeine production as shown in Fig. 1, it is considered that the light irradiation does not cause the direct enhancement of the activity of enzymes involved in caffeine synthesis. It was observed that the *Coffea arabica* cells turned green gradually under light (Fig. 8) and the number of chloroplasts formed was related to the irradiation rate. However, the specific production of caffeine was independent of the irradiation rate in the wide range over 1.0×10^{18} quanta/s/L-medium. Thus, it is suggested that the mechanism of chloroplast formation is not directly related to the enhancement of caffeine production.

As shown in Fig. 6, the activation of caffeine production was saturated at a low rate of irradiation, while the growth was suppressed by increasing the irradiation rate. Therefore, the irradiation rate to maximize the production seemed to be determined by the balance between the light effect to enhance the specific productivity of the cells and the effect to suppress the cell growth.

4.2 Semicontinuous Cultures

It is clear from a comparison of Fig. 9 with Fig. 10 that the transfer from light into darkness is rather effective to enhance the productivity of caffeine in the long-term cultivation of *Coffea arabica* cells using the light irradiation. The main cause of this enhancement effect seems to be the removal of the light-induced suppression of cell growth due to the transfer into darkness. When the cells were transferred into darkness after the 11th day, it was observed that the cells continued to grow even after the 20th day, this is also supported by the fact that the sucrose consumption continued until the 45th day shown in Fig. 10. While, in the case of continuous light irradiation, the cell growth was considerably suppressed approximately after the 20th day as shown in Fig. 9. In the former case, the increase of the productivity was again observed after the 35th day, and the higher productivity would continue for the longer time if sucrose in the medium was not consumed totally. Since it seems difficult to maintain the productivity at higher level in darkness after the transfer from light, the interval of repeated exposure to light would be one of the most important factors affecting the productivity of caffeine for the long-term cultivation.

5 Conclusions

The process of the culture of *Coffea arabica* cells was shown to be influenced by various factors, such as light irradiation rates, irradiation periods, and growth regulators. The caffeine production was greatly increased by light irradiation. The optimal rate of the light irradiation to maximize the production was determined as about 2×10^{18} quanta/s/L-medium in the batch culture for 20 days. Semicontinuous culture was conducted using light for 56 days. Although long-term continuous irradiation greatly suppressed the cell growth, the transfer of the photocultured cells into darkness was effective for recovery of the decreased growth activity and the production rate.

Acknowledgment. This work was partly supported by the Japanese Society for Promoting Sciences (JSPS) for the Japan-US Cooperative Science Program.

6 References

1. Furuya T (1988) Saponin (Ginseng Saponins). In: Cell culture and somatic cells, Academic, New York, p 213.
2. Yamada Y, Fujita Y (1983) Production of useful compounds in culture. In: Evans DA et al. (eds) Handbook of plant cell culture, vol 1, Macmillan, p 717.
3. Ayabe S, Iida K, Furuya T (1986) Plant Cell Reports 3: 186.
4. Sahai OP, Shuler ML (1984) Biotech Bioeng 25: 111.
5. Haldimann D, Brodelius P (1987) Phytochem 26: 1431.
6. Furuya T, Koge K, Orihara Y (1990) Plant Cell Reports 9: 125.
7. Brodelius P (1985) Trends in Biotechnology 3: 280.
8. Frischknecht PM, Baumann TW (1985) Phytochem 24: 2255.
9. Bjoerk L (1986) Found Biotech Ind Ferment Res 4: 177.
10. Frischknecht PM, Baumann TW, Wanner H (1977) Colloq Sci Int Cafe [CR] 8: 139.
11. Hagimori M, Matsumoto T, Obi Y (1982) Plant Physiol 69: 653.
12. Hatchard CG, Parker CA (1956) Proc R Soc London Ser A 235: 518.

5.5 A Model for Embryo Development in Dicotyledonous Plants

Hugo Vits[1], Wei-Shou Hu[1]*, E. John Staba[2] and Todd J. Cooke[3]

[1]Department of Chemical Engineering and Materials Science and [2]Department of Medicinal Chemistry, University of Minnesota, Minneapolis, MN 55455, USA
[3]Department of Botany, University of Maryland, College Park MD 20742, USA

Contents

Somatic embryogenesis is potentially attractive for the large-scale propagation of transgenic plant varieties. Some of its drawbacks are the low efficiency of the proembryo to mature torpedo embryo conversion process and the large number of abnormal embryos associated with it. To further the understanding of the process of embryo development, we have formulated a hypothesis linking the transition from the globular stage to the heart stage during dicotyledonous embryo development, to the combined effect of substrate and hormones on cellular growth rate and differentiation. We propose that limitations in the transport of substrate act as a signal that determines the onset of axial vascularization of the embryo. The mechanical stress fields generated by the growth pattern of the embryo then causes the morphological changes. Differences in growth rates between the central core and the periphery of the embryo result in the formation of cotyledons. We have formulated a continuum mathematical model for the embryo morphogenesis employing conservation equations for substrate and hormone, a low-Reynolds force balance and a biomass balance relating the velocity field to growth rate. The solution strategy, based on a Galerkin-finite-element method coupled to a spine-dependent mesh generator, is briefly described.

1 Introduction

With the employment of genetic engineering techniques to alter the genetic makeup of plants, the ability to regenerate whole plants through tissue culture becomes crucial. Plantlet micropropagation based on shoot and somatic embryo cultures are promising methods for commercial production [1]. However, flask shoot culture is labor intensive since the mature shoots often need to be dissected, transferred to a rooting medium, hardened in a greenhouse, and eventually, transferred to the field. Bioreactor shoot cultures [2] also require

---

* To whom all correspondence should be addressed.

Bioproducts and Bioprocesses 2
Editors: Yoshida, Tanner
© Springer-Verlag Berlin Heidelberg 1993

many of these high cost procedures. Somatic embryogenesis presents definite conceptual advantages for large-scale plantlet production [3]. Millions of embryos can be produced in a laboratory scale bioreactor initiated from a small number of cells [4]. The use of artificial seeds [5] minimizes the necessity of disassembly and renders the automation of large-scale manipulation and planting more feasible. Since the discovery of somatic embryogenesis in carrot (*Daucus carota* (L) [6], this developmental pathway has been demonstrated in many significant crop and woody species.

Morphologically, dicotyledonous somatic and zygotic embryos look remarkably alike. However, the early developmental events differ [7]. In zygotic embryogenesis, the terminal cells convert into a proembryo and the remaining cells originate the suspensor. Further cell divisions result in a spherical organism or globular embryo (Fig. 1). The embryo then elongates in the direction of the suspensor, and morphological bipolarity becomes evident through the formation of both shoot and root meristems. As two cotyledons begin to take form, the embryo acquires a heart-like shape. With full development of the cotyledons and further axis elongation, the embryo enters the torpedo stage and proceeds into a dormant state. Somatic embryos can either arise from single cells [8] or callus cell masses after appropriate changes in the hormonal and nutritional composition of the medium. Morphological polarity becomes manifest with the onset of axis elongation and internal vascularization. A variable number of cotyledons at the apical end marks the beginning of the heart stage. Later, torpedo-like structures can evolve directly into plantlets, undergo secondary embryogenesis (i.e. multiple embryos growing on maturing embryos), or become dormant.

The efficiency of conversion from somatic proembryo to embryo can be low and is strongly affected by culture conditions. The number of normal somatic embryos often decreases in advanced developmental stages; secondary embryogenesis and direct conversion into plantlets are frequent. Understanding the causes of this loss in developmental efficiency is important to assess the potential of somatic embryogenesis for mass vegetative propagation. Here, we propose a

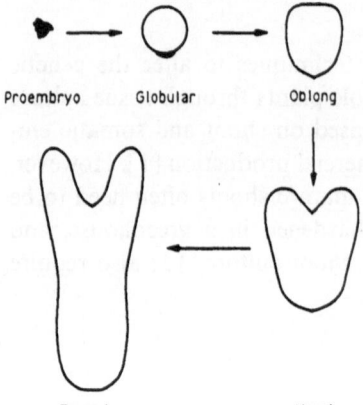

Proembryo   Globular   Oblong

Torpedo   Heart   **Fig. 1.** Developmental stages in dicot embryogenesis hypothesis for embryo development that links mass transfer and environmental effects to the morphological changes of the embryo.

2 Hypothesis

Morphogenesis in the dicotyledonous embryo entails a change from a spherical structure into an elongated torpedo shape. It is generally perceived that the growth pattern in the globular embryo is radially symmetric. It is conceivable that because of nutrient consumption, hormone production and degradation, and mass transfer resistance that concentration gradients of nutrients and hormone develop as the embryo grows. These non-uniform nutrient and hormone distributions may have developmental consequences: the change in the shape of the embryo alters nutrient and hormone patterns which, in turn, affect further form changes and growth patterns through growth-related mechanical effects. Thus, embryo development involves the recurrent interactions of nutrient–hormone–growth–form.

As the globular embryo grows in size, the gradient of the limiting nutrient steepens, and the concentration in the central region can reach a critical level. It is postulated that the embryo senses this metabolic stress and responds by changing the hormone synthetic rate. This new hormone concentration then mediates the enhancement of nutrient transport through the formation of a vascular system. The direction of the vascular system may be specified by the presence of the suspensor in zygotic embryos or the suspensor-like cell appendage in suspension cultures, as suggested by Fry and Wangermann [9]. With a vascular system in place, nutrient is transported through the embryo's exterior surface and vascular system; thus, the nutrient and growth rate distribution patterns change from radially to axially symmetric. Top-to-bottom asymmetry can possibly be a consequence of polar transport along the vascular system, if a preferential flux of limiting substrate is oriented from top to bottom. Alternatively, this asymmetry may be due to a non-uniform top-to-bottom distribution of hormone. The non-uniform distribution of hormone also affects growth rate, resulting in the morphological transition from globular to heart stage. The recurring action of nutrient–hormone–growth–form can be extended to cotyledons formation. One can postulate that the overall growth rate is lower in the near axis region than in the periphery of the embryo, possibly due to hormonal inhibition or cell differentiation. Unequal growth rates in the inner and outer regions could thus lead to the formation of cotyledons.

Figure 2 illustrates possible profiles for substrate, hormone and growth rate at different stages of development. At the early globular stage, the growth field is uniform and not limited by mass transfer. As size increases, the substrate concentration decreases near the center, which stimulates increased hormone synthesis and manifests the onset of vascularization and form change. Gene effects are not explicitly stated in this hypothesis but are embedded, for example, in the vascularization induction mechanism.

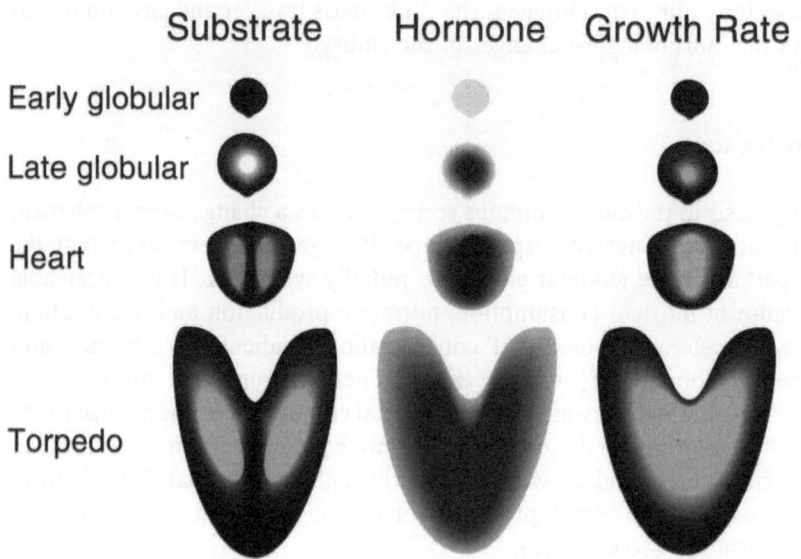

Fig. 2. Hypothetical distributions of nutrient, hormone and growth rate at successive stages of embryo development

The formation of abnormal embryos would result from growth rate distributions that are perturbed from the normal state either by the direct action of environmental factors on growth parameters or by their indirect action on the hormonal balance of the embryo. It is likely that if the perturbations are small enough, the embryo will be able to compensate them via some feedback mechanism on the hormone synthesis/degradation. However, if the perturbation is too large, this response may not be sufficient and the embryo will no longer be structurally stable and may deviate from the expected normal development.

Some experimental evidence is consistent with our hypothesis. Pollock and Jensen [10] found that the cell division pattern was unevenly distributed after the globular stages in *Gossypium* and *Capsella* embryos. Forman and Jensen [11] examined the oxygen consumption by embryos and found an uneven distribution of metabolic activity suggesting a possible oxygen transfer limitation. The work by Fry and Wangermann [9] on mature zygotic embryos, and by Schiavone and Cooke [12] on somatic embryos, suggest that polar transport of auxin exists in the embryo and is essential in the developmental transition from globular to heart stage.

3 Model for Embryo Development

3.1 Conservation Equations

By approximating the heart and torpedo embryos to an axisymmetric shape, the three-dimensional embryo may be represented by a two-dimensional model. In cylindrical coordinates, the model variables become independent of the angular spatial variable. Symmetry boundary conditions can be used and the numerical tractability of the model is greatly improved. A continuum approach has been selected since dealing with average properties of the cells, and not with the cellular microstructure, is less demanding mathematically. This modelling strategy differs from work by Cazzulino et al. [13], where the process of embryo morphological change is represented by a compartment model. Such a model can adequately fit somatic embryo culture data, but it does not provide a mechanistic explanation for development.

Four conservation equations are formulated on the embryo domain: substrate, hormone, biomass and momentum. The local availabilities of substrate, s, and hormone, h, can be described by general mass conservation equations as

$$\frac{\partial s}{\partial t} + s\nabla \cdot v + v \cdot \nabla s = \nabla \cdot J_s + R_s, \tag{1}$$

$$\frac{\partial h}{\partial t} + h\nabla \cdot v + v \cdot \nabla h = \nabla \cdot J_h + R_h, \tag{2}$$

where v is the velocity field, $s\nabla \cdot v$ is the rate of dilution of substrate by growth of biomass and $v \cdot \nabla s$ represents the substrate movement by a convective mechanism. $\nabla \cdot J_s$ is the rate of transport of substrate by non-convective mechanisms, mainly diffusion or carrier mediated transport. R_s is the consumption rate of substrate by the local metabolic activity in the volume element. Similar terms compose the hormone continuity equation.

Biomass can only move convectively by growth of the surrounding cells. The balance equation for the biomass concentration, X, taken on a wet basis, is

$$X\nabla \cdot v = R_x \tag{3}$$

after assuming X to be uniform in space and constant in time. This equation links the local rate of expansion $\nabla \cdot v$ to the local rate of growth R_x.

The momentum balance on the biomass phase is

$$\frac{\partial (Xv)}{\partial t} + \nabla \cdot Xvv = -\nabla \cdot T + F, \tag{4}$$

where T is the total stress tensor and F represents the body forces acting upon the element. Since the embryo growth is slow and the displacements are small, a low-Reynolds approximation can be made [14]. The convective term and the local rate of change of momentum terms can be neglected. By assuming that no body forces are acting on the biomass ($F = 0$), we get

$$\nabla \cdot T = 0. \tag{5}$$

Besides the usual strain, rate-of-strain and pressure dependent terms, the total stress tensor must include a dependence on growth representing the forces on the element of volume due to its own expansion and the interactions with the neighboring elements.

3.2 Boundary and Initial Conditions

Since an axisymmetric domain is being considered, the boundary conditions can be specified on the surface of the embryo and on the symmetry axis. As the limitations to nutrient transport will be assumed to exist only inside the embryo and not in the film surrounding it, the concentration of nutrient at the surface is nearly constant:

$$s = s_0. \tag{6}$$

The hormone is internally synthesized and then released into the maternal tissue or culture medium. The hormone flux at the surface of the embryo given by $[n \cdot J_h]|_\rho$ must be equal to the flux across the liquid film surrounding the embryo,

$$[n \cdot J_h]|_\rho = k_g(h^* - h_0), \tag{7}$$

where h_0 may be taken as constant and low ($h_0 \approx 0.$), if no hormone is externally supplied and the number of embryos per unit of bioreactor volume is low. The hormone concentration h^* represents the value in equilibrium with the intra-embryo hormonal concentration at the surface.

A mechanical boundary condition is established through a momentum balance on the surface while neglecting the stress contributions from the liquid and any gradients in surface tension,

$$- n \cdot T + 2H\sigma n = 0, \tag{8}$$

where n is the normal unit vector at the surface, H is the mean curvature and σ is the surface tension coefficient. The surface tension term accounts for the resistance of surface cells to surface area increases.

Given that the embryo continually changes size and/or shape, the embryo surface is not defined explicitly. It must be tracked throughout the developmental process. Its position at all times can be related to the growth of the embryo body by a boundary equation expressing velocity continuity between the cells and the surface of the embryo.

$$\frac{D}{Dt}(\rho(z,t) - r) = 0, \tag{9}$$

where $\rho(z, t)$ is the time-dependent surface equation, in cylindrical coordinates, and r is the radial cylindrical coordinate.

At the axis of symmetry, the boundary conditions derive from symmetry arguments. For substrate and hormone,

$$n \cdot \nabla s = 0 \tag{10}$$
$$n \cdot \nabla h = 0$$

meaning that, on average, no hormone or substrate crosses the axis of symmetry.

The normal component of the velocity on the axis of symmetry is null, thus any point initially on the axis will remain there for the entire length of the process. However, movement along the axis is not excluded.

$$n \cdot v = 0, \tag{11}$$

$$tn \cdot T = 0.$$

Uniform distributions for substrate and hormone are convenient initial conditions. The initial velocity profiles need to be specified carefully to avoid discontinuities. Appropriate profiles can be obtained by integrating Eq. (3) once the initial growth rate is known.

3.3 Metabolic, Transport Rates and Constitutive Equations

The dependence of growth on substrate and hormone is considered to be of the saturation type for substrate and of the inhibitory type at high hormone concentrations. Substrate consumption is proportional to the growth rate through the yield coefficient Y_{sx}:

$$R_x = -R_s Y_{sx} = \mu_{max} \frac{s}{K_s + s} \frac{h}{K_h + h + \dfrac{h^2}{K_I}}. \tag{12}$$

Changes in hormonal concentration modulate growth via Eq. (12), and induce vascular differentiation via Eq. (14). The postulated net hormone synthesis rate comprises a constitutive synthesis, a first-order degradation rate and an inducible synthesis term,

$$R_h = K_1 - K_2 h + K_3 \frac{h^2}{s} \tag{13}$$

At low substrate concentrations, the last term may increase significantly and act as a concentration signal for differentiation of the transport properties.

The fluxes of substrate and hormone are taken as diffusion-like terms and dependent on the hormone concentration to account for transport-related differentiation. The diffusion coefficient matrices D_s and D_h lump and average the effects of diffusion through the extracellular space, the cytoplasmic membrane and the intracellular space as well as carrier mediated transport. For simplicity, the effective diffusion coefficient is assumed to vary linearly between a minimum and a maximum threshold for each of the principal cartesian directions. Terms of the principal diagonal of the diffusion matrices are assumed to be null.

$$D_{si} = \begin{cases} D_{si\,min} & \text{if } h < h_{min} \\ D_{si\,min} + \dfrac{D_{si\,max} - D_{si\,min}}{h_{max} - h_{min}}(h_{max} - h_{min}) & \text{if } h \in [h_{min}, h_{max}] \\ D_{si\,max} & \text{if } h > h_{max}. \end{cases} \tag{14}$$

with $i = 1, 2, 3$ principal cartesian axes of symmetry.

Biologically, such a dependence could be the result of synthesis of the channel or carrier molecules in response to a hormone concentration change. Anisotropy in transport coefficients may be necessary for the induction of the preferential axis of transport. Expressions similar to Eq. (14) are assumed to be valid for the transport of hormone.

The mechanical constitutive equation includes a term that is dependent on the local growth characteristics of the embryo. Given the large percentage of water in the biomass, we assume it to be an incompressible material. The total stress is then,

$$T = -pI + \tau_{material}(\gamma, \gamma) + f(R_x)I \tag{15}$$

where p is the hydrostatic pressure and $\tau_{material}$ is the stress tensor. The extra stress term due to growth is found to be proportional to $f(R_x)$ if the stress tensor is proportional to the rate-of-strain (viscous limit) and to the growth rate multiplied by a time constant of the material if the stress tensor is proportional to strain (elastic limit).

The direct determination of the parameters in the rate, transport and mechanical constitutive equations is not an easy task. However, the possibility of determining the spatial growth distribution of the embryo [15] is already a major step forward that can be used in the validation and refinement of our model.

4 Numerical Solution Strategy

The moving boundary problem is put in a Galerkin-finite element form as the weighting functions are coincident with the basis functions. The selected basis functions are biquadratic with support on one element. The mapping between the computational and physical domains is performed isoparametrically. Spines are constructed to link the movement of the boundary to the movement of the elements inside the domain [16]. The usual residual equations partial derivatives need to be modified to account for changes in element size and shape with time. While the cylindrical coordinate system is the natural choice for axisymmetric problems, the use of the spine mesh generator could lead to multivalued spines after the heart transition. As a result the domain would have to be re-meshed. The use of a spherical coordinate system may delay this problem considerably.

The resulting system of time-dependent ordinary differential equations and algebraic equations is solved by the implicit trapezoidal rule. Overall, the process amounts to solve a large set on nonlinear algebraic equations by a Newton method at each time step. The structure of the Jacobian matrix is arrow-like as the movement of the boundaries is linked to interior domain variables. This system of linear equations, generated at each Newton iteration can be solved efficiently by Gaussian elimination if the total number of unknowns is kept in the low thousands.

5 Concluding Remarks

Wider application of somatic embryogenesis is hampered by the low efficiency and the high frequency of abnormality. We have hypothesized that the development of embryo is affected by recurring interactions of nutrient–hormone–growth-form. As part of our efforts to test this hypothesis, we have formulated a mathematical model for the development of a single dicot embryo. The numerical testing of model assumptions such as the anisotropic transport mechanism of induction, the selection of constitutive mechanical, metabolic and transport rate equations will certainly increase our understanding of the above cited interrelations. It can also serve as a guide in determining experiments that are relevant to the validation process.

If the hypothesis is correct or even partially correct, it could imply that one way to improve the process of somatic embryogenesis is to exert proper mass transfer conditions while maintaining appropriate control of the environmental variables. Therefore, enhancing our understanding of the interrelationship between the metabolic and morphogenetic processes in the embryo can possibly have a profound and practical effect on plant biotechnology.

Acknowledgements. This study was supported in part by the National Science Foundation through ECE-8552670 and DCB-89-17378 and by a grant from the Minnesota Supercomputer Institute.

6 Nomenclature

D_s, D_h	transport coefficient matrices for substrate and hormone
h	hormone/growth regulator concentration
H	mean surface curvature
I	identity tensor
J_s, J_h	non-convective rates of substrate and hormone transport
k_g	embryo-environment mass transfer coefficient
K_h	hormone saturation constant
K_I	hormone inhibition constant
K_s	substrate saturation constant
K_1, K_2, K_3	hormone synthesis/degradation constants
n	normal vector to surface or line
p	hydrostatic pressure
$r(r, \Theta, z)$	position vector
R_h	local net rate of hormone synthesis/degradation
R_s	local rate of substrate consumption
R_x	local growth rate
s	limiting nutrient or substrate concentration
t	tangential vector to surface or line
T	total stress tensor

v	velocity vector
X	biomass concentration (wet weight basis) or density
Y_{sx}	yield of biomass on substrate coefficient
σ	surface tension coefficient
ρ	embryo surface equation
μ_{max}	maximum specific growth rate coefficient
$\tau_{material}$	stress tensor
0	null tensor

7 References

1. Sluis CJ, Walker KA (1985) IAPTC Newsletter 49: 2.
2. Park JM, Hu W-S, Staba EJ (1989) Biotechnol Bioeng 34: 1209.
3. Ammirato PV (1983) In: Evans DA, Sharp WR, Ammirato PV, Yamada Y (eds) Handbook of plant cell culture (vol 1). Macmillan, New York, p 82.
4. Stuart DA, Strickland SA, Walker KA (1987) Hort Sci 22: 800.
5. Kitto SL, Janick J (1985) J Amer Soc Hort Sci 110: 227.
6. Reinert J (1958) Naturwissenschaft 45: 344.
7. Ateeves TA, Sussex IM (1989) Patterns in plant development. Cambridge University Press, Cambridge, p 6.
8. Nomura K, Komamine A (1985) Plant Physiol 79: 988.
9. Fray SC, Wangermann E (1976) New Phytol 77: 313.
10. Pollock EG, Jensen WA (1964) Amer J Bot 51: 915.
11. Forman J, Jensen WA (1965) Plant Physiol 40: 765.
12. Schiavone FM, Cooke TJ (1987) Cell Differentiation 21: 53.
13. Cazzulino DL, Pedersen H, Chin CK, Styer D (1990) Biotechnol Bioeng 35: 781.
14. Purcell E (1977) Amer J Phys 45: 1.
15. Komamine A, Matsumoto M, Tsukahara M, Fujiwara A, Kawahara R, Ito M, Smith J, Nomurak K, Fujimura T (1990) Prog Plant Cell Molec Biol Kluwer, Dordrecht p 307.
16. Kistler SE, Scriven LE (1983) Int J Numer Methods Eng 4: 207.

5.6 Kinetics of Hybridoma Cell Growth in Continuous Culture

Kelly K. Frame[1] and Wei-Shou Hu*

Department of Chemical Engineering and Materials Science, University of Minnesota, 421 Washington Avenue SE, Minneapolis, MN 55455 USA

Contents

We have performed studies of continuous culture using hybridoma AFP-27 cells. Under glucose limiting conditions the behavior of the biomass can be described by a Monod-type model with a number of modifications. The modifications made to the simple Monod-type model include: (1) a threshold limiting substrate concentration for cell growth; (2) two regions of growth rate with different yield coefficients for the limiting substrate; (3) high death rate associated with low dilution rate. In the course of our study we also observed that non-producing cells arise in a continuous culture and compete with the producing cells. Our preliminary analysis indicates that in the low dilution rate region, producing cells may remain as a dominant population.

1 Introduction

Significant progress has been made in the development of animal cell bioreactors in the last decade. This development has facilitated the practice of industrial animal cell cultivation on a relatively large scale. Lagging behind this development is our understanding of the reaction kinetics in animal cell systems. Associated with this is the lack of suitable kinetic descriptions of animal cell behavior in bioreactors.

The lack of a suitable kinetic model affects not only the control and optimization of batch processing, but also the operation of a perfusion culture. Many of the reactors developed utilize cell retention and continuous perfusion to achieve a high cell concentration [1]. Despite the wide application of perfusion bioreactors, the operation conditions, such as the time profile of the flow rate, are still largely selected by educated guessing. The need for a kinetic description of cell behavior with some predictive value is obvious.

---

[1] Current Address: R.W. Johnson Pharmaceutical Research Institute, U.S. Route 202, Box 300, Raritan, NJ 08869-0602, USA.
* To whom all correspondence should be addressed.

Bioproducts and Bioprocesses 2
Editors: Yoshida, Tanner
© Springer-Verlag Berlin Heidelberg 1993

The development of a kinetic model for animal cell processes has been rather limited because of the complexity of animal cell systems. Saturation type growth models such as the Monod model have been used to describe the growth of microbial systems [2]. The majority of the growth models developed for animal cells have adopted the basic form of the Monod model. Multiplicative Monod-type models have also been proposed for describing the kinetics of hybridoma cell cultures [3]. In such a model, the growth rate is assumed to be dependent on the concentrations of more than one substrate.

Attempts have been made to carry out continuous cultures at steady state to develop kinetic models [4, 5, 3]. Such an approach is tedious, expensive and prone to either contamination or mechanical failure. For hybridoma cultures producing antibodies, the productivity may also change over time in a continuous culture [6]. Nevertheless, the results tend to support the notion that a Monod type unstructured model can be used to correlate the specific growth rate to the limiting substrate concentration with some modifications [4]. In this communication we report our investigation on the kinetic behavior of a hybridoma cell line in continuous culture.

2 The Occurrence of Non-Producing Cells

We have carried out a total of 28,000 h of continuous cultures using hybridoma AFP-27 cells. In our initial continuous culture studies, we observed a decrease in antibody concentration in four different continuous cultures of AFP-27 cells. All cultures were operated at dilution rates of approximately $0.030 \, h^{-1}$; using DMEM with 10% horse serum as the feed medium. Shown in Fig. 1 are results from one of the four cultures.

After 47.5 h of batch growth, dilution was initiated at a rate of $0.029 \, h^{-1}$. The concentration of viable cells reached a steady-state value of approximately 1.5×10^6 cells/ml around 400 h. The residual glucose concentration in the culture stabilized at a concentration of slightly less than 0.02 mg/mL, while the lactate concentration stabilized at a concentration of approximately 0.65 mg/mL.

While all the other measured variables remain constant, the concentrations of antibody decreased over the course of the culture. The antibody concentration, as measured by ELISA, decreased from approximately 40 μg/mL at the beginning of the culture to approximately 20 μg/mL at the end of the culture. Similarly, the antibody concentration, as measured by RIA, decreased from 80 μg/mL to 40 μg/mL over the same period.

We then measured the fractions of non-producing cells in the cultures over the time courses (Fig. 2). The fraction of antibody-producing cells decreased from approximately 98% soon after inoculation to approximately 50%. The specific antibody productivities were then estimated based on only the antibody-producing viable cells present, as opposed to the entire population of viable cells. The apparent specific antibody productivity was divided by the

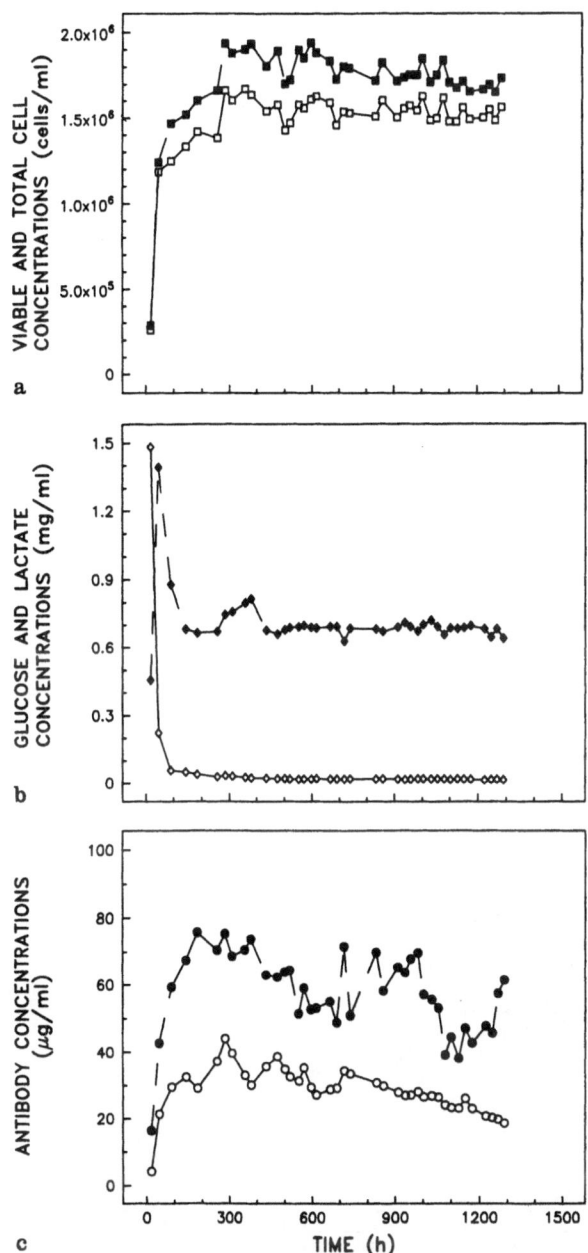

Fig. 1. Continuous cultures of AFP-27 cells: (a) Concentrations of viable (□) and total (■) cells. (b) Concentrations of glucose (◇) and lactate (◆). (c) Concentrations of antibody, as measured by ELISA (○) and RIA (●).

estimated fraction of antibody-producing cells in the various cultures at any time (Fig. 2). These fractions of producing cells were estimated by connecting the data points in Fig. 2 with line segments and assuming the fractions between clonings to fall on these line segments.

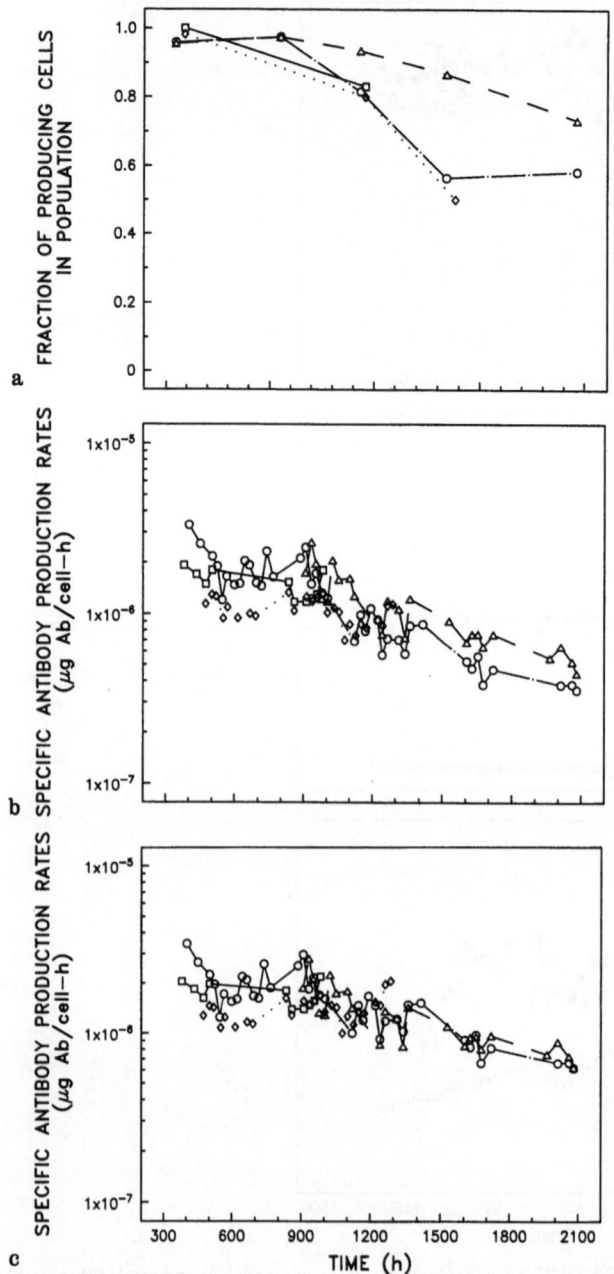

Fig. 2. Continuous cultures of AFP-27 cells. (a) Fraction of producing cells in the cultures. (b) Specific antibody productivities of the cultures, based on all viable cells. (c) Specific antibody productivities of the cultures, based on only antibody-producing viable cells.

The removal of the effect of the non-producing cells completely removed any statistically significant decreases in antibody productivity. We thus attribute this decrease of antibody productivity to the increased population of non-producing cells in the cultures over time [6]. We subsequently isolated the non-producing clones and both producing cells (AFP27-P) and the non-producing cells (AFP27-NP) of AFP-27 were used in this study. The use of the non-producing cells allowed us to circumvent the problem of competition between the producing and non-producing populations in long-term continuous culture. It also enabled us to simplify the mass balance equation for the rate-limiting substrate by neglecting the consumption associated with antibody production.

3 The Kinetic Behavior of Non-Producing Cells

In cell culture studies, the cell concentration (number of cells per unit culture volume) is most commonly used in expressing the dynamics of cell growth. In our investigation, we measured not only cell concentration but also cell volume distributions and, in some cases, dry cell weight. The volume of viable hybridoma cells was significantly larger than that of non-viable cells. During a batch culture the volume of the viable hybridoma cells varied with the growth stages; in continuous cultures it changed with dilution rates. Furthermore, proportionality existed between the volume of the viable cells and their dry weight [7]. This observation allowed us to use the volume data in conjunction with cell concentration data as an indirect measurement of the dry weight (biomass) concentration.

One assumption used in a Monod-type model is that one and only one substrate is growth-rate limiting in the culture. We have carrried out a series of continuous cultures in which step changes in the concentrations of different combination of nutrients were introduced. In the cases where the cell concentration changed afterwards, the system was allowed to reach a new steady state. The results indicated that glucose was the rate-limiting substrate under our experimental conditions.

For non-producing cells (AFP-27-NP), the steady-state viable cell concentration decreased in almost a linear fashion with increasing dilution rates (Fig. 3). The total cell concentration (viable cells plus dead cells) was essentially equal to the viable cell concentration for higher dilution rates but became significantly higher as D was decreased. The mean volume of viable cells increased in a linear fashion with increasing dilution rate. The viable cell concentration data in Fig. 3, when expressed in viable biomass concentration (the product of mean cell volume, viable cell concentration, and dry weight per mean cell volume), demonstrated a lower slope at the lower D values (Fig. 4). We also found that the biomass concentration, as opposed to cell concentration, can be better described by a Monod-type model.

The steady-state residual glucose concentration remained at very low levels (approximately 0.01 mg/mL) until D exceeded approximately 0.045 h^{-1} (Fig. 4).

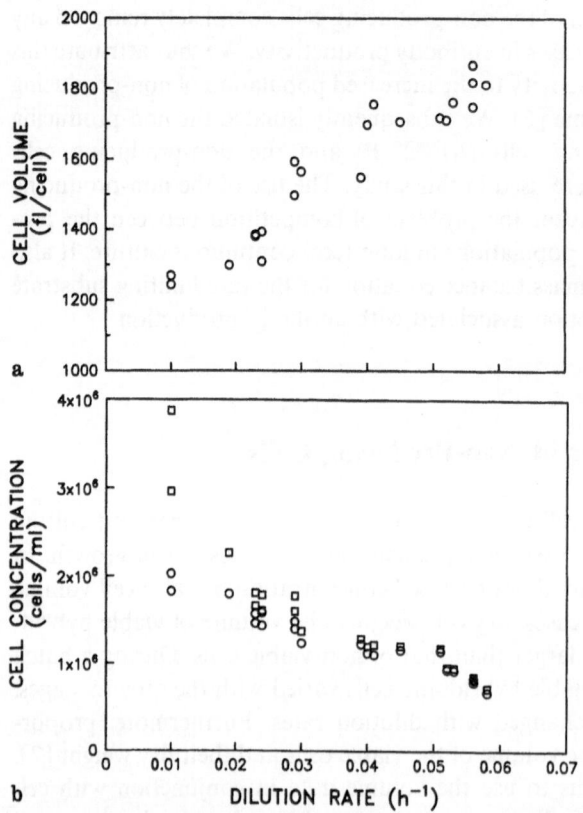

Fig. 3. Steady state cultures (a) viable cell mean volume (b) viable (○) and total (□) cell concentrations.

The lactate concentration increased from 0.15 mg/mL at $D = 0.01\,\mathrm{h}^{-1}$ to 0.6 mg/mL at $0.025\,\mathrm{h}^{-1}$ and remained at approximately 0.6–0.65 mg/mL for the majority of the range of D.

The mass balances on the concentrations of viable biomass, dead biomass and substrate are

$$\frac{dx_v}{dt} = (\mu - \mu_d - D)x_v, \tag{1}$$

$$\frac{dx_d}{dt} = \mu_d x_v - D x_d, \tag{2}$$

$$\frac{ds}{dt} = D(s_f - s) - q_s x_v. \tag{3}$$

The disappearance of dead cells due to lysis was assumed to be negligible and the only removal term considered for dead cells was wash-out. At steady state, we have

$$\mu = \mu_d + D, \tag{4}$$

and

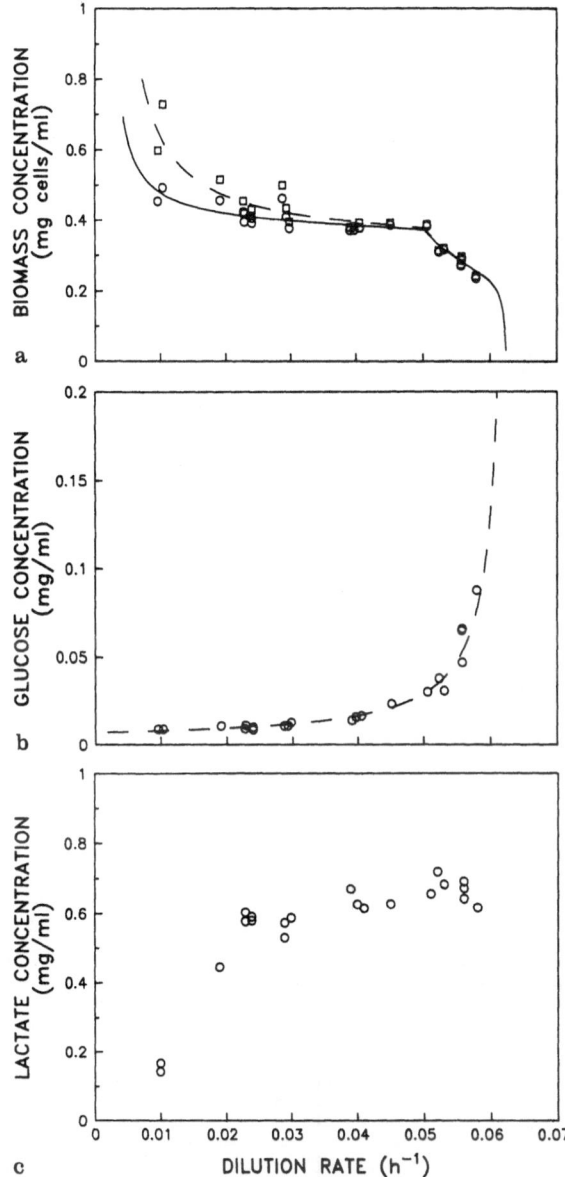

Fig. 4. Steady state cultures (a) viable (○) and total (□) biomass concentrations (b) glucose concentrations (c) lactate concentrations.

$$\mu_d = \frac{x_d}{x_v}D. \tag{5}$$

At low dilution rates the dead cell concentration was relatively high and gave rise to positive μ_d (Fig. 5). Thus, in the continuous culture of AFP cells, μ was not identical to D. When the specific growth rate is plotted against the residual

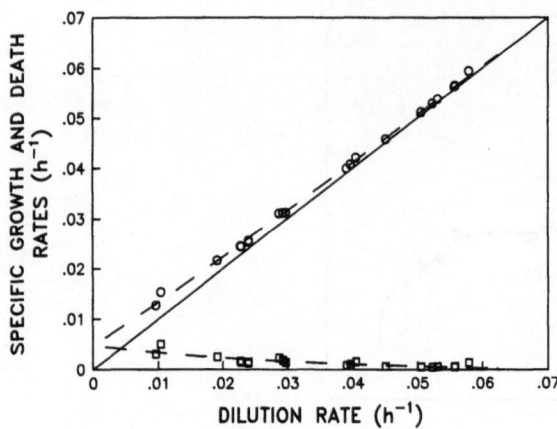

Fig. 5. Specific growth (○) and death (□) rates at various dilution rates

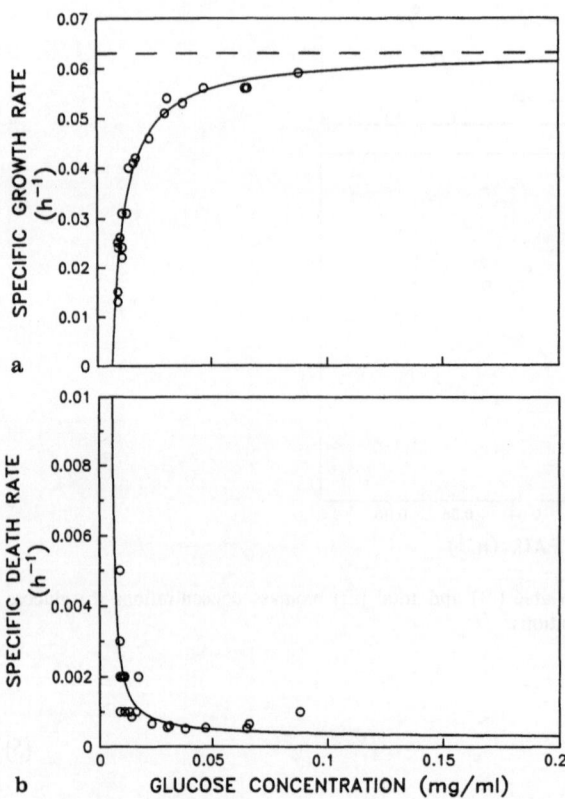

Fig. 6. (a) Specific growth rate as a function of steady state residual glucose concentration. **(b)** Specific death rate as a function of residual glucose concentration

limiting substrate concentration (Fig. 6), the growth rate does not directly approach the origin. Instead, it must either show a sigmoidal approach or have a non-zero intersection with the glucose axis if it continues its linear approach. We adopted the latter for our model and introduced a modification to the Monod model by including a positive "threshold" substrate concentration, s_t, at which point the growth rate goes to a minimum value, μ_{min}. At this minimum value, $\mu = \mu_d$. The modified Monod model is then

$$\mu = \begin{cases} 0, & s < s_t \\ \mu_{min} + \dfrac{(\mu_{max} - \mu_{min})(s - s_t)}{K_s + (s - s_t)}, & s \geq s_t. \end{cases} \tag{6}$$

Since glucose was the only substrate which had an effect on the systems, we simplified our analysis by assuming that μ_d was also a function of s. An inverted Monod model that appears to fit our results is

$$\mu_d = (\mu_{min} - D_{min}) - \frac{\mu_{dmax}(s - s_t)}{K_d + (s - s_t)}, \tag{7}$$

where, D_{min} is the dilution rate corresponding to the point where $\mu = \mu_{min}$ (and where $s = s_t$).

For a system obeying a maintenance model of substrate consumption [8, 9], a plot of the specific substrate consumption rate versus the growth rate is linear, with a slope equal to the reciprocal of the yield coefficient of the biomass on substrate and an intercept equal to the maintenance coefficient of the biomass. The relationship between q_s and μ for AFP-27-NP behaved similarly, but with two interesting deviations (Fig. 7). First of all, the relationship was linear, but was broken into two distinct regions with different slopes. Secondly, the inter-

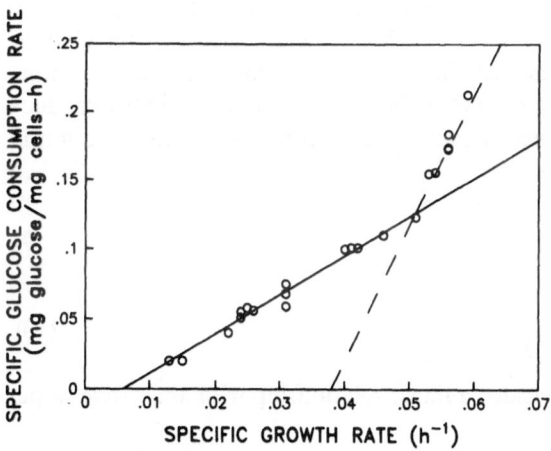

Fig. 7. Specific glucose consumption rate as a function of specific growth rate

cept of either region was negative. In other words, following a maintenance model, the data would lead to the calculation of a negative maintenance coefficient.

Rather than interpreting the lower line segment as having a negative intercept, it was considered as having a zero intercept at the growth rate where $\mu = \mu_{min}$. Such a "zero intercept" may not be true, since cells are still growing at μ_{min}, the specific rate-limiting substrate consumption rate is unlikely to be zero. Nevertheless, due to the lack of data points very close to μ_{min}, it was used in our subsequent analysis. The functional relationship between q_s and μ then follows a linear relationship with a zero maintenance coefficient and a slope equal to the reciprocal of the yield coefficient of viable biomass on substrate:

$$q_s = \frac{1}{Y_{x/s}} (\mu - \mu_{min}) \tag{8}$$

q_s would appear to be a continuous function. As μ increases past some critical growth rate, μ_{crit}, the yield coefficient changes from $Y_{x/s}$ to $Y'_{x/s}$ and the minimum growth rate changes from μ_{min} to μ'_{min}. The intercept of the upper line segment does not necessarily have any physical meaning, but is merely a convenient way of describing the relationship. To ascertain that the change in the yield coefficient was not due to a change in the rate-limiting substrate, we performed rate-limiting substrate studies on both sides of μ_{crit}. In both regions, glucose was the only nutrient that had an effect on biomass concentration.

4 The Kinetic Behavior of Producing Cells

The kinetic behavior of producing cells (AFP-27-P) in continuous culture was very similar to that of non-producing cells. An additional variable measured was, of course, the antibody concentration. It decreased with increasing dilution rate (data not shown). The rate-limiting substrate was identified to be glucose in both regions of dilution rate between which the yield coefficients for glucose were different. The mass balance equations are also similar to those for non-producing cells except that (1) a term for substrate utilized for antibody production is added to substrate balance equation, and (2) an additional equation is needed for antibody concentration.

$$\frac{dp}{dt} = q_p x_v - Dp, \tag{9}$$

$$\frac{ds}{dt} = D(s_f - s) - q_s x_v - \frac{q_p x_v}{Y_{p/s}}. \tag{10}$$

A plot of q_p versus μ (Fig. 8) revealed a linear relationship with a negative slope and a positive intercept.

$$q_p = \alpha - \beta \mu \tag{11}$$

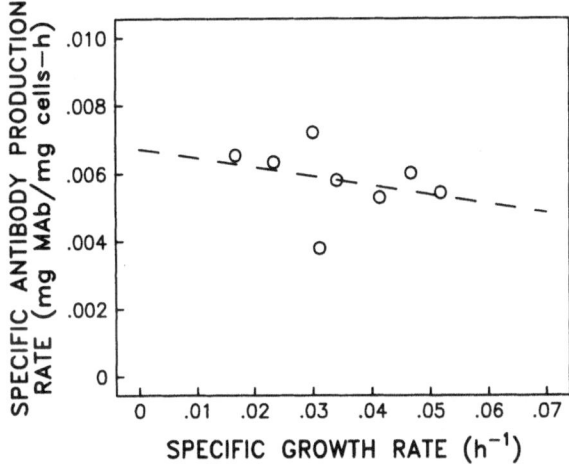

Fig. 8. Antibody concentrations at different dilution rates

This is in contrast to more familiar "mixed" models for product formation in which the growth-associated term is usually positive.

To determine the values of the parameters, we assumed that the yield coefficient of biomass in both regions of dilution rate ($Y_{x/s}$ and $Y'_{x/s}$) were the same as those for the non-producing cells. A summary of the parameter values for both producing and non-producing cells is listed in Table 1. The prediction of the kinetic behavior of AFP-27-NP cells, according to the model, is shown as lines in Figs. 4–8.

5 Competition Between Producing and Non-Producing Cells

The differences in the values of μ_{max}, k_s and other parameters for producing and non-producing cells suggest a complex competitive nature of the two populations in continuous culture. It is conceivable that at some region in the low dilution rate range, the non-producing cells may not have competitive advantage over the producing cells despite their higher maximum specific growth rate, μ_{max}.

As stated above, non-producing cells arose in a continuous culture of hybridoma cells. A majority of these non-producing clones grew at a faster rate than the producing clone in batch culture [10]. The one clone (AFP-27-NP) studied extensively in continuous culture also exhibited higher μ_{max} than the producing clone. Thus, it was expected that competition was present between the two populations (assuming non-producing and producing cells can each be

Table 1. Parameter values for AFP-27-NP and AFP-27-P. The values for $Y_{x/s}$ and $Y'_{x/s}$ were assumed to be the same as for both NP and P cells

Parameter	Value ± Sd		Units
	NP	P	
μ_{min}	0.00630 ± 0.000978	0.0130 ND	h^{-1}
μ_{crit}	0.0508 ND	0.0429 ND	h^{-1}
μ'_{min}	0.0381 ± 0.00174	0.0344 ND	h^{-1}
μ_{max}	0.0630 ± 0.00273	0.0526 ± 0.00711	h^{-1}
K_s	0.00613 0.00145	0.00451 ± 0.00165	mg glucose/ml
s_t	0.00712 ± 0.000559	0.00545 ± 0.00491	mg glucose/ml
D_{min}	0.00183 ± 0.00275	0.00636 ± 0.00236	h^{-1}
$\mu_{d,max}$	0.00420 ± 0.00249	0.00628 ± 0.00217	h^{-1}
K_d	0.00241 ± 0.00337	0.00433 ± 0.00567	mg glucose/ml
α		0.00672 ± 0.00121	mg Ab/mg cells-h
β		0.0266 ± 0.0336	mg Ab/mg cells
$Y_{p/s}$		0.408 ND	mg Ab/mg glucose
$Y_{x/s}$	0.355 ± 0.0143		mg cells/mg glucose
$Y'_{x/s}$	0.101 ± 0.0111		mg cells/mg glucose

treated as a single population). The non-producing cells presumably arose from their own growth as well as from the conversion of producers to non-producers. The mass balances of a system with both P and NP are as follows:

$$\frac{dx_v^{NP}}{dt} = (\mu^{NP} - \mu_d^{NP} - D)x_v^{NP} + r_c x_v^P \tag{12}$$

$$\frac{dx_v^{P}}{dt} = (\mu^{P} - \mu_d^{P} - D)x_v^{P} - r_c x_v^P \tag{13}$$

$$\frac{dx_d^{NP}}{dt} = \mu_d^{NP} x_v^{NP} - Dx_d^{NP} \tag{14}$$

$$\frac{dx_d^{P}}{dt} = \mu_d^{P} x_v^{P} - Dx_d^{P} \tag{15}$$

$$\frac{ds}{dt} = D(s_f - s) - q_s^{NP} x_v^{NP} - q_s^{P} x_v^{P} - \frac{q_p x_v^{P}}{Y_{p/s}}. \tag{16}$$

For a pure, simple competition with no conversion rate and no viability concerns, the population with the lower μ value at a given s value would wash out; while the other population would be stable, as long as D was below the wash-out dilution rate for the remaining population. Figure 9 indicates that the AFP-27-NP population would be expected to wash out at low residual glucose concentrations (i.e. low D values), while the AFP-27-P population would be expected to wash out at high residual concentrations.

Preliminary analysis was done by numerically integrating the above equations to determine the locations of the steady-states for the high and low estimates of r_c (Fig. 10). For positive values of r_c at D values well below $D = 0.038\ h^{-1}$, the AFP-27-NP population is essentially negligible; however, its steady-state level increases as r_c increases. As D approaches $0.038\ h^{-1}$, positive values of r_c enable the coexistence of significant populations of both AFP-27-P and AFP-27-NP cells. For example, at $D = 0.38\ h^{-1}$ and $r_c = 2 \times 10^{-3}$ AFP-27-NP cells formed per AFP-27-P cell per h, the biomass concentrations of AFP-27-P and AFP-27-NP are almost equal. With tight convergence criteria (i.e. criteria for the amount of change in the viable biomass concentrations over given time periods), these coexistence points do not appear to converge, so they are perhaps not true steady-states.

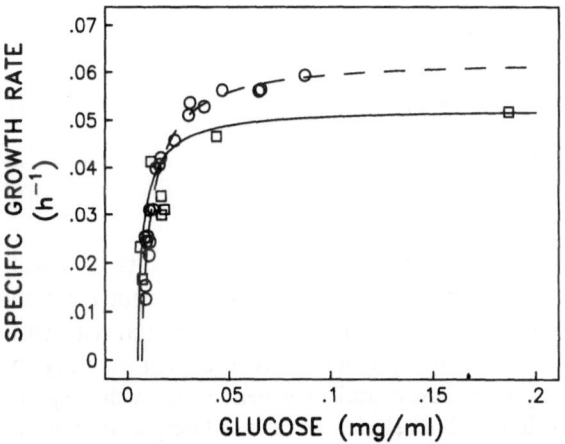

Fig. 9. Prediction of steady state cell concentrations of AFP-27-P (*solid line*) and AFP-27-NP (*dotted line*)

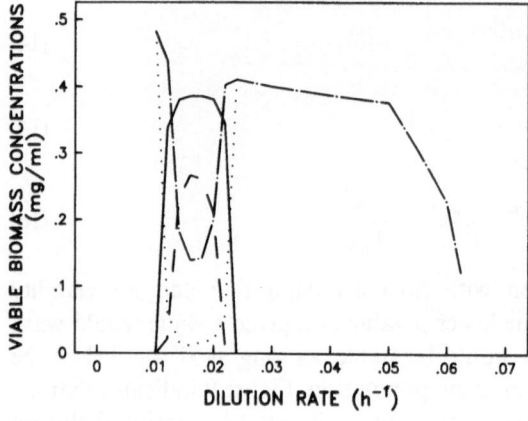

Fig. 10. Steady state biomass concentrations of AFP-27-P (– – – –, ———) and NP ($\cdots$, —·—). The two sets of curves are for $r_c = 2 \times 10^{-4}$ (– – –, $\cdots$) and $r_c = 2 \times 10^{-3}$ (———, —·—).

6 Discussion

Our continuous culture studies of hybridoma cell growth kinetics present a few peculiar phenomena of particular interest to biochemical engineering: (1) a threshold glucose concentration, s_t, for cell growth and a corresponding minimum growth rate, μ_{min}; (2) two regions of growth between which the yield coefficients for glucose are different; (3) specific death rate is a function of dilution rate; (4) the potential competitive nature of antibody producing and non-producing cells.

In the low dilution rate region, it may well be that a sigmoidal type of behavior is more suitable than a minimum growth rate and a threshold glucose concentration. Alternatively, a model with three parameters, such as the one shown below, can possibly describe the growth adequately.

$$\mu = \begin{cases} 0, & s \leq s_t \\ \dfrac{\mu_{max} s}{s_t \left(\dfrac{\mu_{max} - \mu_{min}}{\mu_{min}} + s \right)}, & s > s_t. \end{cases} \tag{17}$$

Since in our model μ_d is related to μ, a change in the functionality of μ will certainly affect that of μ_d. We have shown previously that the spent medium harvested from a batch culture in which cell death has occurred contains growth inhibitory substances [11]. Exponentially growing cells lose their viability in a first-order fashion after being centrifuged and resuspended in such a spent medium. We have also shown that such a cell death was not likely to be caused by nutrient limitation or the accumulation of lactate and ammonia. The growth inhibitory substance(s) are most likely to have a relatively small molecular weight because it is dialyzable through a membrane with a molecular weight cut-off of 2000. At a low dilution rate where the dead cell concentration is high, the potential effect of such an inhibitory substance cannot be discounted. It is possible that μ_d can be expressed as a function of such an inhibitory substance, once it is identified.

Compared to conventional Monod type models for growth with a single limiting substrate, our model is certainly more complicated. However, this increased complexity not only fits our experimental data better, but also predicts a competition advantage for producing cells at low dilution rates. The notion that the loss of antibody productivity can be prevented at low dilution rates is very appealing from a processing perspective and certainly warrants further studies.

Acknowledgements. This research was supported by grants from the National Science Foundation (ECE-8512427, ECE-8552670) and from Ecolab, Inc., St. Paul, MN. KKF was the recipient of a fellowship from the Amoco Foundation. The authors thank A.G. Fredrickson for valuable discussions.

7 Nomenclature

D	dilution rate of reactor
p	antibody concentration
q_p	specific antibody production rate
q_s	specific consumption rate of the substrate by the viable biomass
r_c	conversion rate of producer to non-producer
s	concentration of substrate
s_f	concentration of substrate in feed medium
x_d	concentration of dead biomass
x_v	concentration of viable biomass
$Y_{p/s}$	yield of antibody on substrate
μ	specific growth rate of viable biomass
μ_d	specific death rate of viable biomass
Superscript P	producing cells
Superscript NP	non-producing cells

8 References

1. Hu W-S, Peshwa MV (1991) Can J Chem Engr 69: 409.
2. Fredrickson AG, Tsuchyia HM (1977) In: Amundson NR, Lapidus L (eds) Microbial kinetics and dynamics in chemical reactor theory: A review. Prentice Hall, Englewood, NJ, p 405.
3. Miller WM, Blanch HW, Wilke CR (1988) Biotechnol Bioeng 32: 947.
4. Frame KK, Hu W-S (1991) Biotechnol Bioeng 37: 55.

308   Kelly K. Frame and Wei-Shou Hu

5. Tovey M, Brouty-Boye D (1976) Exptl Cell Res 101: 346.
6. Frame KK and Hu W-S (1990) Biotechnol Bioeng 35: 469.
7. Frame KK, Hu W-S (1990) Biotechnol Bioeng 36: 191.
8. Marr AG, Nilson EH, Clark DJ (1963) Ann NY Acad Sci 102: 536.
9. Pirt LJ (1965) Proc Roy Soc B 163: 224.
10. Frame KK, Hu W-S (1991) Enz Microbiol Technol 13: 690.
11. Dodge TC, Ji G-Y, Hu W-S (1987) Enz Microb Technol 9: 607.

5.7 Development of New Immobilization Method and Continuous Production of Bioproducts in Immobilized Mammalian Cell System

Takeshi Kobayashi and Shinji Iijima,

Department of Biotechnology, Faculty of Engineering, Nagoya University, Furo-cho, Chikusa-ku, Nagoya 464-01, Japan

A novel immobilization method for animal cells, in which the alginate gel was covered with urethane polymer was developed. By this method, gel particles became resistant to physical stress, cell leakage was minimized and a high concentration of cells could be obtained. Use of the immobilized hybridoma cells in a fluidized-bed reactor improved the production of the monoclonal antibody. The production of monoclonal antibody increased with culture time and finally reached 300 mg/ml gel per day on day 17, which was eight times higher than the value obtained by the repeated-batch culture in spinner flasks. Using this method, we were able to introduce air by bubbling, which is the most effective oxygen transfer method. This bubbling caused free cells to die. In the case of the alginate gels, cell leakage was evident but the free cells soon died. In contrast, cells did not leak from the polymer layer and thus gave very good results which were obtained for hybridoma growth and monoclonal antibody production. In the immobilized cell culture with gel entrapment, diffusion of oxygen seems to be poor compared with the suspension culture. Therefore, we attempted to improve the oxygen supply by using pure oxygen gas instead of air. After reaching a stationary phase, oxygen gas containing 0–5% carbon dioxide was introduced at 10 d. By this change, although glucose concentration in the bioreactor was almost the same the lactate concentration decreased to about two-thirds of that under air. Furthermore, the antibody production rate increased 2-fold. It became feasible to obtain a high concentration of monoclonal antibody continuously for a long period.

Bioproducts and Bioprocesses 2
Editors: Yoshida, Tanner
© Springer-Verlag Berlin Heidelberg 1993